T0093068

Intelligent Fractal-Based Image Analysis
Applications in Pattern Recognition and Machine Vision

Cognitive Data Science in
Sustainable Computing

Intelligent Fractal-Based
Image Analysis

Applications in Pattern Recognition and
Machine Vision

Volume Editors

Soumya Ranjan Nayak
School of Computer Engineering, KIIT Deemed to be University,
Bhubaneswar, Odisha, India

Janmenjoy Nayak
Department of Computer Science, Maharaja Sriram Chandra Bhanja Deo
University, Baripada, Odisha, India

Khan Muhammad
Department of Applied Artificial Intelligence, Sungkyunkwan University,
Seoul, South Korea

Yeliz Karaca
University of Massachusetts Chan Medical School, Worcester, MA,
United States

Series Editor

Arun Kumar Sangaiah
National Yunlin University of Science and Technology, Taiwan (RoC)

ACADEMIC PRESS
An imprint of Elsevier

ELSEVIER

Academic Press is an imprint of Elsevier
125 London Wall, London EC2Y 5AS, United Kingdom
525 B Street, Suite 1650, San Diego, CA 92101, United States
50 Hampshire Street, 5th Floor, Cambridge, MA 02139, United States

Copyright © 2024 Elsevier Inc. All rights are reserved, including those for text and data mining, AI training, and similar technologies.

Publisher's note: Elsevier takes a neutral position with respect to territorial disputes or jurisdictional claims in its published content, including in maps and institutional affiliations.

No part of this publication may be reproduced or transmitted in any form or by any means, electronic or mechanical, including photocopying, recording, or any information storage and retrieval system, without permission in writing from the publisher. Details on how to seek permission, further information about the Publisher's permissions policies and our arrangements with organizations such as the Copyright Clearance Center and the Copyright Licensing Agency, can be found at our website: www.elsevier.com/permissions.

This book and the individual contributions contained in it are protected under copyright by the Publisher (other than as may be noted herein).

Notices

Knowledge and best practice in this field are constantly changing. As new research and experience broaden our understanding, changes in research methods, professional practices, or medical treatment may become necessary.

Practitioners and researchers must always rely on their own experience and knowledge in evaluating and using any information, methods, compounds, or experiments described herein. In using such information or methods they should be mindful of their own safety and the safety of others, including parties for whom they have a professional responsibility.

To the fullest extent of the law, neither the Publisher nor the authors, contributors, or editors, assume any liability for any injury and/or damage to persons or property as a matter of products liability, negligence or otherwise, or from any use or operation of any methods, products, instructions, or ideas contained in the material herein.

ISBN: 978-0-443-18468-0

For information on all Academic Press publications
visit our website at https://www.elsevier.com/books-and-journals

Publisher: Mara Conner
Editorial Project Manager: Suchita Gera
Production Project Manager: Swapna Srinivasan
Cover Designer: Vicky Pearson Esser

Typeset by VTeX

Contents

Part I
Intelligent fractal-based image analysis

1. A deep insight into intelligent fractal-based image
   analysis with pattern recognition

   *H. Swapnarekha, Janmenjoy Nayak, Bighnaraj Naik, and
   Danilo Pelusi*

Part II
Recognition model using fractal features

8. Comparative analysis of approaches to optimize fractal image compression

Rakesh Garg and Richa Gupta

Part III
Fractals in disease identification and control

9. Alzheimer disease (AD) medical image analysis with convolutional neural networks

Ayesha Sohail, Muddassar Fiaz, Alessandro Nutini, and M. Sohail Iqbal

10. An intelligent fractal-dimension-based model for brain-tumor MRI analysis

Rakesh Garg, Richa Gupta, and Neha Agarwal

11. Fractal dimension analysis using hybrid RDBC and IDBC for gray scale images

Surbhi Vijh, Sumit Kumar, and Mukesh Saraswat

12. Preliminary analysis and survey of retinal disease diagnosis through identification and segmentation of bright and dark lesions

Jaskirat Kaur, Deepti Mittal, Ramanpreet Kaur, and Gagandeep

Contributors

Neha Agarwal, Amity University, Noida, Uttar Pradesh, India

Riaz Ahmad, Faculty of Science, Yibin University, Yibin, Sichuan, China

Belkacem-Toufik Badeche, Collaborating Academics - University of Montpellier, Montpellier, France

Sung Wook Baik, Sejong University, Seoul, South Korea

Ghulam Bary, Faculty of Science, Yibin University, Yibin, Sichuan, China
Key Laboratory of Computational Physics of Sichuan Province, Yibin University, Yibin, China

Anouar Ben Mabrouk, Laboratory of Algebra, Number Theory and Nonlinear Analysis Lab UR11ES50, Department of Mathematics, Faculty of Sciences, University of Monastir, Monastir, Tunisia
Department of Mathematics, Higher Institute of Applied Mathematics and Computer Science, University of Kairouan, Kairouan, Tunisia
Department of Mathematics, Faculty of Sciences, University of Tabuk, Tabuk, Saudi Arabia

Mourad Ben Slimane, Department of Mathematics, College of Sciences, King Saud University, Riyadh, Saudi Arabia

Sourav Kumar Bhoi, Department of Computer Science and Engineering, Parala Maharaja Engineering College, Berhampur, India

Carlo Cattani, Department of Economics, Engineering, Society and Business Organization - DEIM, Tuscia University, Viterbo, Italy

Ünal Çavuşoğlu, Sakarya University, Software Engineering Department, Sakarya, Türkiye

Muddassar Fiaz, Comsats University Islamabad, Islamabad, Pakistan

Gagandeep, Department of Computer Science Engineering, Chandigarh Engineering College, Mohali, Punjab, India

Rakesh Garg, Amity University, Noida, Uttar Pradesh, India

Richa Gupta, Amity University, Noida, Uttar Pradesh, India

M. Sohail Iqbal, Department of General Medicine, Shahdara Hospital, Lahore, Pakistan

Hadi Jahanshahi, Department of Mechanical Engineering, University of Manitoba, Winnipeg, MB, Canada

Kalyan Kumar Jena, Department of Computer Science and Engineering, Parala Maharaja Engineering College, Berhampur, India

Sezgin Kaçar, Sakarya University of Applied Sciences, Electrical and Electronics Engineering Department, Sakarya, Türkiye

Yeliz Karaca, University of Massachusetts Chan Medical School, Worcester, MA, United States

Jaskirat Kaur, Department of Electronics and Communication Engineering, Punjab Engineering College (Deemed to be University), Chandigarh, India

Ramanpreet Kaur, Department of Electronics and Communication Engineering, Chandigarh Engineering College, Mohali, Punjab, India

Sumit Kumar, ASET, Amity University, Noida, Uttar Pradesh, India

Deepti Mittal, Department of Electrical and Instrumentation Engineering, Thapar Institute of Engineering and Technology, Patiala, Punjab, India

Khan Muhammad, Department of Applied Artificial Intelligence, Sungkyunkwan University, Seoul, South Korea

Bighnaraj Naik, Department of Computer Application, Veer Surendar Sai University of Technology, Burla, Odisha, India

Janmenjoy Nayak, Department of Computer Science, Maharaja Sriram Chandra Bhanja Deo University, Baripada, Odisha, India

Soumya Ranjan Nayak, School of Computer Engineering, KIIT Deemed to be University, Bhubaneswar, Odisha, India

Alessandro Nutini, Centro Studi Attività Motorie – Biology and Biomechanics Dept., Lucca, Italy

Danilo Pelusi, Faculty of Communications Sciences, University of Teramo, Teramo, Italy

Mukesh Saraswat, Jaypee Institute of Information Technology, Noida, Uttar Pradesh, India

Utkarsh Sinha, Amity School of Engineering and Technology, Amity University, Noida, Uttar Pradesh, India

Ayesha Sohail, School of Mathematics and Statistics, The University of Sydney, Camperdown, NSW, Australia

H. Swapnarekha, Department of Information Technology, Aditya Institute of Technology and Management, Tekkali, Andhra Pradesh, India

Fath U Min Ullah, School of Engineering and Computing, University of Central Lancashire (UCLan), Preston, United Kingdom

Surbhi Vijh, School of Engineering and Technology, Sharda University, Greater Noida, India

Foreword

Nature is inherently full of patterns at different scales. From image to sound, we humans are accustomed to perceiving different stimuli from our surroundings through our organs, which have learned to process and classify such stimuli disregarding whether they occur at different spatial resolutions, playing speeds or sampling frequencies. The evidence that the human brain is capable of accommodating varying scales in their processed information is captivating, posing exciting challenges when it comes to endowing machine learning algorithms with similar modeling capabilities.

From a practical perspective, spatial invariance is of utmost importance in several fields of application, ranging from medical imaging to vehicular perception. Indeed, when deployed to tackle modeling tasks in these fields, the actionability of machine learning models – namely, their capability to support decision making effectively and reliably – is stringently subject to their capability to process input data at different spatial depths or flowing at varying speeds over time (as in, e.g., object detection from vehicular cameras). Possibilities expand sharply when multi-modal information enters the picture, as the distinct modalities from which the models learn (for instance, LIDAR, camera, and sound sensors in a vehicle) may evolve differently over time.

This noted importance of scale invariance in the construction of machine learning models is the main rationale for the advent and progressive maturity of fractal-based computer vision witnessed over the years. This research area aims to leverage the mathematical concept of fractals to analyze and process visual data. Fractals are, in essence, complex geometric patterns that display self-similarity at different scales. Since they exhibit similar patterns when zoomed in or out, fractals can be considered a powerful tool for understanding and characterizing the intricate structures found in natural images and scenes. By representing an image as a fractal, it is possible to extract more meaningful visuals from certain images, such as texture and patterns, in a multi-scale fashion. Fractal-based algorithms can also identify self-similar patterns within the image, which can support modeling tasks such as texture synthesis, image compression, and in general, pattern recognition problems requiring the aforementioned scale invariance.

This book provides a thorough reference material for students, researchers, practitioners, and newcomers interested in the mathematical principles of fractal geometry, emphasizing how fractal geometry can be synergized with image analysis techniques that rely on machine learning. The book also overviews avantgarde techniques and real-world applications where this discipline has been shown to provide large performance benefits when compared to traditional computer vision counterparts. Overall, the sequence of chapters and their purposed arrangement, from theory to applications through techniques, excel at engaging the reader around fractal geometry, and at clearly exposing the enormous potential and sensational future faced by this research area.

Together with the growing momentum of geometric machine learning, it is my firm belief that the publication of this book is extremely opportune, and that it will undoubtedly enthrone fractal-based image analysis as a subject of intense research in the years to come.

Javier (Javi) Del Ser
Research Professor in Artificial Intelligence,
TECNALIA, Basque Research & Technology Alliance (BRTA),
Derio, Bizkaia, Spain

Preface

The fascinating world of intelligent fractal-based image analysis is an intersection where the realms of pattern recognition, machine vision, and fractal geometry converge. Accordingly, we intend to embark on a comprehensive journey through the theory, techniques, and applications of fractal-based image analysis within the domain of intelligent systems through our book entitled "Intelligent Fractal-Based Image Analysis: Applications in Pattern Recognition and Machine Vision". Fractal geometry, introduced to the world by the visionary mathematician Benoit Mandelbrot, has revolutionized our understanding of the complexity and beauty of natural phenomena. Fractals, with their self-similar patterns recurring at different scales, are ubiquitous in the universe, from the branching structures of trees to the intricate designs found in snowflakes. Harnessing the power of fractals, researchers have developed innovative methods for analyzing images, leading to the emergence of fractal-based image analysis as a promising field.

Fractal graphics (FG) is an interdisciplinary field that deals with how to utilize computers to attain a high level of comprehension from digital images by analyzing its own structural properties, like self-similar and self-affine patterns. The concept of fractal geometry, dealing with shapes found in nature having non-integer or fractal dimensions, is closely linked to scale invariance, and thus, it provides a framework for the analysis of natural phenomena in various scientific and engineering domains. The current research on Fractal Dimension (FD) has gained more popularity among various researchers due to the ease and compatibility in any field of application supported by the literature concerning engineering applications, the healthcare domain, among many others. The concept of FD principally comes under the concept of FG for measuring the dimensions (surface roughness) of complex objects found in nature like clouds, coastlines, borderlines, landscapes, mountains, and so forth, which failed to be analyzed by Euclidean geometry (EG). This fractal dimension is the branch of FG (namely computer graphics) that adopts the mathematical set exhibiting a repeating pattern displayed at every scale.

Fractals, as geometric shapes, and patterns that can describe the roughness, or irregularity, present in almost every object in nature. Repeating and iterating their geometry at smaller or larger scales, fractals are infinite, complex patterns used in modeling physical and dynamic systems in numerous fields of applications, like engineering science, health science, social sciences, and many more. The infinite complexity and aesthetically astounding beauty of fractals are endowed with theoretical, mathematical, historical, and technological aspects. Inspired by these exceptional characteristics of fractals, recent literature reveals the vital role that fractals play in digital image analysis, specifically in fractal information, fractal design, fractal arts, texture characterization, color characterization, 3D modeling, pattern matching, pattern recognition, remote sensing, image fractal stenography, porous material analysis, etc. Moreover, in health science applications, it delivers more recent research analytics in terms of brain tumor analysis, breast cancer analysis, heart rate analysis, oncopathology, and many more by interpreting health care data.

As advancements in artificial intelligence and machine learning continue to reshape the landscape of image analysis, fractal-based approaches offer a unique perspective on tackling the challenges posed by complex visual data. By exploiting the self-similarity and information-rich nature of fractals, intelligent algorithms have been developed to extract meaningful features, identify patterns, and perform recognition tasks across a wide range of applications. However, a recent study demonstrated that the integration of artificial intelligence (AI) with fractal characteristics has resulted in new interdisciplinary research in the fields of pattern recognition and machine vision by interpreting the buzzword called fractal feature in invariant scales. This book provides insights into the current strengths and weaknesses of different applications as well as research findings on fractal graphics in engineering and health science applications.

Our book is structured with a total of 12 chapters, each addressing different aspects of fractal-based image analysis. The chapters provide in-depth coverage of key concepts, techniques, and applications, offering readers a holistic understanding of this multidisciplinary field. Throughout the book, we strike a balance between technical rigor and accessibility, ensuring that both novice learners and experienced practitioners can grasp the underlying principles and gain practical insights. The main objective of this edited book is to represent the most recent core concepts and methodologies and to discuss future directions and emerging opportunities in detail. Apart from the core concepts, we also focus on two major real time applications: engineering science and health science applications. In view of these elements, this book mainly focuses on stepwise discussions, exhaustive literature reviews, detailed analyses, and rigorous experimentation results, demonstrating and encouraging novel application-oriented approaches. More importantly, it elaborates on the most recent trends in detailed knowledge (core concept) of fractal image analysis especially box-counting analysis, multi-fractal analysis, 3D fractal analysis, chaos theory, color imaging, and pattern recognition through the use and accordance of fractal theory. Subsequently, the book also describes the complete understanding of two main different applications, which are engineering science and health science. Thus the book aims to provide the readers, who are willing to understand, study, and contribute to the different application areas, with detailed information and thought-provoking ideas to enable better understanding of fractal analysis in imaging science. Finally, this book also discusses the recent pattern recognition applications in bio-medical image analysis in terms of retinopathy image analysis, chest X-ray image analysis, MRI Image analysis, cancer cell analysis, and so forth.

This book contains in-depth fundamental and advanced research contributions from a methodological perspective to deal with various applications of fractal-based approaches for engineering and healthcare imaging problems. The current challenges in handling the complex objects' analysis problems in terms of self-similarity properties are some of the major highlights of the book. The motivational aspects and outcomes obtained from the analyses included in the book can provide directions to improve the solutions and rationality among different AI-fractal integrated based models that investigate advanced fractal theories spanning the areas of neural networks, fuzzy logic, machine learning, deep learning, and hybrid intelligent systems in solving pattern recognition problems, for the purpose of exploring the more efficient application of fractal theories to a wide range of medical image processing modalities. The presentation of case studies that illustrate the application and integration of fractal theories to intelligent computing in the resolution of important pattern recognition and machine vision problems for end-user applications is another aspect of this book, which can be appealing to researchers, data scientists, industry professionals, practitioners, and graduate students in the fields of fractal graphics and its related applications. This volume includes 3 Parts and 12 Chapters, and is organized as follows.

Part I: Intelligent fractal-based image analysis

Chapter 1 presents a detailed review of intelligent fractal-based image analysis with pattern recognition by description of the gene expression in digital as well as micro array form for disease diagnosis. Moreover, a broad literature study has been carried out related to the feature prediction of high dimensional microarray cancer data using various deep learning techniques, and subsequently observed that the need for developing intelligent pattern recognition systems has been rising rapidly due to the advancement in various pattern recognition and digital image processing approaches. Furthermore, the study also provides a critical investigation of various pattern recognition and image processing approaches used in the fractal analysis of images that may assist the researchers in developing more sophisticated systems in various application domains. Lastly, the chapter is concluded by giving a critical analysis with several challenges and future research scopes.

Chapter 2 deals with the processing of Mandelbrot set fractal images for several applications. Fractal analysis is considered as a crucial research domain in the current scenario. This chapter focuses on the machine learning based approaches for the processing and analysis of several Mandelbrot set fractal images. The machine learning based approach mainly focuses on the clustering techniques such as hierarchical clustering, distance mapping, multidimensional scaling (MDS), and distance matrix for the processing of fractal images. These clustering techniques are mainly used to segregate the Mandelbrot set fractal images into several clusters for further processing. Afterwards, the performance of each of these clustering techniques is analyzed. The simulation of this work is carried out using Orange3-3.24.1.

Chapter 3 addresses the chaos-based image encryption applications that have been rendered beneficial owing to the randomness and unpredictable dynamic behavior of chaotic systems. The authors of the chapter discuss the continuous and discrete-time chaotic systems while going through the chaos-based image encryption studies conducted over the recent years. Moreover, they also deal with integer-order, fractional-order or variable-order systems as well as their use in image encryption applications. The chapter further handles the related methods employed in encryption applications besides the analysis methods utilized in chaos-based image encryption applications. Regarding the steps of these applications, numerical solutions of chaotic systems, random number generation operations, and randomness tests and encryption operations using the XOR operation are included in the chapter, which also presents the security evaluation of chaos-based encryption algorithms. The chapter is concluded with a case study on the way medical images are handled by a chaos-based image encryption algorithm. The chapter provides contributions for chaos-based encryption algorithms to become more efficient and secure, indicating the need that cryptanalysis studies in the literature be increased and general attack scenarios applied to the designs that will and should be developed as well as standardized.

Chapter 4 comprehensively demonstrates the efficacy of fractal features for image classification, with a particular emphasis on brain MRI image classification. This research investigates the potential of fractal dimensions as a robust feature extractor, considering a wide range of parameters, such as Smoothness, Hausdorff Dimension, and various Gray-Level Co-occurrence Matrix (GLCM) features, including Contrast, Correlation, Homogeneity, Energy, and Entropy. To ensure effective feature extraction and optimized data analysis, numerous preprocessing techniques have been applied. These techniques include offline data augmentation and data normalization methods that have been pivotal in enhancing the reliability of the experiment. The calculated feature vectors have been aggregated and various exploratory data analysis techniques have been presented to investigate the effects of the FD on the accuracy of classification. Subsequently, to achieve better results and reduce computer time, efficient techniques such as feature selection and extraction, feature engineering, normalization, and offline data augmentation have been deployed. The study also includes a deep dive into the classification process. Utilizing well-established machine learning (ML) based classifiers such as Support Vector Machine (SVM), K-Nearest Neighbors (KNN), Random Forests, and Artificial Neural Networks, the binary dataset of Normal and Tumor images was accurately classified. The findings were carefully validated using diverse evaluation metrics and the results highlights the overall importance of fractal dimension based feature extraction of images.

Part II: Recognition model using fractal features

Chapter 5 highlights the main feature of image processing, obtaining results that are smooth and qualitative. The chapter shows the importance of fractal analysis for the purpose of deciphering and analyzing. In addition, the study shows that fractal analysis gives a comprehensive keenness and the normalized image parameter, which is evaluated to ascertain the analytical strength of quantum interferences. To these ends, the authors of the chapter evaluate the produced image characteristics by the normalized fractal-based technique to deploy in scientific accomplishment with its application. In addition, they calculate and explore the performance of the hybrid system by differential equations to analyze the peculiarities of the image excreted sources. By exploring several quantum correlations for the sources image analysis, the chapter provides contributions to the related fields, while harnessing the particle femtoscopy in the analysis to explore various special correlations for the investigation of the particulars regarding chaotic-coherent dynamics and the intrinsic morphology of the particle-generating sources for pattern recognition purposes.

Chapter 6 discusses the concept of learning the sequence for VD using CNN, LSTM, and its variants including multi-layer, GRU, etc. Similarly, the CNN and LSTM based methods are surveyed with their main contributions and drawbacks. Next, the chapter explains their working flow along with their frameworks, experimental performance, and their visual results. The authors shed light on the working of RNN and their limitations explaining why LSTM is a better performer than RNN. Lastly, the authors have concluded the chapter with a discussion and the recommendations in terms of future research directions that may assist the forthcoming challenges of VD. Soon, it is aimed to investigate VD methods in terms of the Internet of Things (IoT) in consideration with multi-view surveillance scenarios.

Chapter 7 is concerned with the development of a wavelet-based study to comprehend the anisotropic, or directional, behavior of images, particularly nano images, and oscillations in nanomaterials. Wavelets have significance as being sophisticated tools in image processing, which allows the explanation of various intricate cases of images, like the nano- and bio- ones, as well as mixed cases of both. One of the main aims of the chapter is to adapt the notion of oscillating singularity to directional cases when the focuses will be on the local directional-wise oscillations. The role of wavelets in fractal/multifractal structures understanding is indicated by reviewing the essential wavelet characterizations of irregularities and singularities in one-dimensional and two-dimensional cases. Another aim is the revision of the notion of regularity or oscillating singularity in higher dimensions, by showing that the cases can be extended to non-isotropic cases as is the case with classical Hölder regularity. Thus various methods are applied to the recognition problems like neural networks, fuzzy logic, morphological methods, genetic algorithms, and so forth. The chapter provides a direction in that mathematical models such as wavelets for AI objects recognition are critical, paving the way for further investigation and applications in the related domains.

Chapter 8 presents a comparative study of strategies for faster fractal image compression (FIC) as addressed accordingly. Several fields have widely accepted FIC as a commending compression technique. The fractal compressor has an asymmetric encoder and decoder concerning the time spent in each. The encoding process includes searching for a large image pool. The quantity of time elapsed in searching is a matter of significant concern. Enormous mean square error (MSE) computations lead to a delay in the encoder. Over the years, many methodologies have been suggested to improve the performance of the FIC algorithm. This chapter presents a comprehensive investigation of thirteen broadly coordinated approaches, i.e., Block Classification, Partitioning of Image Blocks, Nearest Neighbor Search, Clustering, Vector Quantization, Discrete Cosine Transforms, Variance-Based FIC, Spatial Correlation, Iteration free Convergence, Local Search or No Search Method, Parallel Algorithm, Wavelet Fractal Hybrid Encoder, and Genetic Algorithm along with their strengths and weaknesses. Each approach was deliberated chronologically for a superior understanding of the development of FIC. The frequency domains have been reported and combined with FIC to improve the algorithm's performance. It is observed that the high threshold value results in the image's poor quality, while a low threshold leads to a long encoding time in search of less FIC. The underlying objective of this chapter is to recognize the methods that allow the expansion of new strategies to speed up the encoding procedure.

Part III: Fractals in disease identification and control

Chapter 9 presents the use of Convolutional Neural Networks (CNN), which provides a new deep learning system ensuring better disease classification, broader use of data as well as improved classification of tissue pathological features. Moreover, the authors of the chapter provide the definition and application of machine learning and deep learning systems towards diagnostic imaging related to Alzheimer's disease. The chapter includes information on certain diagnostic techniques that involve medical imaging such as structural, functional, and molecular. By considering these application-related aspects, the chapter demonstrates the importance of the integration of fractals, imaging, and AI in disease diagnostics in medical and other related areas. The chapter provides contributions in terms of showing the importance of integrating the data obtainable from the various diagnostic systems so that a better discretionary capacity can be achieved for defining the tissue that is affected by pathologies.

Chapter 10 is concerned with the implementation of an intelligent fractal dimension-based model for brain tumor analysis by using MRI images. An entirely automated brain tumor segmentation technique based on Convolutional Neural Networks (CNN) is also proposed. The intended networks are designed to different brain tumor types like glioma, pituitary, and meningioma present in the dataset. A segmentation-based feature extraction algorithm (SFTA) is exercised on the MRI of the brain to determine the region of interest (ROI). Texture features solve semantic gap issues, and it provides the required description of the organ's texture in an MRI Scan. The chapter exercises the fractal dimension as a feature to analyze the MRI scan of the brain. Two methods of fractal dimension are explored to find an optimized method for analysis such as Box counting and the Differential box-counting method, used to observe the structure of the MRI scan. Further, a SFTA-based feature extraction with fractal dimension is proposed to detect ROI. In the result section, an MRI image is analyzed to observe the structure of the tumor using fractal dimension and segmentation. The automated technique is developed using the Flask framework based on Python by implementing several convolutional neural network models such as InceptionV3, VGG-16, RESNET-50, Xception VGG-19, and Efficient NetB0.

Chapter 11 demonstrates the fractal dimension analysis using Hybrid Relative Differential Box Counting (RDBC) and Improved Differential Box Counting (IDBC) for gray scale images. Fractal dimension (FD) plays a significant role in determining surface quality in digital images considering essential aspects of fractal geometry for pattern recognition. An effective and enhanced Differential Box-Counting technique (DBC) is presented to achieve a precise and accurate evaluation of the fractal dimension of images. In this chapter, the authors discuss a modified DBC (RIDBC) approach to overcome three issues in existing DBC, i.e., computational error, least friction surface, and homogeneous fractal dimension determined either by increasing or decreasing the constant value to each intensity point. The modified DBC is composed of hybrid RDBC and IDBC. The RIDBC method is experimented and validated on four different images, i.e., Brodatz images, smooth texture using a linear transformation, smooth synthetic textured images, and rough synthetic textured images. For accurate computation, the adjustment coefficient factor is formulated for fractal dimension analysis. The chapter is concluded by presenting the promising results in comparison to existing state-of-the-art methods such as DBC, RDBC, and IDBC respectively.

Chapter 12 aims to provide a general background to design a computer-aided retinal disease diagnostic method. Detailed understanding of various retinal lesions related to systemic retinal diseases and their characteristic visual appearance is also provided in this chapter. This preliminary analysis and survey are designed to understand the mathematical interpretation of various retinal anatomical structures (landmarks) and lesions to detect various retinal diseases and grade the severity level of non-proliferative and proliferative retinal diseases from retinal fundus images. Additionally, to design a robust automated retinal disease diagnostic method, a retinal image dataset comprising of fundus images having varying attributes is essential. Taking these facts into consideration, the detailed preliminary study of various retinal lesions and their classification is carried out in this chapter to design an effective computer-aided diagnostic method for the detection of various retinal diseases. Furthermore, the need for computer-aided retinal disease detection, retinal image enhancement, characterization of landmark (healthy) structures, characterization of retinal lesions, and computer-aided retinal disease classification and grading method have also been discussed.

We anticipate that our edited book will provide new horizons towards fractal-based image analysis for everyone who is or could be interested in the theory, applications, simulations, and modeling of intelligent systems in different applied areas. Hence, we would like to thank all the contributors and the reviewers for their contributions and dedicated efforts for the successful completion of this edited book. We would particularly thank the editorial team of Elsevier for their valuable technical support and efforts. We hope that the work reported in this volume will motivate further research and encourage development efforts in the performance evaluation of various cancer diseases and allied domains. In our exploration, we begin by introducing the fundamental principles of fractal geometry and its relevance to image analysis. We delve into the mathematical representations and estimation techniques that enable the characterization of fractal properties in digital images. Building upon this foundation, we examine traditional and contemporary approaches to image analysis, including feature extraction, segmentation, object detection, and classification.

To address the growing demand for intelligent systems, we delve into the integration of fractal-based image analysis with advanced machine learning techniques. Deep learning and convolutional neural networks tailored for fractal data are explored, providing readers with the tools necessary to leverage the power of intelligent algorithms in their own research and applications. The book further explores the practical applications of fractal-based image analysis across a diverse range of domains. From medical imaging and remote sensing to industrial inspection and multimedia analysis, we showcase real-world scenarios where fractal-based techniques have demonstrated their efficacy and versatility. The chapters in this book have been authored by leading experts in the field, each bringing their unique perspectives and insights. Their contributions ensure a comprehensive coverage of the subject matter and provide readers with a wealth of knowledge and practical guidance.

We would like to express our gratitude to the contributors, reviewers, and editors who have dedicated their time and expertise to make this book possible. Their invaluable contributions have shaped this work into a valuable resource for the research community. We invite our readers to embark on this enlightening journey into the realm of intelligent fractal-based image analysis. By immersing yourself in the pages that follow, you will gain a deeper appreciation regarding the intricate relationship between fractal geometry and intelligent systems. We hope this book inspires the readers to explore new frontiers and unleash the full potential of fractal-based image analysis in their own research and applications.

Enjoy the exploration that will spark a new interest in the secrets of the universe!

Acknowledgments

We, as the editors, present this volume of carefully curated chapters on *Intelligent Fractal-Based Image Analysis: Applications in Pattern Recognition and Machine Vision*. This volume is a collective effort of many prolific researchers, whose research contributions are really helpful to make a full shaped book. They have been urged to submit chapters of state-of-the-art and to demonstrate the original findings of fractal analysis and the imparting of information to both national and international cooperatives in various fields and angles of investigation.

We would like to extend our sincere gratitude to the reviewers for their meticulous reviewing efforts and important suggestions for improvement of the quality of the chapters. The reviewers have diligently shown their interest to review the chapters and develop critical suggestions for the authors to enhance their research work. We would also like to extend our cordial thanks to all the contributors involved in this process. We would like to express our profound gratitude and genuine appreciation to the editorial members of Elsevier Publishing for all their support and cooperation to complete our edited book.

Editors

Soumya Ranjan Nayak
School of Computer Engineering,
KIIT Deemed to be University,
Bhubaneswar, Odisha, India

Janmenjoy Nayak
Department of Computer Science,
Maharaja Sriram Chandra Bhanja Deo University,
Baripada, Odisha, India

Khan Muhammad
Department of Applied Artificial Intelligence,
Sungkyunkwan University,
Seoul, South Korea

Yeliz Karaca
University of Massachusetts Chan Medical School,
Worcester, MA, United States

Part I

Intelligent fractal-based image analysis

Chapter 1

A deep insight into intelligent fractal-based image analysis with pattern recognition

H. Swapnarekha[a], Janmenjoy Nayak[b], Bighnaraj Naik[c], and Danilo Pelusi[d]

[a]Department of Information Technology, Aditya Institute of Technology and Management, Tekkali, Andhra Pradesh, India, [b]Department of Computer Science, Maharaja Sriram Chandra Bhanja Deo University, Baripada, Odisha, India, [c]Department of Computer Application, Veer Surendar Sai University of Technology, Burla, Odisha, India, [d]Faculty of Communications Sciences, University of Teramo, Teramo, Italy

1.1 Introduction

Pattern recognition is the activity of identifying patterns in data and then utilizing those patterns to make proper decisions. In general, human beings excel in the activity of pattern recognition. Indeed, most of the daily activities of humans involves various pattern recognition tasks. Due to the rapid growth of large amounts of high-dimensional and unstructured data, it becomes impossible for humans to interpret the data. This drives the need for developing intelligent systems for identifying patterns promptly and accurately. Therefore the goal of pattern recognition is to develop various algorithms and methodologies for automatically identifying the patterns that humans normally perform. In the present era of artificial intelligence, pattern recognition has been widely used in a variety of applications such as text pattern recognition, speech recognition, facial recognition, recognizing patterns in medical images, and so on. In most of the methodologies developed for solving pattern recognition problems, patterns are represented as a set of measurements called features. The selection of too many features increases the complexity of the classifier, while the selection of weakly informative features reduces the accuracy of recognition algorithm. Therefore the selection of features plays a vital role in the development of efficient classifiers. Generally, two approaches have been used in the selection of features. One approach is feature selection that involves the selection of a small subset of features that is exquisitely required to determine the results of the recognition system, whereas the second approach is feature extraction that involves the re-

Copyright © 2024 Elsevier Inc. All rights reserved, including those for text and data mining, AI training, and similar technologies.

duction in the dimension of information features while preserving the original information required for classification.

As the daily activities of humans depend heavily on digital images, the field of digital image processing is always a fascinating area among researchers, as it refers to the process of removing any type of irregularities or noise present in the image in order to provide enhanced pictorial information for interpretation of images by human beings. Due to the advancement of computing technology, image processing systems have been widely used in a variety of applications, such as remote sensing [1], medical imaging [2], agriculture [3], forensic studies [4], document processing [5], and so on. As a high number of digital images are uploaded every minute, development of various image processing techniques, such as image segmentation, feature extraction, and image classification are of higher importance to provide automatic analysis of digital images. Image segmentation refers to the process of dividing a digital image into several segments having similar features. The primary goal of image segmentation is simplification of a digital image so that it can be analyzed easily. Image classification refers to the process of assigning all pixels in the image to specific classes that aims at the accurate detection of features present in an image. Feature extraction is the process of extracting a set of relevant features that define the shape of object uniquely and precisely. As feature extraction aims at enhancing the efficacy of the recognition system, the selection of an appropriate feature extraction approach in accordance with the input, needs to be performed with utmost care.

In the field of mathematics, Fractals is a new area of knowledge and is one of the most significant discoveries of the present era as it offers more powerful tools for human beings to analyze the complexity. The fractal dimension technique has drawn the focus of many researchers in various disciplines. This is mainly due to the fact that fractal geometry can be used to describe the irregular patterns and complex objects of nature that cannot be determined by the conventional Euclidean geometry [6]. Moreover, the fractal technique can be used efficiently to model much time series data as well as images as fractals are basically mathematical sets having a high level of geometrical intricacy. One of the significant characteristics of fractals is fractal dimension because it holds the data about their geometric structure. Due to this feature, the fractal dimension technique has been widely utilized in various applications such as pattern recognition, segmentation of image, texture analysis, medical signal analysis, and so on [7–10].

The expeditious advancement in image acquisition devices has resulted in the availability of large amounts of image data due to the rise in the throughput and installation of a large number of image acquisition devices. The processing of these images manually by human experts is an extensive and tedious efforts as it is susceptible to human error. In the past few decades, artificial intelligence approaches have seen vast advancement in various application of computer vision and pattern recognition. Machine learning approaches have been extensively utilized in the research of image processing as they obtain knowledge from data representation [11–13]. Kusumo et al. [14] analyzed various image processing based features for efficient identification of corn disease. The performance of features such as color, local features on image, SURF (speeded up robust features), objector detector, and ORB (oriented fast and rotated BRIEF) have been analyzed using distinct machine learning approaches. The experimental results show that color is the most informative feature in the efficient identification of corn disease as it obtained the best accuracy in most of the classifiers evaluated. Nageswaran et al. [15] have used machine learning and image processing approaches for the efficient detection and classification of lung cancer. During preprocessing, the authors used a geometric mean filter for enhancing the quality of image. After preprocessing, images were segmented using the k-mean approach. Then, the classification of the images was performed using various machine learning approaches such as ANN (Artificial Neural Network), KNN (K Nearest Neighbor), and random forest. The experiments were conducted on 83 CT scans collected from 70 patients and the results indicate that ANN attained better results in the detection and classification of lung cancer. However, machine learning approaches demand significant domain knowledge and sophisticated engineering as these approaches depend on the data representation. Learning higher-level features from the data is considered as a possible way to overcome the limitations of traditional machine learning approaches. In recent years, Deep learning approaches has shown better efficacy than traditional machine learning methods for processing extensive quantities of images as they automatically learn higher-level features from the data representations. Deep learning approaches have been widely applied in many application fields, including speech recognition [16], computer-aided diagnosis [17], text identification [18], drug discovery [19], face recognition [20], and other areas, due to their advantages. This study will discuss the importance of the fractal analysis of digital images using intelligent approaches. Initially, this survey will provide a detailed analysis of various approaches used in the development of an efficient pattern recognition system. Moreover, this study also addresses distinct image processing approaches and the importance of fractal features in the accurate analysis of digital images.

The rest of the paper is organized as follows. Section 1.2 presents various approaches used in the development of a pattern recognition system and its applications in various domains. The different image processing approaches used in the accurate interpretation of images are presented in Section 1.3. The importance of fractal features and analysis of fractal-based images using intelligent approaches is discussed in Section 1.4. Further, Section 1.5 presents a critical analysis of the research performed on fractal-based image analysis with pattern recognition using intelligent approaches. Finally, the chapter concludes in Section 1.6.

1.2 Approaches and applications of pattern recognition

The advancement in computing technology has not only facilitated the processing of large amounts of data but also assisted in the usage of distinct and elaborated approaches for the analysis and classification of data. Simultaneously, the need for developing automatic pattern recognition systems is increasing tremendously due to the availability of vast databases and demanding performance requirements such as higher accuracy and faster recognition speed at low cost. According to Jain et al. [21], four major approaches are used in the development of efficient pattern recognition systems. Fig. 1.1 represents the different approaches used in the pattern recognition systems.

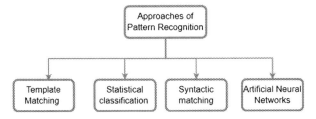

FIGURE 1.1   Different approaches used in pattern recognition systems.

- **Template matching**

   This is one of the simplest and earliest approach that is used to determine the likeliness between two entities of similar type. In this approach, the pattern to be identified is compared against the stored template by considering all translation, rotation, and scale changes. Even though template matching is effective in some application domains, it still results in failure if the patterns are misinterpreted due to huge intraclass variations among patterns, change in viewpoint or imaging process.

- **Statistical classification**

   In this classification, each pattern is characterized by d features and is considered as a point in d-dimensional space. This approach aims at selecting those measures so that the pattern vectors associated with distinct categories can hold disjoint and solid regions in d-dimensional feature space. The way the patterns belonging to distinct classes can be separated determines the efficacy of the feature set.

- **Syntactic matching**

   In complex pattern recognition systems, it is better to view patterns from a hierarchical perspective in which patterns are made up of simple subpatterns that in turn are made from other simple subpatterns [22,23]. In the hierarchical perspective, the simplest subpatterns are known as primitives and a complex pattern is represented as an association between those primitives. In syntactic matching, patterns are recognized as sentences of a language that are generated as per the rules of grammar and primitives are recognized as alphabets of a language. Therefore complex patterns in a syntactic pattern are represented by a few primitives and grammatical rules. As syntactic matching provides information about the construction of a pattern from the primitives in addition to classification, it is considered as one of the most appealing approaches of pattern recognition systems. Although syntactic matching is an efficient approach, it still leads to many difficulties when dealing with segmentation of noisy patterns and deduction of grammatical rules from training data.

- **Artificial neural networks**

A neural network is a network of neurons that are used to mimic the operations of the brain in order to represent the relationship between enormous amount of data. The fundamental characteristic of a neural network is the ability to learn complicated non-linear relationships between input and output and acquaint themselves with the data. A feed-forward network, which consists of Radial-Basis Function (RBF) and multilayer perceptrons, is one of the most commonly used neural networks for the task of pattern classification [24]. The other network that is used in pattern classification task is the Self-organizing Map [25]. The growing popularity of neural networks in pattern recognition problems is mainly due to the accessibility of adequate learning algorithms and their low reliance on domain-specific knowledge.

1.2.1 Applications of pattern recognition

Nowadays, pattern recognition has been widely used in many areas of applications that involves the observation of structure. The following are some of the areas where pattern recognition has been frequently used.

1.2.1.1 Healthcare

The advancement in data acquisition and monitoring approaches in the healthcare sector has resulted in the generation of large volumes of medical data. Therefore analysis of large volumes of medical data to provide enhanced healthcare is one of the key challenges in today's medical field. Over the last few decades, Computer-aided diagnosis (CAD) has become one of the major areas of research in the field of medical imaging and diagnosis due to the advancement in various pattern recognition algorithms [26–28]. To overcome the limitations of manual diagnosis of coronary artery disease using ECG, an approach for automatic diagnosis of coronary artery disease has been suggested by Desai et al. [29]. The authors used HOS (Higher-order statistics) cumulant features for the management of risk factors associated with coronary artery disease. Further, the dimensionality of cumulant coefficients has been diminished using the PCA (Principal component analysis) approach. Then, the medically substantial features whose p-value is less than 0.05 are subjected to classification using Rotation Forest (ROF) and Random Forest (RF) ensemble classifiers and results reveal that the ROF ensemble approach obtained better accuracy than the RF approach. Gong et al. [30] have suggested three algorithms known as SA-PNN (Simulated Annealing-non-parallel probabilistic neural network), LSA-PPNN (Linear SA-parallel probabilistic neural network), and SA-PPNN for efficient pattern recognition of epilepsy from iEEG (intracranial electroencephalogram) readings. These algorithms were developed based on the principles of DWT (discrete wavelet transforms) and PPNN. The network parameter of PNN is optimized using SA and LSA approaches. Then, the performance of the three algorithms has been evaluated and results indicate the running time of SA-PPNN has been reduced by 2.18 times when compared with SA-PNN. Moreover, the LSA-PPNN average operating time has been decreased by 9.97 times when compared with SA-PPNN. Table 1.1 represents various pattern recognition studies performed in the healthcare sector.

1.2.1.2 Speech recognition

Since speech is the most vital and essential means of communication among humans, recognition of human speech has become an interesting issue among researches. The advancement in computing technologies and the progress of hidden Markov modeling has triggered more effective statistical approaches that further assist in developing more dynamic speech-recognition systems. Deng et al. [36] have defined deep generative and discriminative approaches that have been used in the recognition of speech and its associated related pattern recognition problems. The deep generative model is used to characterize the distribution of data and the related targets, while the deep discriminative approach is used to define the distribution of targets accustomed with data. Moreover, the authors have also specified the ways of integrating generative approaches with discriminative approaches for effective recognition of speech. For efficient automatic recognition of speech words, a novel deep learning based speech recognition model has been suggested by Jermsittiparsert et al. [37]. In this approach, speech sound that is the input source has a direct effect on the accuracy of the classifier. The suggested approach has been evaluated on a Berlin database consisting of 500 utterances of both males and females and the results reveal that the suggested approach has attained an accuracy of 94.21% for MFCC, 83.54% for prosodic, 83.65% for LSP, and 78.13% for LPC features when compared with other conventional approaches. Analysis of other works carried out in speech recognition is presented in Table 1.2.

1.2.1.3 Agriculture

The agriculture sector is one of the most important sectors that plays a vital role in the global economy. Due to the continuous growth of population in the world, the demand for agricultural products is also increasing. Accurate diagnosis and providing timely solutions to the problems in agriculture can result in increased productivity. The advancement in image processing and pattern recognition techniques has led to the development of autonomous systems that can be used

TABLE 1.1 Analysis of other pattern recognitions studies performed in healthcare sector.

Author & Year	Objective	Dataset	Approach	Results	Observations	Ref
Yao et al. 2022	For automatic identification and classification of distinct patterns of calcifications dissemination in mammographic images	Collected 581 mammographic images from 292 breast cancer cases	Graph Convolutional Network approach	Accuracy= 64.325 ± 1.694, Sensitivity= 0.643 ± 0.017, and Specificity= 0.847 ± 0.009	Does not include larger representations of the linear calcification distribution pattern in the analysis	[31]
Javed et al. 2021	To identify the interaction between drugs	DrugBank dataset	Random forest	Accuracy= 95.4%	Only a few biomedical datasets are considered for analysis	[32]
Dominguez et al. 2020	For efficient identification of prostate cancer	156 patients with prostate cancer	Pattern recognition Artificial neural network	Sensitivity= 97%, specificity= 88%	Only a few samples were used for analysis	[33]
Kim et al. 2019	For efficient classification and recognition of respiratory patterns from signal data	15 000 data items collected from the UWB Radar that generates data at 25 frames per second	Ultra-wideband Radar and 1D CNN	Average recognition rate of five respiration patterns is 93.9%	Did not consider different breathing pattern when person is sleeping	[34]
Dakhly et al. 2019	Compares the efficiency of IOTA simple rules versus pattern recognition approaches in determining ovarian cancer as benign and malignant	396 women with ovarian masses from Kasr El Aini Hospital, Cairo	IOTA (International Ovarian Tumor Analysis) simple rules and pattern recognition	Accuracy is 91.7%	Did not include expert sonographer in the analysis of results	[35]

TABLE 1.2 Analysis of other pattern recognitions studies performed in speech recognition.

Author & Year	Objective	Dataset	Approach	Results	Observations	Ref
Rocha et al. 2016	To classify patterns of speech signals	EPUSP, INATEL and IFMA voice banks	Multilayer Perceptron and Learning vector quantization	Results in selection of best config-uration neural network to perform speech recognition	Did not consider speech signals from different languages	[38]
Kong et al. 2020	Recognition of audio patterns from Audioset	AudioSet dataset extracted from YouTube videos consisting of 527 classes	Pretrained audio neural networks	Mean average precision is 0.439 and AUC is 0.973	Only a few audio patterns were considered for analysis	[39]
Khan et al. 2016	Analysis of subvocal Hindi phonemes using EMG signals	Sub vocal speech samples from EMG (elec-tromyogra-phy) signals	Hidden Markov model (HMM) classifier	Accuracy= 85.56%	Considered smaller dataset	[40]
Petridis et al. 2020	To recognize speech from the mouth images	OuluVS2, CUAVE, AVLetters, and AVLetters2 database	Bidirec-tional Long-Short Memory networks (LSTM)	Attains clas-sification rate of 0.6% on OuluVS2, 3.4% on CUAVE, 3.9% on AVLetters, 11.4% on AVLetters database, respectively	Data imbalance between different datasets	[41]
Tailor et al. 2018	To recognize speech of Gujarati language	Speech dataset consisting of 650 routine Gujarat utterances	HMM Model	Accuracy= 87.23%	Considered smaller dataset	[42]

to tackle various challenges of the agricultural sector. For vigorous recognition and detection of grape berries and grape bunches in vineyards, a novel approach that made use of visible spectrum cameras has been proposed by Pérez-Zavala et al. [43]. In order to split regions of clustered pixels into grape bunches, the proposed approach makes use of a procedure along with the shape and consistency information of grape bunches. The proposed approach utilizes HOG (Histogram of oriented gradients) and LBP (Linear binary patterns) methods to obtain shape and texture information of grape bunches. Further, the suggested approach has been validated using datasets containing 163 images of various grapevine types obtained from four different countries. The experimental results reveal that the suggested approach detects grape bunches with an average precision and recall of 88.61% and 80.34%, respectively, whereas single berries are recognized with a precision and recall of 99% and 84%, respectively. A semi-automatic approach has been proposed by Liu et al. [44] for the efficient detection of agricultural parcels. The proposed approach makes use of a pixel-based approach and SAM (spectral angle mapper) properties to derive the agricultural parcels accurately. The proposed approach has been evaluated by considering aerial images collected from Gaofen-1 wide field of view (GF-1 WFV), Resource 1-02C (ZY1-02C), and Gaofen-2 (GF-2). The empirical outcomes show that the proposed approach has obtained an accuracy of 99.09%, 96.51%, and 84.42% on GF-1 WFV, GF-2, and ZY1-02C respectively. Table 1.3 presents the analysis of other studies carried out in agriculture.

1.3 Various approaches of image analysis

Image analysis refers to the process of removing noise and any sort of irregularities existing in the image utilizing a digital computer. The noise or irregularity may exist in the image either at the time of image formation or during image transformation. In mathematical terms, an image is represented as a two-dimensional function $f(x, y)$ y), where x and y represent the spatial coordinates and the amplitude f at any pair of coordinates (x, y) represents the intensity of the image at that point. The image is called a digital image when all the intensity values of f and (x, y) coordinates are finite and discrete quantities. In general, a digital image is made up of a finite number of elements known as pixels in which every pixel has a specific value and location. Nowadays, image processing systems are gaining more popularity in a variety of application domains such as industrial automation, remote sensing, medical image analysis, robotics, forensics, and so on due to the availability of advanced computing technology [50]. The primary goal of image processing is to provide a description and interpretation of the scene by extracting essential features from the image. Basically, the process of image analysis involves the study of feature extraction, segmentation, and classification approaches, as shown in Fig. 1.2 [51].

TABLE 1.3 Analysis of other pattern recognitions studies performed in agriculture.

Author & Year	Objective	Dataset	Approach	Results	Observations	Ref
Bhagat et al. 2022	To obtain segmentation of agriculture patterns from aerial images	Agriculture vision dataset consisting of 21 061 aerial farmland images	Efficient-NetB7	Attains mean dice score of 74.78 for RGB, 68.11 for NIR, and 84.23 for RGB + NIR images	Only a few patterns were considered	[45]
Ekramirad et al. 2021	For the detection of apples infested by codling moth pest	Sample apples processed in Department of Entomology, University of Kentucky, USA	Vibro-acoustic signal monitoring system	Accuracy of match filtering and ensemble approach is 100%	Did not consider the duration of signal	[46]
Tenhunen et al. 2019	For automatic recognition of cereal rows from a set of aerial photographs	Set of real-time color photographs taken from crop field using drones	Clustering based approach	Accuracy= 94%	Limited to finding nearly parallel, roughly straight rows of plants, occurring in a window of restricted size	[47]
Rothe et al. 2015	For the detection of cotton leaf disease	Images captured from the fields at Central Institute of Cotton Research Nagpur, and the cotton fields in Buldana and Wardha district	Active contour model	Accuracy= 85%	Requires longer training and testing time	[48]
Ewida et al. 2014	For automatic grading of palm date fruit	570 date samples (305 collected from local market and 265 samples dried by the researchers)	Fuzzy logic algorithm	Accuracy= 99.8%	Few samples were considered	[49]

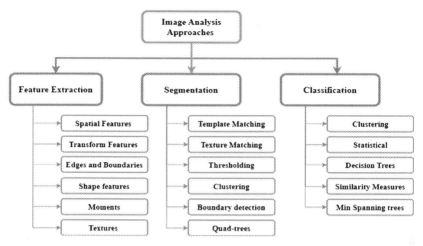

FIGURE 1.2   Various approaches of image analysis.

1.3.1   Feature extraction

In image processing systems, features play a critical role in the identification of related information. Feature extraction is the process of determining the set of parameters that describe the shape of an object accurately and uniquely. The primary objective of the feature extraction approach is to acquire the most pertinent information that characterizes the object from the raw data and present that information in a reduced dimensionality area. Each feature extracted in the feature extraction process forms a feature vector that is then utilized by the classifiers to identify the input unit with the intended output unit. Several studies show that much work has been done by researchers to design the best feature extraction approaches in order to acquire more relevant information from the objects [52–54]. To classify plants into relevant taxonomies, an automatic plant recognition system that is capable of recognizing plants from their leaf images has been developed by Munisami et al. [55]. Initially, the images of leaves were collected and uploaded in the server using a mobile application. Before obtaining potential matches of the image using the pattern matcher, the server applies preprocessing and feature extraction approaches on the image. The distinct features that are obtained using feature processing approaches are length, width, area, and perimeter of leaf, color histogram, a distance map, a centroid based radial distance map, hull perimeter, and hull area. After extracting these features, the authors conducted experiments on 640 leaves associated with 32 species and the results reveal that the suggested system has attained an accuracy of 83.5%. Further, the recognition accuracy of the system has been enhanced to 87.3% by utilizing the information acquired from the color histogram. Chetoui et al. [56] proposed a novel approach that makes use of a SVM (Support Vector Machine) and two distinct texture feature extraction approaches known as LTP (Local

Ternary Pattern) and LESH (Local Energy-based Shape Histogram) for the automatic detection of diabetic retinopathy. The features that were extracted using LTP and LESH approaches have been classified using SVM. The authors used a histogram binning scheme for representing features. Further, experiments were conducted on 1200 retinal fundus images and outcomes indicate that LESH approach obtained better accuracy and AUC of 90.4% and 93.1%, respectively, using SVM with RBF (radial basis function) kernel. The review of other studies carried out using feature extraction approaches is represented in Table 1.4.

1.3.2 Segmentation

In computer vision, image segmentation is a crucial phase as it is used to represent a digital image in a more recognizable form. Image segmentation is the process of partitioning an image into essential regions in order to obtain more relevant information regarding that region. Due to its importance in digital image processing, researchers are constantly working to optimize image segmentation approaches in order to make images smoother and noise free. For efficient detection of brain tumors, a novel approach that makes use of an independent component analysis-linear discriminate analysis algorithm (ICA-LDA) with an Adaptive region-based histogram enhancement (ARHE) approach has been suggested by Saravanan et al. [62]. Further, the authors utilized a weighted average approach for the image fusion approach. In the pre-processing stage, an adaptive median filter and the ARHE approach are used for reducing the noise and enhancing the image. Then, the ICA approach is used for extracting the features from the image. Finally, for the classification of images as normal and abnormal, the LDA approach was utilized by the authors. The simulation results indicate that the ICA-LDA approach has obtained better performance in terms of accuracy, specificity, and sensitivity. For efficient analysis of two-dimensional medical images, a novel approach known as CE-Net (context encoder network) has been suggested by Gu et al. [63]. The suggested CE-Net approach captures more prominent features and retains more special information to perform segmentation of two-dimensional medical images. A feature encoder, context extractor, and feature decoder are the three major component of the proposed framework. In the proposed architecture, a pretrained ResNet block was used as the feature extractor and a DAC (dense atrous convolution) block and a RMP (residual multi-kernel pooling) block have been used as the context extractor model. Further, experiments have been performed using benchmark datasets and the results show that the suggested framework attains better segmentation in various tasks such as vessel detection, optic disc segmentation, lung segmentation, retinal optical coherence tomography layer segmentation, and cell contour segmentation when compared with the existing U-Net approach. Table 1.5 represents the analysis of other studies carried out using the segmentation approach.

TABLE 1.4 Analysis of other studies performed using the feature extraction approach.

Author & Year	Objective	Dataset	Approach	Results	Ref
Yang et al. 2019	To distinguish lesions of tumor images	500 ultrasound images from the network CANCER-CAPTAC-GBM database	Xception and DenseNet	Accuracy of Xception=99.01% and DenseNet= 99.16%	[57]
Berbar et al. 2018	For detection and classification of malignant masses in mammograms	54 normal and 237 malignant mass images	WaveletCT1, WaveletCT2 and ST-GLCM	Accuracy, sensitivity and AUC of ST-GLCM is 97.89%, 96.12%, and 0.8769 and WaveletCT2 is 97.84%, 97%, and 0.89331	[58]
Amitrano et al. 2018	To extract data from time-series images	50 COSMOSkyMed stripmap three meter resolution images acquired in HH polarization between 2010 and 2016	Object-based image analysis (OBIA) and self-organizing maps (SOM	Median of the object-based false alarm=0	[59]
Bahadure et al. 2017	Detection of Brain tumor from Magnetic Resonance images (MRI)	Digital Imaging and Communications in Medicine (DICOM) dataset and Brain Web dataset	Berkeley wavelet transformation (BWT) and Support vector Machine (SVM)	Accuracy=96.5% Specificity=94.2% Sensitivity=97.72%	[60]
Weimer et al. 2016	To detect defects in manufacturing scenarios	Dataset consisting of weakly supervised learning examples for industrial optical inspection	Convolutional neural network	Average accuracy= 99.2%	[61]

1.3.3 Classification

The classification task assists in taking appropriate decisions in our daily life. The requirement for classification emerges whenever an object is placed in a particular group based on the attributes identical to that object. Due to the advancement in digital imaging, numerous images are produced in day-to-day life. These images need to be classified into various groups in order to ac-

TABLE 1.5 Other studies performed using the segmentation approach.

Author & Year	Objective	Dataset	Approach	Results	Ref
Huang et al. 2020	For efficient representation of lesion region and to obtain optimal features	103 patients MRI images from Department of Radiology, Guangzhou Panyu Central Hospital, Guangzhou, China	DSFR (Deep segmentation feature-based Radiomics) approach + SVM classifier	Accuracy on clinical task of deep venous thrombosis (DVT) is 97.6% and pancreatic neuroendocrine neoplasms (pNENs) is 92.9%	[64]
Wang et al. 2019	For analyzing root images from complex soil background	3824 soybean root images from greenhouse in the Department of Botany and Plant Pathology, Purdue University	SegRoot approach based on CNN	R^2 value=0.9791	[65]
Jamil et al. 2019	For early identification of melanoma from dermoscopy images	100 dermoscopy images from a European dataset	HSV color approach	True detection rate is 97.25%, False Positive rate is 3.51%, and error probability is 3.01%	[66]
Skourt et al. 2018	To perform segmentation of Lung CT images for early detection of lung cancer	LIDC (Lung Image Database Consortium) image dataset	U-net architecture	Dice coefficient index=0.952	[67]
Chen et al. 2018	To perform multi-class segmentation of brain cerebrospinal fluid (CSF) in CT images	781 2D CT image slices were collected from 133 stroke patients from local hospitals	Dense-Res-Inception Net	Dice coefficient of Pancreas, Kidney, Liver and spleen dataset is 83.42%, 95.96%, 96.57%, and 95.64%, respectively	[68]

cess them easily and quickly. The major objective of the classification task is the accurate finding of features in an image. The classification becomes more complex if the image contains noisy and hazy content. If an image includes multiple objects, then classification is quite a challenging task. Over the last few decades, several researchers have been continuously working on designing advanced classification approaches for enhancing the classification accuracy [69,70]. Several studies have also been carried out in selecting appropriate classification approaches in image analysis [71,72]. For accurate identification of cervical cancer, a framework based on a deep learning approach has been proposed by Zhang et al. [73] for the classification of cervical lesions. The proposed framework consists of two steps. In the first step, lesions in the cervical images were segmented using a convolutional neural network. In the second step, CapsNet was utilized for the accurate identification and classification of cervical images. The proposed framework has been evaluated on a dataset consisting of 6692 cervical lesion images and the outcomes indicate that the developed CapsNet approach has attained a better testing accuracy of 80.1% when compared with other standard approaches. For the efficient classification of hyperspectral images, a generative adversarial network (GAN) has been proposed by Zhu et al. [74]. In the proposed GAN, two convolutional neural networks (CNN) are used. One CNN is used for discriminating the inputs and another is used for generating fake inputs. In addition, the proposed GAN consists of 1D-GAN and 3D-GAN as a spectral and a spatial-spectral classifier. Then, the proposed approach was evaluated using three standard hyperspectral datasets and the results indicate that the proposed GAN model attains better results in comparison with state-of-the-art approaches. The analysis of other studies carried out in image processing using the classification approach is represented in Table 1.6.

1.4 Fractal based image analysis

In recent years, fractal-based pattern recognition has been gaining significant importance among researchers because of its capability to handle feature information in a complex environment. In 1975, the fractal theory was introduced by B.B. Mandelbrot as a new area of modern mathematical theory [80,81]. The fractal dimension that is used to describe the complexity of a fractal set is considered as the vital parameter of fractal theory. In the field of image processing and pattern recognition, fractal dimensions are basically utilized for the purpose of image compression, segmentation of textures, and extraction of features.

1.4.1 Image compression using fractal features

In recent years, the concept of fractal image compression has gained more popularity and became one of the popular approaches among various modern image coding approaches as it results in high compression ratio and fast decoding of digital images [82,83]. In this approach, image compression has been performed using the Collage theorem and a recursive iterated function system [84]. Several research studies have been performed to show the effectiveness of fractal image compression [85–87]. To extract fractal features, an approach that makes use of various edge detection approaches along with the box-counting fractal image compression approach has been proposed by Basharan et al. [88]. The features that were extracted were given as an input vector to the classifiers for the recognition of partial discharge (PD) patterns. Then, the authors simulated artificially multiple PD sources in a HV laboratory to assess the performance

TABLE 1.6 Review of other studies performed using the classification approach.

Author & Year	Objective	Dataset	Approach	Results	Ref
Ragab et al. 2021	For accurate classification of breast cancer lesions in mammograms	6560 samples from CBIS-DDSM (curated breast imaging subset of DDSM) and (MIAS) the mammographic image analysis society digital mammogram database	Pre-trained deep convolutional neural network and SVM classifier	Accuracy of CBIS-DDSM and MIAS dataset are 97.90% and 97.40%, respectively	[75]
Yan et al. 2020	For efficient classification of breast histopathological images	3771 breast cancer histopathological images	Hybrid convolutional and recurrent deep neural network	Average accuracy= 91.3%	[76]
Dornik et al. 2018	Classification of soils using digital maps of topography	171 soil profiles from Romanian Soil Taxonomy System and World Reference Base for Soil Resources	Object-based image and Random Forest classifier	Attains overall accuracy of 58% that is 10% higher than optimized pixel-based map	[77]
Jain et al. 2018	To classify the hyperspectral images collected from satellites	Images collected from hyperspectral remote sensing datasets of Indian Pines and Pavia University	Support vector machine optimization using self-organizing maps	Overall accuracy= 95.46%	[78]
Kuo et al. 2016	To categorize 30 varieties of rice grains	Samples were collected from Genetic Stocks Oryza germplasm collection, Agricultural Research Service, United States Department of Agriculture	Sparse representation-based classification	Overall accuracy= 89.1%	[79]

of the suggested approach and the results indicate that the suggested sigmoidal kernel-based multi-class SVM shows better recognition results when compared with other approaches. To overcome the drawback of fractal image compression, an approach known as the Tree Seed Algorithm was suggested by Muneeswaran et al. [89]. This approach is particularly utilized for the optimization of encoding process in order to make efficient utilization of the resources. The suggested approach has been evaluated and the empirical results indicate the enhancement of the algorithm under various facets, such as pressure proportion, encoding time, and MSE (Mean Square Error).

1.4.2 Texture segmentation using fractal features

In the applications of image processing, texture plays a vital role in the configuration and analysis of natural images. The application of image processing includes segmentation of texture regions, automatic classification of remote sensing data and segregation and analysis of medical images. As traditional approaches in texture segmentations depend on individual pixel values, these approaches result in over segmented images. Nowadays, fractal dimension has become one of the most popular tools in texture analysis because of its robustness in image scaling. Sarafrazi et al. [90] proposed a novel feature to perform efficient texture segmentation in remote sensing image processing. This approach makes use of the Contourlet transform and Fractal analysis to perform texture segmentation. Further, the performance of the suggested feature has been compared with other statistical, power spectrum, fuzzy, and fractal features and the results indicate that the suggested feature has obtained better performance in the presence of noise. A novel approach based on a non-Euclidean method has been suggested by Pant [91] for determining high dimensional objects. In the proposed approach, the structure of the object has been inferred using fractal features as the fractal dimension varies in the range 2.0 to 3.0. Then, the image texture attained from the fractal map has been utilized for the segregation of the objects. Then, the evaluation of segmentation has been performed and the outcomes show that the fractal dimension values have resulted in the efficient segmentation of texture images. Yousif et al. [92], have proposed a set of five fractal features for the analysis of texture images. These features make use of two clustering algorithms known as k-means and DBSCAN (Density Based Scan) to cluster the obtained fractal features into a relevant group. In the k-means algorithm, the reasonable value of k is obtained by using two statistical algorithms known as Elbow and Silhouette. Further, the proposed algorithms have been evaluated and the experimental results show that k-means obtained better performance than the DBSCAN.

1.4.3 Feature extraction using fractal dimensions

Feature extraction is the process of simplifying data representation by decreasing its dimensionality while obtaining relevant features that are appropriate for the task. Feature extraction is a crucial process as it has a large impact on the speed and classification accuracy. As fractal dimensions are used to describe irregular shapes and complex objects of nature, it has been used in various domains to characterize such complex objects [93,94]. To enhance the performance of two distinct speech recognition systems, Ezeiza et al. [95] combined the fractal dimensions of the time series with the standard Mel frequency cepstral coefficients (MFCCs) in the feature vector. To enhance sleep identification performance, Finotello et al. [96] proposed a novel approach that makes use of fractal dimension for extracting relevant features from the EEG (electroencephalography) signals. For accurate classification of heart sound signals, two feature extraction approaches were proposed by Hamidi et al. [97]. The first approach makes use of curve fitting to obtain the information present in the heart sound signal sequence, while the second approach combines the features extracted using Mel frequency cepstral coefficients with the fractal features. Then, the proposed approach was evaluated using six different datasets and the results reveal that the suggested approach obtains better performance than Filter banks and wavelet transforms.

1.4.4 Intelligent approaches for analysis of fractal features

As vision is the most important sense in human beings, images play an essential role in human life. Nowadays, a vast number of images are generated in day-to-day life because of the advancement in digital image processing approaches. In order to provide accurate identification and analysis of image, the selection of appropriate features needs to be carried out with utmost care. As traditional image processing approaches are not enough to cope with high throughput image data, machine learning approaches have emerged as a new research area in image processing to cope with the complex problems. Shi et al. [98] have performed a systematic research study on signal modulation using fractal dimension. The authors have extracted Box, Higuchi, Petrosian, Sevcik, and Katz fractal dimensions from eight distinct signal modulations for recognizing signal patterns. Simultaneously, noise robustness and computational intricacy of five distinct fractals has been evaluated using an anti-noise function and running time, respectively. Finally, the classification of distinct modulation signals depending on their fractal features has been performed using random forest, grey relation analysis, k-nearest neighbor, and a back propagation neural network and the results indicate that random forest obtained a better recognition performance when compared with other standard approaches. To perform effective classification of Chinese medical plants, analysis of a machine learning approach based on fractal dimension, visible/near infrared spectroscopy, and a morpho-colorimetric method has been performed by Xue et al. [99]. In this approach, automated image analysis and visible/nearby spectroscopy features acquired from leaves are given as inputs to the artificial neural network (ANN) for classification of medical plants. Then, the performance of the model has been validated using 20 medicinal plants collected from Chinese medicinal plant collection site in the Northwest A&F University. The results reveal that the ANN approach based on the morpho-colorimetric approach has obtained an accuracy of 98.3%, whereas the ANN approach that makes use of visible/near infrared spectroscopy has obtained an accuracy of 92.5%. From the results, it can be observed that ANN based on the morpho-colorimetric approach can be used for the non-destructive evaluation of leaves, while the ANN approach based on visible/near infrared spectroscopy can be used for the non-intrusive detection of Chinese medicinal plants.

Even though machine learning approaches have been widely used, these approaches still require substantial domain expertise and sophisticated engineering knowledge. Due to the availability of a sufficient number of annotated images, deep learning approaches have exhibited excellence performance over machine learning approaches in the analysis of images using fractal features. To expand the invariance of fractal dimension to filtering operations, a novel FDIF (fractal dimension invariant filtering) approach has been suggested by Xu et al. [100]. Initially, a local fractal model for images has been developed by the authors. In addition to anisotropic filter banks, a nonlinear postprocessing step has been added to retain the invariance of the fractal dimension of an image. Simultaneously, the model has been reimplemented using a convolutional neural network where anisotropic structures of image are extracted using convolutional layer and the structure has been enhanced using a nonlinear layer. The model has been evaluated and the experimental results indicate that the suggested approach can efficiently detect intricate curves from the texture-like images when compared with the standard approaches. To enhance the assessment of vascular complexity in cine-angiography images of patients affected by PAOD (peripheral artery occlusive diseases), an automatic deep learning segmentation approach based on fractal dimensions has been proposed by Bruno et al. [101]. The suggested approach comprises of three major steps. The first step involves changing of cine-angiographies to single static images with a broader field of view. The automatic segmentation of vascular tissues using a deep learning approach has been done in the second step and the third step involves the assessment and calculation of fractional dimensions of an image. The suggested approach has been evaluated and results indicate that the assessment of vascular complexity has been enhanced by the segmentation of a vascular tree from images of cine-angiography. In addition, the approach is also used to minimize the interobserver variability in assessing the vascular complexity in cine-angiography images. Moreover, the results also reveal the efficacy of fractal dimension in assessing the vascular tree complexity. The analysis of other studies carried on the analysis of fractal features using intelligent approaches is depicted in Table 1.7.

1.5 Critical analysis

A systematic analysis of research performed on fractal-based image analysis with pattern recognition using intelligent approaches has been presented in this study. In this section, a systematic analysis of a number of articles published

TABLE 1.7 Other studies performed on fractal image analysis using intelligent approaches.

Author & Year	Objective	Dataset	Approach	Results	Ref
Moldovanu et al. 2021	For efficient recognition and classification of skin cancer based on fractal dimensions and statistical color cluster features	248 nevi and 407 melanoma images collected from 7-point, PH2 and Med-Node datasets	RBFNN (Radial basis function neural network) classifier	RBFNN obtained an accuracy of 95.42%, 94.88%, and 94.71% on 7-point, PH2 and Med-Node datasets, respectively	[102]
Ahmadi et al. 2021	For the detection of brain tumor using fractal features	200 MRI images	Fuzzy logic and Wavelet based neural network	Accuracy=100%	[103]
Hu et al. 2021	For efficient prediction of dry weight, fresh weight, and plant height of rice using fractal dimensions	424 RGB images from potting farm of Huazhong Agricultural University	Multiple regression Model	Multiple regression model obtained R^2 value of 0.8697, 0.8631, and 0.9196 for dry weight, fresh weight, and plant height prediction model, respectively	[104]
Lahmiri et al. 2019	For early diagnosis of Alzheimer's disease	70 images from ADNI (Alzheimer's Disease Neuroimaging Initiative) dataset	SVM	100% Accuracy, sensitivity, and specificity	[105]
Fuentes et al. 2018	For automatic categorization of leaves of 16 grapevine cultivars	16 different grapevine cultivars from vineyard located in Palma de Mallorca	ANN based on visible/near infrared spectroscopy and morpho-colorimetric	Accuracy of ANN with morpho-colorimetric approach=94%	[106]
Yazgaç et al. 2022	For the classification of sound signals	UrbanSound8k dataset consisting of 8723 labels	CNN + data augmentation approaches based on fractional-order-calculus	Enhanced classification accuracy by 8.5%	[107]

continued on next page

TABLE 1.7 (*continued*)

Author & Year	Objective	Dataset	Approach	Results	Ref
Roberto et al. 2021	For the classification of histological images	Histological tissues of breast, colorectal, non-Hodgkin lymphoma and liver from UCSB, CR, NHL, LG, and LA datasets, respectively	Fractal geometry and convolutional neural network	Attained accuracy of 95.55%, 99.39%, 89.66%, 99.62%, and 99.62% on NHL, CR, UCSB, LG, and LA datasets, respectively	[108]
Nisa et al. 2020	To classify malware images efficiently	Malimg dataset	SFTA (segmentation-based fractal texture analysis) + AlexNet and Inception V3	Obtained 99.3% accuracy	[109]
Chatra et al. 2019	Used for the classification of texture images	Texture images from textured surfaces and KTH-TIPS datasets	Binary Dragon Fly with Deep Neural Network (BDADNN)	BDADNN attained better accuracy in all k cross validations on both textured surfaces and KTH-TIPS datasets	[110]
Ning et al. 2017	Estimation of human pose	Images from MPII Human Pose and extended Leeds Sports Poses (LSP) dataset	Deep fractal neural networks	Attained 91.2 and 92.3 percentage correct keyscore (PCK)@0.5 score on MPII and LSP datasets	[111]

in distinct applications of pattern recognition using various approaches of pattern recognition and image processing is presented to show the significance of pattern recognition approaches in different application domains. Further, a systematic analysis of the growth of publications in pattern recognition using fractal features is also presented. It is observed from the analysis that the need for developing automatic pattern recognition systems has been rising rapidly due to the advancement in digital image processing approaches. Moreover, the development of pattern recognition systems using fractal features has been gaining significant importance among researchers because of its capability of handling feature information in complex environment.

1.5.1 Percentage of work carried out using various pattern recognition approaches

Fig. 1.3 depicts the distinct approaches used in the development of efficient pattern recognition systems. It is observed from the figure that 48% of the work has been carried out in the development of pattern recognition systems using ANN technology. Next, 43% of the work has been performed in pattern recognition using a statistical classification approach. It is also observed from the figure that the least amount of work i.e., 6% and 3% in pattern recognition system has been accomplished using syntactic and template matching approaches.

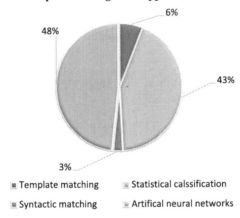

Percentage of research work carried out using various pattern recognition approaches

- Template matching
- Syntactic matching
- Statistical calssification
- Artifical neural networks

FIGURE 1.3 Percentage of work carried out using various pattern recognition approaches.

1.5.2 Application of pattern recognition approaches in various application domains

The application of pattern recognition approaches in various application domains is depicted in Fig. 1.4. From the figure, it is observed that the majority of work, i.e., 28% was published in the healthcare sector. Next, 25% of the work has been performed in character recognition systems. It is also observed from the figure that 18%, 14%, and 10% of the work has been carried out in speech recognition, agriculture, and manufacturing domains, respectively. Finally, least work, i.e., 2% has been performed in fingerprint identification and industrial automation systems.

1.5.3 Contribution of work using distinct approaches of image analysis

The advancement in computing technology has resulted in the development of more sophisticated image processing systems in various application domains

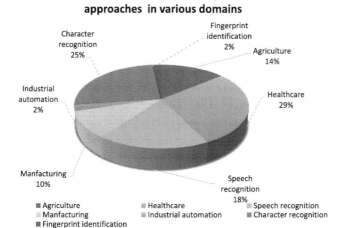

FIGURE 1.4 Application of pattern recognition in various domains.

such as medical image analysis, remote sensing, forensics, robotics, and so on. In general, image processing involves the study of feature extraction, segmentation, and classification approaches in order to provide an automatic analysis of the image. Fig. 1.5 depicts the work carried out using various approaches of image processing. From the figure, it is observed that the majority of the work, i.e., 42% has been performed on the classification of images. Next, 29% of the work has been performed in both feature extraction and segmentation of images.

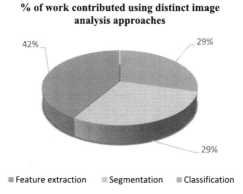

FIGURE 1.5 Percentage of work contributed using various image analysis approaches.

1.5.4 Growth of publications in fractal based image analysis

Fig. 1.6 displays the growth of publications in fractal based image analysis from 2015 to October, 2022. From the figure, it is observed that the number of publications is increasing from year 2015 to 2021. As we considered publications up to October 2022 only, a slight decrease in the number of publications has been observed in that year. The growth of publications is mainly due to fractal features that allows structure quantification in both spatial and temporal scales. Due to this feature, fractal based image analysis has been extensively used in distinct applications.

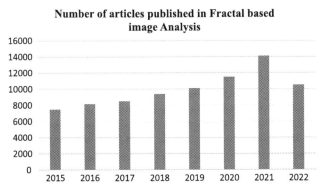

FIGURE 1.6 Growth of articles in fractal based image analysis.

1.6 Conclusion

In the present era of artificial intelligence, the advancement in digital image processing approaches has resulted in the generation of a large number of images in day-to-day human life. This drives the demand for developing intelligent pattern recognition systems in order to provide accurate identification and analysis of images. In recent years, fractal-based pattern recognition has been gaining significant importance among researchers as fractal features can be used to describe irregular patterns and complex objects of nature. This study presents a detailed review of intelligent fractal based image analysis with pattern recognition. From the study it is observed that the need for developing intelligent pattern recognition systems has been rising rapidly due to the advancement in various pattern recognition and digital image processing approaches. Moreover, the study also provides a critical investigation of various pattern recognition and image processing approaches used in the fractal analysis of images that may assist researchers in developing more sophisticated systems in various application domains.

References
[1] Yanfei Zhong, et al., Computational intelligence in optical remote sensing image processing, Applied Soft Computing 64 (2018) 75–93.
[2] Muhammad Imran Razzak, Saeeda Naz, Ahmad Zaib, Deep learning for medical image processing: overview, challenges and the future, in: Classification in BioApps, 2018, pp. 323–350.

[3] Shivam Thakur, et al., Autonomous farming—visualization of image processing in agriculture, in: Inventive Communication and Computational Technologies, Springer, Singapore, 2020, pp. 345–351.

[4] Komal Saini, Shabnampreet Kaur, Forensic examination of computer-manipulated documents using image processing techniques, Egyptian Journal of Forensic Sciences 6 (3) (2016) 317–322.

[5] Menbere Kina Tekleyohannes, et al., i DocChip: a configurable hardware accelerator for an end-to-end historical document image processing, Journal of Imaging 7 (9) (2021) 175.

[6] B.B. Mandelbrot, Ebookstore Release Benoit B Mandelbrot Fractals: Form, Chance and Dimension Prc, Freeman, 1977, p. 365.

[7] Matteo Baldoni, et al., Towards automatic fractal feature extraction for image recognition, in: Feature Extraction, Construction and Selection, Springer, Boston, MA, 1998, pp. 357–373.

[8] Taoi Hsu, Kuo-Jui Hu, Multi-resolution texture segmentation using fractal dimension, in: 2008 International Conference on Computer Science and Software Engineering, vol. 6, IEEE, 2008.

[9] Ting Cao, et al., Crack image detection based on fractional differential and fractal dimension, IET Computer Vision 13 (1) (2019) 79–85.

[10] C. Khotimah, D. Juniati, Iris recognition using feature extraction of box counting fractal dimension, Journal of Physics. Conference Series 947 (1) (2018).

[11] Xiao Tian, Hugh Daigle, Han Jiang, Feature detection for digital images using machine learning algorithms and image processing, in: SPE/AAPG/SEG Unconventional Resources Technology Conference, OnePetro, 2018.

[12] Mustafa Merchant, et al., Mango Leaf Deficiency Detection Using Digital Image Processing and Machine Learning, 2018 3rd International Conference for Convergence in Technology (I2CT) IEEE, 2018.

[13] J. Rodellar, et al., Image processing and machine learning in the morphological analysis of blood cells, International Journal of Laboratory Hematology 40 (2018) 46–53.

[14] Budiarianto Suryo Kusumo, et al., Machine learning-based for automatic detection of corn-plant diseases using image processing, in: 2018 International Conference on Computer, Control, Informatics and Its Applications (IC3INA), IEEE, 2018.

[15] Sharmila Nageswaran, et al., Lung cancer classification and prediction using machine learning and image processing, BioMed Research International 2022 (2022).

[16] Kuniaki Noda, et al., Audio-visual speech recognition using deep learning, Applied Intelligence 42 (4) (2015) 722–737.

[17] Heang-Ping Chan, Lubomir M. Hadjiiski, Ravi K. Samala, Computer-aided diagnosis in the era of deep learning, Medical Physics 47 (5) (2020) e218–e227.

[18] Batuhan Balci, Dan Saadati, Dan Shiferaw, Handwritten text recognition using deep learning, in: CS231n: Convolutional Neural Networks for Visual Recognition, Stanford University, Course Project Report, Spring, 2017, pp. 752–759.

[19] Yankang Jing, et al., Deep learning for drug design: an artificial intelligence paradigm for drug discovery in the big data era, The AAPS Journal 20 (3) (2018) 1–10.

[20] Puja S. Prasad, et al., Deep learning based representation for face recognition, in: ICCCE 2019, Springer, Singapore, 2020, pp. 419–424.

[21] Anil K. Jain, Robert P.W. Duin, Jianchang Mao, Statistical pattern recognition: a review, IEEE Transactions on Pattern Analysis and Machine Intelligence 22 (1) (2000) 4–37.

[22] Hsi-Ho Liu, K.S. Fu, A syntactic pattern recognition approach to seismic discrimination, Geoexploration 20 (1–2) (1982) 183–196.

[23] John Edward Albus, et al., Syntactic Pattern Recognition, Applications, vol. 14, Springer Science & Business Media, 2012.

[24] Rajan Dharwal, Loveneet Kaur, Applications of artificial neural networks: a review, Indian Journal of Science and Technology 9 (47) (2016) 1–8.

[25] Umut Asan, Secil Ercan, An introduction to self-organizing maps, in: Computational Intelligence Systems in Industrial Engineering, Atlantis Press, Paris, 2012, pp. 295–315.

[26] Enas M.F. El Houby, Framework of computer aided diagnosis systems for cancer classification based on medical images, Journal of Medical Systems 42 (8) (2018) 1–11.

[27] Ronald M. Summers, Deep learning and computer-aided diagnosis for medical image processing: a personal perspective, in: Deep Learning and Convolutional Neural Networks for Medical Image Computing, Springer, Cham, 2017, pp. 3–10.

[28] Shihui Chen, et al., Research progress of computer-aided diagnosis in cancer based on deep learning and medical imaging, Sheng Wu Yi Xue Gong Cheng Xue Za Zhi = Journal of Biomedical Engineering = Shengwu yixue gongchengxue zazhi 34 (2) (2017) 314–319.

[29] Usha Desai, et al., Automated diagnosis of coronary artery disease using pattern recognition approach, in: 2017 39th Annual International Conference of the IEEE Engineering in Medicine and Biology Society (EMBC), IEEE, 2017.

[30] Chen Gong, Xingchen Zhou, Yunyun Niu, Pattern recognition of epilepsy using parallel probabilistic neural network, Applied Intelligence 52 (2) (2022) 2001–2012.

[31] Melissa Min-Szu Yao, et al., End-to-end calcification distribution pattern recognition for mammograms: an interpretable approach with GNN, Diagnostics 12 (6) (2022) 1376.

[32] Rabia Javed, et al., An efficient pattern recognition based method for drug-drug interaction diagnosis, in: 2021 1st International Conference on Artificial Intelligence and Data Analytics (CAIDA), IEEE, 2021.

[33] George A. Dominguez, et al., Detecting prostate cancer using pattern recognition neural networks with flow cytometry-based immunophenotyping in At-risk men, Biomarker Insights 15 (2020) 1177271920913320.

[34] Seong-Hoon Kim, Zong Woo Geem, Gi-Tae Han, A novel human respiration pattern recognition using signals of ultra-wideband radar sensor, Sensors 19 (15) (2019) 3340.

[35] Dina M.R. Dakhly, et al., Diagnostic value of the International Ovarian Tumor Analysis (IOTA) simple rules versus pattern recognition to differentiate between malignant and benign ovarian masses, International Journal of Gynaecology and Obstetrics 147 (3) (2019) 344–349.

[36] Li Deng, Navdeep Jaitly, Deep discriminative and generative models for speech pattern recognition, in: Handbook of Pattern Recognition and Computer Vision, 2016, pp. 27–52.

[37] Kittisak Jermsittiparsert, et al., Pattern recognition and features selection for speech emotion recognition model using deep learning, International Journal of Speech Technology 23 (4) (2020) 799–806.

[38] Priscila Lima Rocha, Washington Luis Santos Silva, Artificial neural networks used for pattern recognition of speech signal based on DCT parametric models of low order, in: 2016 IEEE 14th International Conference on Industrial Informatics (INDIN), IEEE, 2016.

[39] Qiuqiang Kong, et al., Panns: large-scale pretrained audio neural networks for audio pattern recognition, IEEE/ACM Transactions on Audio, Speech and Language Processing 28 (2020) 2880–2894.

[40] Munna Khan, Mosarrat Jahan, Sub-vocal speech pattern recognition of Hindi alphabet with surface electromyography signal, Perspectives in Science 8 (2016) 558–560.

[41] Stavros Petridis, et al., End-to-end visual speech recognition for small-scale datasets, Pattern Recognition Letters 131 (2020) 421–427.

[42] Jinal H. Tailor, Dipti B. Shah, HMM-based lightweight speech recognition system for gujarati language, in: Information and Communication Technology for Sustainable Development, Springer, Singapore, 2018, pp. 451–461.

[43] Rodrigo Pérez-Zavala, et al., A pattern recognition strategy for visual grape bunch detection in vineyards, Computers and Electronics in Agriculture 151 (2018) 136–149.

[44] Dongsheng Liu, Ling Han, Semi-automatic extraction and mapping of farmlands based on high-resolution remote sensing images, International Journal of Pattern Recognition and Artificial Intelligence 36 (01) (2022) 2254002.

[45] Sandesh Bhagat, et al., MS-Net: a CNN architecture for agriculture pattern segmentation in aerial images, in: International Conference on Computer Vision and Image Processing, Springer, Cham, 2022.

[46] Nader Ekramirad, et al., Development of pattern recognition and classification models for the detection of vibro-acoustic emissions from codling moth infested apples, Postharvest Biology and Technology 181 (2021) 111633.

[47] Henri Tenhunen, et al., Automatic detection of cereal rows by means of pattern recognition techniques, Computers and Electronics in Agriculture 162 (2019) 677–688.

[48] P.R. Rothe, R.V. Kshirsagar, Cotton leaf disease identification using pattern recognition techniques, in: 2015 International Conference on Pervasive Computing (ICPC), IEEE, 2015.

[49] E.H. Ewida, et al., Computer application on pattern recognition for palm-date grading, Misr Journal of Agricultural Engineering 31 (2) (2014) 619–630.

[50] B. Chitradevi, P. Srimathi, An overview on image processing techniques, International Journal of Innovative Research in Computer and Communication Engineering 2 (11) (2014) 6466–6472.

[51] Aparna Vyas, Soohwan Yu, Joonki Paik, Fundamentals of digital image processing, in: Multiscale Transforms with Application to Image Processing, Springer, Singapore, 2018, pp. 3–11.

[52] Moacir Ponti, Tiago S. Nazaré, Gabriela S. Thumé, Image quantization as a dimensionality reduction procedure in color and texture feature extraction, Neurocomputing 173 (2016) 385–396.

[53] Jyotismita Chaki, Ranjan Parekh, Samar Bhattacharya, Plant leaf recognition using texture and shape features with neural classifiers, Pattern Recognition Letters 58 (2015) 61–68.

[54] Li-sheng Wei, Quan Gan, Tao Ji, Skin disease recognition method based on image color and texture features, Computational & Mathematical Methods in Medicine 2018 (2018).

[55] Trishen Munisami, et al., Plant leaf recognition using shape features and colour histogram with K-nearest neighbour classifiers, Procedia Computer Science 58 (2015) 740–747.

[56] Mohamed Chetoui, Moulav A. Akhloufi, Mustanha Kardouchi, Diabetic retinopathy detection using machine learning and texture features, in: 2018 IEEE Canadian Conference on Electrical & Computer Engineering (CCECE), IEEE, 2018.

[57] Aimin Yang, et al., Research on feature extraction of tumor image based on convolutional neural network, IEEE Access 7 (2019) 24204–24213.

[58] Mohamed A. Berbar, Hybrid methods for feature extraction for breast masses classification, Egyptian Informatics Journal 19 (1) (2018) 63–73.

[59] Donato Amitrano, et al., Feature extraction from multitemporal SAR images using selforganizing map clustering and object-based image analysis, IEEE Journal of Selected Topics in Applied Earth Observations and Remote Sensing 11 (5) (2018) 1556–1570.

[60] Nilesh Bhaskarrao Bahadure, Arun Kumar Ray, Har Pal Thethi, Image analysis for MRI based brain tumor detection and feature extraction using biologically inspired BWT and SVM, International Journal of Biomedical Imaging (2017) 2017.

[61] Daniel Weimer, Bernd Scholz-Reiter, Moshe Shpitalni, Design of deep convolutional neural network architectures for automated feature extraction in industrial inspection, CIRP Annals 65 (1) (2016) 417–420.

[62] S. Saravanan, R. Karthigaivel, V. Magudeeswaran, A brain tumor image segmentation technique in image processing using ICA-LDA algorithm with ARHE model, Journal of Ambient Intelligence and Humanized Computing 12 (5) (2021) 4727–4735.

[63] Zaiwang Gu, et al., Ce-net: context encoder network for 2d medical image segmentation, IEEE Transactions on Medical Imaging 38 (10) (2019) 2281–2292.

[64] Bingsheng Huang, et al., Deep semantic segmentation feature-based radiomics for the classification tasks in medical image analysis, IEEE Journal of Biomedical and Health Informatics 25 (7) (2020) 2655–2664.

[65] Tao Wang, et al., SegRoot: a high throughput segmentation method for root image analysis, Computers and Electronics in Agriculture 162 (2019) 845–854.

[66] Uzma Jamil, et al., Melanoma segmentation using bio-medical image analysis for smarter mobile healthcare, Journal of Ambient Intelligence and Humanized Computing 10 (10) (2019) 4099–4120.

[67] Brahim Ait Skourt, Abdelhamid El Hassani, Aicha Majda, Lung CT image segmentation using deep neural networks, Procedia Computer Science 127 (2018) 109–113.

[68] Liang Chen, et al., DRINet for medical image segmentation, IEEE Transactions on Medical Imaging 37 (11) (2018) 2453–2462.

[69] Yongming Rao, et al., Global filter networks for image classification, Advances in Neural Information Processing Systems 34 (2021) 980–993.

[70] Zeynep Akata, et al., Label-embedding for image classification, IEEE Transactions on Pattern Analysis and Machine Intelligence 38 (7) (2015) 1425–1438.

[71] Laleh Armi, Shervan Fekri-Ershad, Texture image classification based on improved local quinary patterns, Multimedia Tools and Applications 78 (14) (2019) 18995–19018.

[72] Adriana Romero, Carlo Gatta, Gustau Camps-Valls, Unsupervised deep feature extraction for remote sensing image classification, IEEE Transactions on Geoscience and Remote Sensing 54 (3) (2015) 1349–1362.

[73] XiaoQing Zhang, Shu-Guang Zhao, Cervical image classification based on image segmentation preprocessing and a CapsNet network model, International Journal of Imaging Systems and Technology 29 (1) (2019) 19–28.

[74] Lin Zhu, et al., Generative adversarial networks for hyperspectral image classification, IEEE Transactions on Geoscience and Remote Sensing 56 (9) (2018) 5046–5063.

[75] Dina A. Ragab, et al., A framework for breast cancer classification using multi-DCNNs, Computers in Biology and Medicine 131 (2021) 104245.

[76] Rui Yan, et al., Breast cancer histopathological image classification using a hybrid deep neural network, Methods 173 (2020) 52–60.

[77] Andrei Dornik, Lucian Drăguţ, Petru Urdea, Classification of soil types using geographic object-based image analysis and random forests, Pedosphere 28 (6) (2018) 913–925.

[78] Tzu-Yi Kuo, et al., Identifying rice grains using image analysis and sparse-representation-based classification, Computers and Electronics in Agriculture 127 (2016) 716–725.

[79] Deepak Kumar Jain, et al., An approach for hyperspectral image classification by optimizing SVM using self organizing map, Journal of Computational Science 25 (2018) 252–259.

[80] Yu-Dong Zhang, et al., Fractal dimension estimation for developing pathological brain detection system based on Minkowski-Bouligand method, IEEE Access 4 (2016) 5937–5947.

[81] Juan Pablo Amezquita Sanchez, et al., Detection of ULF geomagnetic anomalies associated to seismic activity using EMD method and fractal dimension theory, IEEE Latin America Transactions 15 (2) (2017) 197–205.

[82] Erjun Zhao, Dan Liu, Fractal image compression methods: a review, in: Third International Conference on Information Technology and Applications (ICITA'05), vol. 1, IEEE, 2005.

[83] Manish Joshi, Ambuj Kumar Agarwal, Bhumika Gupta, Fractal image compression and its techniques: a review, in: Soft Computing: Theories and Applications, 2019, pp. 235–243.

[84] M.F. Barnsley, L.P. Hurd, Fractal Image Compression, AK Peters. Ltd, Wellesley, 1993.

[85] R.A. Zubko, Image compression by fractal method, Eastern European Journal of Advanced Technology: Collection of Scientific Works 6 (2014) 23–28.

[86] Swalpa Kumar Roy, et al., Fractal image compression using upper bound on scaling parameter, Chaos, Solitons and Fractals 106 (2018) 16–22.

[87] Rafik Menassel, Idriss Gaba, Khalil Titi, Introducing BAT inspired algorithm to improve fractal image compression, International Journal of Computers & Applications 42 (7) (2020) 697–704.

[88] Vigneshwaran Basharan, Willjuice Iruthayarajan Maria Siluvairaj, Maheswari Ramasamy Velayutham, Recognition of multiple partial discharge patterns by multi-class support vector machine using fractal image processing technique, IET Science, Measurement & Technology 12 (8) (2018) 1031–1038.

[89] V. Muneeswaran, et al., Enhanced image compression using fractal and tree seed-bio inspired algorithm, in: 2021 Second International Conference on Electronics and Sustainable Communication Systems (ICESC), IEEE, 2021.

[90] Katayoon Sarafrazi, Mehran Yazdi, Mohammad Javad Abedini, A new image texture segmentation based on contourlet fractal features, Arabian Journal for Science and Engineering 38 (12) (2013) 3437–3449.

[91] T. Pant, Implementation of fractal dimension for finding 3D objects: a texture segmentation and evaluation approach, in: International Conference on Intelligent Interactive Technologies and Multimedia, Springer, Berlin, Heidelberg, 2013.

[92] S.A. Yousif, Arkan Jassim Mohammed, Nadia Mohammed Ghanim Al-Saidi, Texture images analysis using fractal extracted attributes, International Journal of Innovative Computing, Information & Control 16 (4) (2020).

[93] Yoshio Nakamura, Yoshiharu Yamamoto, Isao Muraoka, Autonomic control of heart rate during physical exercise and fractal dimension of heart rate variability, Journal of Applied Physiology 74 (2) (1993) 875–881.

[94] Umut Güçlü, Yağmur Güçlütürk, Chu Kiong Loo, Evaluation of fractal dimension estimation methods for feature extraction in motor imagery based brain computer interface, Procedia Computer Science 3 (2011) 589–594.

[95] Aitzol Ezeiza, et al., Enhancing the feature extraction process for automatic speech recognition with fractal dimensions, Cognitive Computation 5 (4) (2013) 545–550.

[96] Francesca Finotello, Fabio Scarpa, Mattia Zanon, EEG signal features extraction based on fractal dimension, in: 2015 37th Annual International Conference of the IEEE Engineering in Medicine and Biology Society (EMBC), IEEE, 2015.

[97] Maryam Hamidi, Hassan Ghassemian, Maryam Imani, Classification of heart sound signal using curve fitting and fractal dimension, Biomedical Signal Processing and Control 39 (2018) 351–359.

[98] Chang-Ting Shi, Signal pattern recognition based on fractal features and machine learning, Applied Sciences 8 (8) (2018) 1327.

[99] Jinru Xue, et al., Automated Chinese medicinal plants classification based on machine learning using leaf morpho-colorimetry, fractal dimension and visible/near infrared spectroscopy, International Journal of Agricultural and Biological Engineering 12 (2) (2019) 123–131.

[100] Hongteng Xu, et al., Fractal dimension invariant filtering and its CNN-based implementation, in: Proceedings of the IEEE Conference on Computer Vision and Pattern Recognition, 2017.

[101] Pierangela Bruno, et al., Assessing vascular complexity of PAOD patients by deep learning-based segmentation and fractal dimension, Neural Computing & Applications (2022) 1–8.

[102] Simona Moldovanu, et al., Skin lesion classification based on surface fractal dimensions and statistical color cluster features using an ensemble of machine learning techniques, Cancers 13 (21) (2021) 5256.

[103] Mohsen Ahmadi, et al., FWNNet: presentation of a new classifier of brain tumor diagnosis based on fuzzy logic and the wavelet-based neural network using machine-learning methods, Computational Intelligence and Neuroscience (2021) 2021.

[104] Yijun Hu, Jingfang Shen, Yonghao Qi, Estimation of rice biomass at different growth stages by using fractal dimension in image processing, Applied Sciences 11 (15) (2021) 7151.

[105] Salim Lahmiri, Amir Shmuel, Performance of machine learning methods applied to structural MRI and ADAS cognitive scores in diagnosing Alzheimer's disease, Biomedical Signal Processing and Control 52 (2019) 414–419.

[106] Sigfredo Fuentes, et al., Automated grapevine cultivar classification based on machine learning using leaf morpho-colorimetry, fractal dimension and near-infrared spectroscopy parameters, Computers and Electronics in Agriculture 151 (2018) 311–318.

[107] Bilgi Görkem Yazgaç, Mürvet Kırcı, Fractional-order calculus-based data augmentation methods for environmental sound classification with deep learning, Fractal and Fractional 6 (10) (2022) 555.

[108] Guilherme Freire Roberto, et al., Fractal neural network: a new ensemble of fractal geometry and convolutional neural networks for the classification of histology images, Expert Systems with Applications 166 (2021) 114103.

[109] Maryam Nisa, et al., Hybrid malware classification method using segmentation-based fractal texture analysis and deep convolution neural network features, Applied Sciences 10 (14) (2020) 4966.

[110] Kaveri Chatra, Venkatanareshbabu Kuppili, Damodar Reddy Edla, Texture image classification using deep neural network and binary dragon fly optimization with a novel fitness function, Wireless Personal Communications 108 (3) (2019) 1513–1528.

[111] Guanghan Ning, Zhi Zhang, Zhiquan He, Knowledge-guided deep fractal neural networks for human pose estimation, IEEE Transactions on Multimedia 20 (5) (2017) 1246–1259.

Chapter 2

Analysis of Mandelbrot set fractal images using a machine learning based approach

Kalyan Kumar Jena[a], Sourav Kumar Bhoi[a], and Soumya Ranjan Nayak[b]

[a]*Department of Computer Science and Engineering, Parala Maharaja Engineering College, Berhampur, India,* [b]*School of Computer Engineering, KIIT Deemed to be University, Bhubaneswar, Odisha, India*

2.1 Introduction

Fractal image processing [1–12,19,27,29–41,43,45,47–50,52–54,56,57] is an important research area for scientists and researchers. It has a wide variety of applications. A fractal is considered as a never-ending pattern and it has a non-regular geometric shape. Mathematically, a fractal is an everywhere continuous and nowhere differentiable curve. Different natural objects such as mountains, clouds, vegetables, etc., can be approximated using fractals. Fractals can be used to describe the behavior of a group of objects that act similar to each other, and they can also be used to model several structures. The fractals can be classified as self-similar, self-affine, and invariant fractals. ML [16,17,23–26,28] is an important application of artificial intelligence. It helps the systems to learn and improve from experience automatically without being programmed explicitly and it mainly focuses on the development of computer programs that can access data and use it to learn themselves. ML deals with supervised as well as unsupervised learning techniques. Classification techniques belong to supervised learning techniques. Different ML based classification techniques are adaboost, logistic regression, KNN, neural network, tree, random forest, naïve Bayes, etc. Clustering techniques [13–15,18,20–22] belong to unsupervised learning techniques. Different ML based clustering techniques are hierarchical clustering, distance map, MDS, distance matrix, t-SNE, DBSCAN, manifold learning, k-means, Louvain clustering, etc. Clustering techniques are mainly used to group data points with similar features. These techniques are used to group similar objects into a set that is referred to as a cluster. It is a challenging problem to accurately group all the data points with similar features using several clustering techniques. ML based techniques play an important role for the processing of several fractal images. The ML based clustering techniques can be used to

Copyright © 2024 Elsevier Inc. All rights reserved, including those for text and data mining, AI training, and similar technologies.

generate several clusters by processing different fractal images. In this work, several ML based clustering techniques such as hierarchical clustering, distance map, MDS, and distance matrix are discussed to process different MSFIs.

The main contributions of this work are stated as follows:

- An ML based approach is used for the processing of several MSFIs.
- An ML based approach focuses on several clustering techniques such as hierarchical clustering, distance map, MDS, and distance matrix for such processing.
- These clustering techniques are used to generate the MSFIs with several clusters.
- The performance of this work is carried out using Orange3-3.24.1.

The rest of the chapter is organized as follows. Section 2.2 describes related works, Section 2.3 describes the methodology for the processing of fractal images, Section 2.4 describes the results and discussion, and Section 2.5 outlines the conclusions.

2.2 Related works

Different studies have been carried out by several researchers and scientists on fractals as well as other images [1–12,19,27,29–57] and ML [13–18,20–26,28] for a wide variety of applications. Some of the studies are described as follows. Dhal et al. [1] focused on a stochastic fractal search for the segmentation of acute lymphoblastic leukemia image. Biswas et al. [3] focused on the fractal compression of several medical images by the help of classification methods based on fractal dimension. Kumar et al. [4] focused on a lossless image compression mechanism by the help of parallel fractal texture identification methodology. Shahrezaei et al. [10] focused on the fractal analysis as well as texture classification of synthetic aperture radar (SAR) sea-ice images by considering high frequency multiplicative noise on the basis of a transform domain image decomposition mechanism. Rashmi et al. [13] focused on the clustering of multiplane satellite images by using multithreading mechanism. Govender et al. [15] focused on a review on analysis of air pollution by applying k-means as well as hierarchical clustering techniques. Verbeeck et al. [17] focused on the exploratory data analysis in the case of imaging mass spectrometry by using an unsupervised ML approach. Karanja et al. [24] focused on the analysis of the internet of things (IoT) malware by the help of features of image texture as well as ML techniques. Reddy et al. [25] focused on the survey on food recognition and calorie measurement by the help of image processing as well as ML techniques. Diykh et al. [27] focused on a fractal dimension undirected correlation graph based support vector machine model to identify focal as well as non-focal electroencephalography signals. Joardar et al. [33] focused on the recognition of a thermal face by using an enhanced fractal dimension approach that is based on a feature extraction mechanism. Yin et al. [35] focused on the fractal dimension based analysis for seismicity, spatial as well as the temporal distribution, in the case of the circum pacific seismic belt. Joshi et al. [36] focused on a review of fractal image compression and its several techniques. Padmavati et al. [38] focused on the hardware implementation of fractal quadtree compression for several medical images. The review of some studies related to fractal, clustering, and ML is summarized in Table 2.1.

TABLE 2.1 Review of some studies related to fractal, clustering and ML.

Author	Key focus
Dhal et al. [1]	Acute lymphoblastic leukemia image segmentation using fractals
Biswas et al. [3]	Medical image compression using fractals
Kumar et al. [4]	Lossless image compression using fractals
Rashmi et al. [13]	Multiplane satellite image clustering using multithreading
Verbeeck et al. [17]	Imaging mass spectrometry analysis using ML
Karanja et al. [24]	IoT malware analysis using ML
Diykh et al. [27]	Identification of focal and non-focal electroencephalography signals using fractals
Joardar et al. [33]	Thermal face recognition using fractals
Yin et al. [35]	Seismicity spatial analysis using fractals
Padmavati et al. [38]	Fractal quadtree compression

2.3 Methodology

In this work, four ML based clustering techniques [16,17,23–26,28] such as hierarchical clustering, distance map, MDS, and distance matrix are used to carry out clustering mechanisms on MSFIs. Hierarchical clustering is used to generate the hierarchy of clusters. Distance map is used to display the embedding of cluster centers in two dimensions with the distance to other centers preserved. MDS is used to translate the information about the pair wise among a set of k objects into a configuration of k points mapped into an abstract Cartesian space. Distance matrix is used to show the distance between pairs of objects in a matrix format.

In this work, at first the MSFIs are given as input to the Orange3-3.24.1 [58]. Afterwards, an image embedding mechanism is carried out by taking input MSFIs as inputs to generate embeddings or skipped MSFIs as outputs. For image embedding, several embedders such as Inception v3, SqueezeNet (local), VGG-16, VGG-19, Painters, DeepLoc, or Openface can be used. However, in this work, SqueezeNet (local) is considered as embedded for image embedding purposes. Then, the distances will be calculated by considering image embedding data. For distance calculations, several distance metrics such as Cosine, Euclidean, Manhattan, Jaccard, Spearman, Absolute Spearman, Pearson, Absolute Pearson, Hamming, Mahalanobis, and Bhattacharyya can be used. In this work, Cosine is considered as a distance metric and the distance between rows is considered. Hierarchical clustering, distance map, MDS, and distance matrix techniques are used to generate clustered MSFIs by considering distance values. Hierarchical clustering takes distances as inputs and selected data or data as outputs. Distance map takes distances as inputs and selected data or data or features as outputs. The MDS technique takes data or a data subset or distances as inputs and selected data or data as outputs. Distance matrix takes distances as inputs and distances or selected data as outputs. The methodology is described in Fig. 2.1. The steps involved in the methodology of this work are summarized as follows:

Step 1: Input MSFIs.
Step 2: Perform image embedding mechanism by considering MSFIs.
Step 3: Calculate distances by considering image embedding mechanism.
Step 4: Apply hierarchical clustering, distance map, MDS, and distance matrix techniques to generate clustered MSFIs by considering distances values.

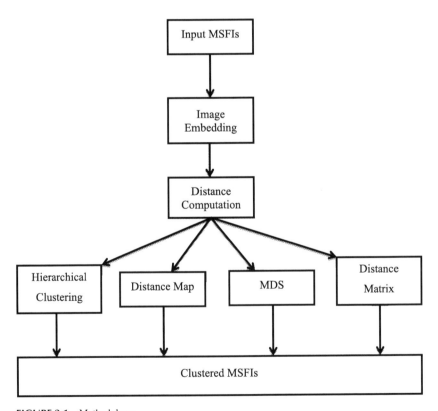

FIGURE 2.1 Methodology.

2.4   Results and discussion

Orange3-3.24.1 [58] is used for the simulation of this work. Several MSFIs with different sizes are taken from the source [59–79]. In this work, 20 MSFIs were taken for testing purposes that are mentioned in Figs. 2.2–2.6. The MSFIs are processed using ML based clustering techniques such as hierarchical clustering,

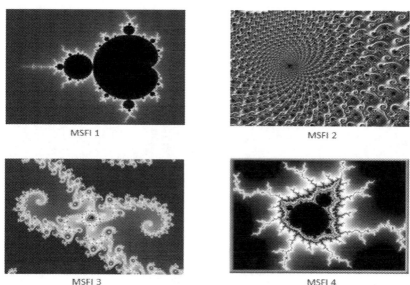

FIGURE 2.2   MSFI 1, MSFI 2, MSFI 3, MSFI 4 with size 259×194, 259×195, 259×194, 259×194, respectively.

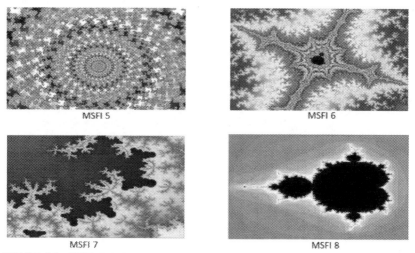

FIGURE 2.3   MSFI 5, MSFI 6, MSFI 7, MSFI 8 with size 275×183, 259×194, 247×204, 225×225, respectively.

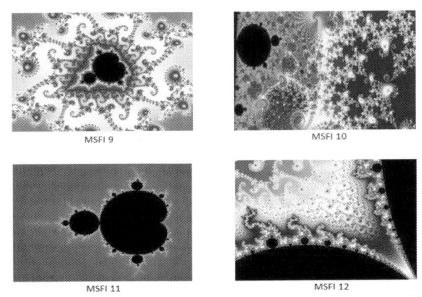

FIGURE 2.4 MSFI 9, MSFI 10, MSFI 11, MSFI 12 with size 259×194, 300×168, 261×193, 259×194, respectively.

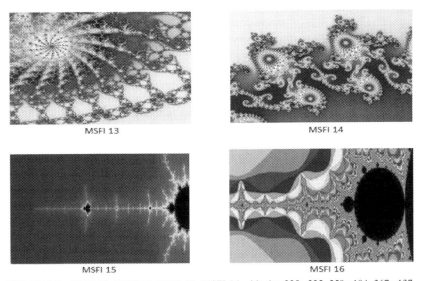

FIGURE 2.5 MSFI 13, MSFI 14, MSFI 15, MSFI 16 with size 225×225, 259×194, 367×137, 344×147, respectively.

distance map, MDS, and distance matrix. The output of hierarchical clustering, distance map, MDS, and distance matrix are shown in Figs. 2.7–2.10, respectively.

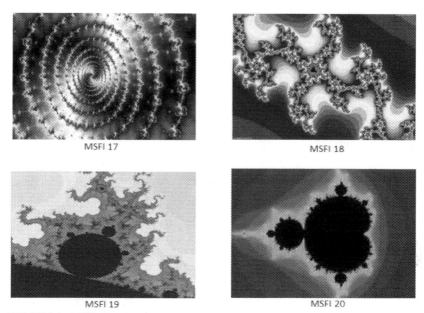

MSFI 17

MSFI 18

MSFI 19

MSFI 20

FIGURE 2.6 MSFI 17, MSFI 18, MSFI 19, MSFI 20 with size 300×168, 275×183, 247×204, 269×188, respectively.

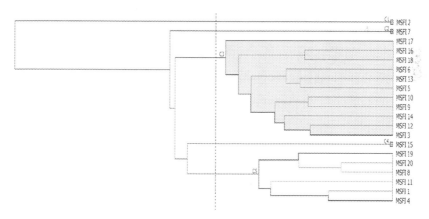

FIGURE 2.7 Output by applying the hierarchical clustering technique.

In this work, for hierarchical clustering the linkage is considered as average, for pruning the maximum depth is considered as 10 and the selection procedure is considered as a height ratio of 46.8%. For distance map, the element sorting is considered as clustering. For MDS and distance matrix, the default setting of Orange3-3.24.1 [58] is considered and for all these techniques image name is considered for annotations and labels purpose. Fig. 2.7 shows the hierarchy

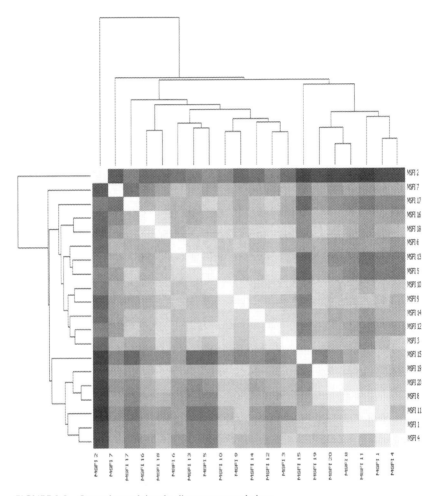

FIGURE 2.8 Output by applying the distance map technique.

of clusters of MSFIs (MSFI 1–MSFI 20) by applying hierarchical clustering technique. Fig. 2.8 displays the embedding of cluster centers in two dimensions with the distance to other centers preserved using the distance map technique. MDS is used to translate the information about those pair wise among a set of objects into a configuration of points mapped into an abstract Cartesian space that is shown in Fig. 2.9. Fig. 2.10 shows the distance between pairs of objects in a matrix format using the distance matrix technique. From Figs. 2.7–2.10, it is observed that the MSFIs are broadly classified into five clusters such C1, C2, C3, C4, and C5. C1 MSFI 2 belongs to cluster C1, MSFI 7 belongs to cluster C2, MSFI 17, MSFI 16, MSFI 18, MSFI 6, MSFI 13, MSFI 5, MSFI 10, MSFI 9, MSFI 14, MSFI 12, MSFI 12, and MSFI 3 belong to cluster C3, MSFI 15

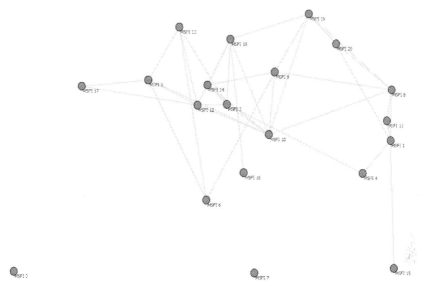

FIGURE 2.9 Output by applying the MDS technique.

	MSFI 1	MSFI 10	MSFI 11	MSFI 12	MSFI 13	MSFI 14	MSFI 15	MSFI 16	MSFI 17	MSFI 18	MSFI 19	MSFI 2	MSFI 20	MSFI 3	MSFI 4	MSFI 5	MSFI 6	MSFI 7	MSFI 8	MSFI 9
MSFI 1		0.118	0.058	0.193	0.256	0.160	0.116	0.186	0.248	0.159	0.124	0.399	0.103	0.136	0.056	0.257	0.194	0.167	0.067	0.129
MSFI 10	0.118		0.153	0.085	0.121	0.087	0.227	0.130	0.132	0.093	0.093	0.258	0.121	0.085	0.128	0.107	0.130	0.162	0.112	0.075
MSFI 11	0.058	0.153		0.236	0.283	0.211	0.160	0.210	0.273	0.209	0.153	0.421	0.126	0.224	0.156	0.280	0.196	0.215	0.089	0.117
MSFI 12	0.193	0.085	0.236		0.085	0.092	0.256	0.140	0.109	0.088	0.148	0.233	0.174	0.073	0.198	0.280	0.132	0.201	0.181	0.123
MSFI 13	0.256	0.121	0.283	0.085		0.128	0.308	0.175	0.157	0.132	0.168	0.266	0.233	0.108	0.257	0.082	0.091	0.167	0.223	0.138
MSFI 14	0.160	0.087	0.211	0.092	0.128		0.236	0.144	0.132	0.094	0.137	0.304	0.184	0.099	0.165	0.126	0.145	0.188	0.166	0.111
MSFI 15	0.116	0.227	0.160	0.256	0.308	0.236		0.231	0.317	0.233	0.255	0.405	0.224	0.228	0.141	0.311	0.204	0.244	0.185	0.255
MSFI 16	0.186	0.130	0.210	0.140	0.175	0.144	0.231		0.166	0.073	0.140	0.311	0.170	0.180	0.201	0.157	0.158	0.234	0.173	0.165
MSFI 17	0.248	0.132	0.273	0.109	0.157	0.132	0.317	0.166		0.142	0.206	0.268	0.230	0.175	0.249	0.104	0.166	0.291	0.242	0.184
MSFI 18	0.159	0.093	0.209	0.088	0.132	0.094	0.233	0.078	0.142		0.096	0.310	0.128	0.094	0.148	0.130	0.161	0.205	0.150	0.127
MSFI 19	0.124	0.093	0.153	0.148	0.168	0.137	0.255	0.140	0.206	0.096		0.387	0.079	0.118	0.148	0.162	0.158	0.176	0.087	0.093
MSFI 2	0.399	0.258	0.421	0.253	0.266	0.304	0.405	0.311	0.268	0.310	0.387		0.402	0.306	0.394	0.240	0.296	0.364	0.404	0.328
MSFI 20	0.103	0.121	0.126	0.174	0.233	0.184	0.224	0.170	0.230	0.128	0.079	0.402		0.147	0.130	0.217	0.202	0.213	0.045	0.122
MSFI 3	0.136	0.085	0.224	0.073	0.108	0.099	0.228	0.180	0.175	0.094	0.118	0.306	0.147		0.114	0.137	0.169	0.143	0.137	0.131
MSFI 4	0.056	0.128	0.156	0.198	0.257	0.165	0.141	0.201	0.249	0.148	0.148	0.394	0.130	0.114		0.262	0.218	0.178	0.117	0.163
MSFI 5	0.257	0.107	0.280	0.080	0.082	0.126	0.311	0.157	0.104	0.130	0.162	0.240	0.217	0.137	0.262		0.097	0.197	0.232	0.147
MSFI 6	0.194	0.130	0.196	0.132	0.091	0.145	0.204	0.158	0.166	0.161	0.158	0.296	0.202	0.169	0.218	0.097		0.144	0.199	0.176
MSFI 7	0.187	0.162	0.215	0.201	0.167	0.188	0.244	0.234	0.291	0.205	0.176	0.364	0.213	0.143	0.178	0.197	0.144		0.192	0.182
MSFI 8	0.067	0.112	0.089	0.181	0.223	0.166	0.185	0.173	0.242	0.150	0.087	0.404	0.045	0.137	0.117	0.232	0.199	0.190		0.113
MSFI 9	0.129	0.075	0.117	0.123	0.138	0.111	0.255	0.165	0.184	0.127	0.093	0.328	0.122	0.131	0.163	0.147	0.176	0.182	0.113	

FIGURE 2.10 Output by applying the distance matrix technique.

belongs to cluster C4, MSFI 19, MSFI 20, MSFI 8, MSFI 11, MSFI 1, and MSFI 4 belongs to cluster C5.

2.5 Conclusions

This chapter focuses on the processing of several MSFIs. The MSFIs are analyzed by applying ML based clustering techniques such as hierarchical clustering, distance map, MDS, and distance matrix. Each technique carries out the processing on MSFIs to generate the output with MSFIs that leads to different clusters. From the analysis of results, it is concluded that these clustering techniques provide similar outputs in terms of MSFIs with different clusters. This work can be extended to analyze the performance of these ML based clustering techniques along with other clustering techniques by focusing on several fractals and other images.

References

[1] K.G. Dhal, J. Gálvez, S. Ray, A. Das, S. Das, Acute lymphoblastic leukemia image segmentation driven by stochastic fractal search, Multimedia Tools and Applications (2020) 1–29.

[2] K. Jang, C. Russo, A. Di Ieva, Radiomics in gliomas: clinical implications of computational modeling and fractal-based analysis, Neuroradiology (2020) 1–20.

[3] A.K. Biswas, S. Karmakar, S. Sharma, Effectiveness of the fractal dimension based classification methods for fractal compression of medical images, in: 2020 First International Conference on Power, Control and Computing Technologies (ICPC2T), IEEE, 2020, pp. 155–160.

[4] R.S. Kumar, P. Manimegalai, Near lossless image compression using parallel fractal texture identification, Biomedical Signal Processing and Control 58 (2020) 101862.

[5] L. Ma, H. Zhang, M. Lu, Building's fractal dimension trend and its application in visual complexity map, Building and Environment (2020) 106925.

[6] D.B.V. Jagannadham, G.V.S. Raju, D.V.S. Narayana, Novel performance analysis of DCT, DWT and fractal coding in image compression, in: Data Engineering and Communication Technology, Springer, Singapore, 2020, pp. 611–622.

[7] M.J.S. Goh, Y.S. Chiew, J.J. Foo, A method for 3D reconstruction of net undulation for fluid structure interaction of fractal induced turbulence, IEEE Sensors Journal (2020).

[8] S.M. Abdullahi, H. Wang, T. Li, Fractal coding-based robust and alignment-free fingerprint image hashing, IEEE Transactions on Information Forensics and Security 15 (2020) 2587–2601.

[9] A.M.H. Saad, M.Z. Abdullah, N.A.M. Alduais, H.H.Y. Sa'ad, Impact of spatial dynamic search with matching threshold strategy on fractal image compression algorithm performance: study, IEEE Access 8 (2020) 52687–52699.

[10] I.H. Shahrezaei, H.C. Kim, Fractal analysis and texture classification of high-frequency multiplicative noise in SAR sea-ice images based on a transform-domain image decomposition method, IEEE Access 8 (2020) 40198–40223.

[11] H.Y. Li, H.W. Chai, X.H. Xiao, J.Y. Huang, S.N. Luo, Fractal breakage of porous carbonate sand particles: microstructures and mechanisms, Powder Technology (2020).

[12] S. Balakrishna, M. Thirumaran, Semantics and clustering techniques for IoT sensor data analysis: a comprehensive survey, in: Principles of Internet of Things (IoT) Ecosystem: Insight Paradigm, Springer, Cham, 2020, pp. 103–125.

[13] C. Rashmi, G.H. Kumar, Multithreading approach for clustering of multiplane satellite images, in: Artificial Intelligence Techniques for Satellite Image Analysis, Springer, Cham, 2020, pp. 25–47.

[14] A. Ghosal, A. Nandy, A.K. Das, S. Goswami, M. Panday, A short review on different clustering techniques and their applications, in: Emerging Technology in Modelling and Graphics, Springer, Singapore, 2020, pp. 69–83.

[15] P. Govender, V. Sivakumar, Application of k-means and hierarchical clustering techniques for analysis of air pollution: a review (1980–2019), Atmospheric Pollution Research 11 (1) (2020) 40–56.

[16] T. Thomas, A.P. Vijayaraghavan, S. Emmanuel, Machine learning and cybersecurity, in: Machine Learning Approaches in Cyber Security Analytics, Springer, Singapore, 2020, pp. 37–47.

[17] N. Verbeeck, R.M. Caprioli, R. Van de Plas, Unsupervised machine learning for exploratory data analysis in imaging mass spectrometry, Mass Spectrometry Reviews 39 (3) (2020) 245–291.

[18] P. D'Urso, V. Vitale, A robust hierarchical clustering for georeferenced data, Spatial Statistics (2020) 100407.

[19] J.T. Machado, A.M. Lopes, Multidimensional scaling locus of memristor and fractional order elements, Journal of Advanced Research (2020).

[20] P.G. Vieira, M.M. de Melo, A. Şen, M.M. Simões, I. Portugal, H. Pereira, C.M. Silva, Quercus cerris extracts obtained by distinct separation methods and solvents: total and friedelin extraction yields, and chemical similarity analysis by multidimensional scaling, Separation and Purification Technology 232 (2020) 115924.

[21] T. Thomas, A.P. Vijayaraghavan, S. Emmanuel, Clustering and malware classification, in: Machine Learning Approaches in Cyber Security Analytics, Springer, Singapore, 2020, pp. 73–106.

[22] K. Jayashree, R. Chithambaramani, Big data and clustering techniques, in: Handbook of Research on Big Data Clustering and Machine Learning, IGI Global, 2020, pp. 1–9.

[23] S. Mirjalili, H. Faris, I. Aljarah, Introduction to evolutionary machine learning techniques, in: Evolutionary Machine Learning Techniques, Springer, Singapore, 2020, pp. 1–7.

[24] E.M. Karanja, S. Masupe, M.G. Jeffrey, Analysis of internet of things malware using image texture features and machine learning techniques, Internet of Things 9 (2020) 100153.

[25] V.H. Reddy, S. Kumari, V. Muralidharan, K. Gigoo, B.S. Thakare, Literature survey—food recognition and calorie measurement using image processing and machine learning techniques, in: ICCCE 2019, Springer, Singapore, 2020, pp. 23–37.

[26] P. Lopez-Exposito, C. Negro, A. Blanco, Direct estimation of microalgal flocs fractal dimension through laser reflectance and machine learning, Algal Research 37 (2019) 240–247.

[27] M. Diykh, S. Abdulla, K. Saleh, R.C. Deo, Fractal dimension undirected correlation graph-based support vector machine model for identification of focal and non-focal electroencephalography signals, Biomedical Signal Processing and Control 54 (2019) 101611.

[28] K.A. Kumar, B.M. Kumar, A. Veeramuthu, V.S. Mynavathi, Unsupervised machine learning for clustering the infected leaves based on the leaf-colors, in: Data Science and Big Data Analytics, Springer, Singapore, 2019, pp. 303–312.

[29] R. Ghezelbash, A. Maghsoudi, E.J.M. Carranza, Mapping of single-and multi-element geochemical indicators based on catchment basin analysis: application of fractal method and unsupervised clustering models, Journal of Geochemical Exploration 199 (2019) 90–104.

[30] L. Liu, S. Li, X. Li, Y. Jiang, W. Wei, Z. Wang, Y. Bai, An integrated approach for landslide susceptibility mapping by considering spatial correlation and fractal distribution of clustered landslide data, Landslides 16 (4) (2019) 715–728.

[31] G. Bhatnagar, Q.J. Wu, A fractal dimension based framework for night vision fusion, IEEE/CAA Journal of Automatica Sinica 6 (1) (2019) 220–227.

[32] K. Uemura, H. Toyama, S. Baba, Y. Kimura, M. Senda, A. Uchiyama, Generation of fractal dimension images and its application to automatic edge detection in brain MRI, Computerized Medical Imaging and Graphics 24 (2) (2000) 73–85.

[33] S. Joardar, A. Sanyal, D. Sen, D. Sen, A. Chatterjee, An enhanced fractal dimension based feature extraction for thermal face recognition, in: Decision Science in Action, Springer, Singapore, 2019, pp. 217–226.

[34] S. Kadam, V.R. Rathod, Medical image compression using wavelet-based fractal quad tree combined with Huffman coding, in: Third International Congress on Information and Communication Technology, Springer, Singapore, 2019, pp. 929–936.

[35] L. Yin, X. Li, W. Zheng, Z. Yin, L. Song, L. Ge, Q. Zeng, Fractal dimension analysis for seismicity spatial and temporal distribution in the circum-Pacific seismic belt, Journal of Earth System Science 128 (1) (2019) 22.

[36] M. Joshi, A.K. Agarwal, B. Gupta, Fractal image compression and its techniques: a review, in: Soft Computing: Theories and Applications, Springer, Singapore, 2019, pp. 235–243.

[37] X.X. Li, D. Tian, C.H. He, J.H. He, A fractal modification of the surface coverage model for an electrochemical arsenic sensor, Electrochimica Acta 296 (2019) 491–493.

[38] S. Padmavati, V. Meshram, A hardware implementation of fractal quadtree compression for medical images, in: Integrated Intelligent Computing, Communication and Security, Springer, Singapore, 2019, pp. 547–555.

[39] S.R. Nayak, J. Mishra, A. Khandual, G. Palai, Fractal dimension of RGB color images, Optik 162 (2018) 196–205.

[40] S.R. Nayak, J. Mishra, Analysis of medical images using fractal geometry, in: Histopathological Image Analysis in Medical Decision Making, IGI Global, 2019, pp. 181–201.

[41] S.R. Nayak, J. Mishra, G. Palai, Analysing roughness of surface through fractal dimension: a review, Image and Vision Computing 89 (2019) 21–34.

[42] K.K. Jena, S. Mishra, S.N. Mishra, S.K. Bhoi, S.R. Nayak, MRI brain tumor image analysis using fuzzy rule based approach, Journal of Research on the Lepidoptera 50 (2019) 98–112.

[43] S.R. Nayak, J. Mishra, A modified triangle box-counting with precision in error fit, Journal of Information & Optimization Sciences 39 (2018) 113–128.

[44] K.K. Jena, S. Mishra, S.N. Mishra, S.K. Bhoi, 2L-ESB: a two level security scheme for edge based image steganography, International Journal on Emerging Technologies 10 (2019) 29–38.

[45] S.R. Nayak, J. Mishra, G. Palai, An extended DBC approach by using maximum Euclidian distance for fractal dimension of color images, Optik 166 (2018) 110–115.

[46] K.K. Jena, S. Mishra, S. Mishra, S.K. Bhoi, Unmanned aerial vehicle assisted bridge crack severity inspection using edge detection methods, in: 2019 Third International Conference on I-SMAC (IoT in Social, Mobile, Analytics and Cloud) (I-SMAC), IEEE, 2019, pp. 284–289.

[47] S.R. Nayak, A. Ranganath, J. Mishra, Analysing fractal dimension of color images, in: IEEE International Conference on Computational Intelligence and Networks, 2015, pp. 156–159.

[48] K.K. Jena, S. Mishra, S.N. Mishra, An edge detection approach for fractal image processing, in: Examining Fractal Image Processing and Analysis, IGI Global, 2020, pp. 1–22.

[49] S.R. Nayak, J. Mishra, R. Padhy, A new extended differential box-counting method by adopting unequal partitioning of grid for estimation of fractal dimension of grayscale images, in: Computational Signal Processing and Analysis, Springer, Singapore, 2018, pp. 45–57.

[50] S.K. Das, S.R. Nayak, J. Mishra, Fractal geometry: the beauty of computer graphics, Journal of Advanced Research in Dynamical and Control Systems 9 (10) (2017) 76–82.

[51] K.K. Jena, S. Mishra, S.N. Mishra, An algorithmic approach based on CMS edge detection technique for the processing of digital images, in: Examining Fractal Image Processing and Analysis, IGI Global, 2020, pp. 252–272.

[52] S.R. Nayak, A. Khandual, J. Mishra, Ground truth study on fractal dimension of color images of similar texture, Journal of the Textile Institute 109 (2018) 1159–1167.

[53] S.R. Nayak, J. Mishra, P.M. Jena, Fractal analysis of image sets using differential box counting techniques, International Journal of Information Technology 10 (2018) 39–47.

[54] S.R. Nayak, J. Mishra, G. Palai, A modified approach to estimate fractal dimension of gray scale images, Optik 161 (2018) 136–145.

[55] K.K. Jena, S.R. Nayak, S. Mishra, S.N. Mishra, Vehicle number plate detection: an edge image based approach, in: 4th Springer International Conference on Advanced Computing and Intelligent Engineering, Advances in Intelligent Systems and Computing, 2019.

[56] S.R. Nayak, J. Mishra, R. Padhy, An improved algorithm to estimate the fractal dimension of gray scale images, in: International Conference on Signal Processing, Communication, Power and Embedded System, IEEE, 2016, pp. 1109–1114.

[57] S.R. Nayak, J. Mishra, P.M. Jena, Fractal dimension of grayscale images, in: Progress in Computing, Analytics and Networking, Springer, Singapore, 2018, pp. 225–234.

[58] https://orange.biolab.si/download/#windows. (Accessed 11 April 2020).

[59] https://www.google.com/search?q=Mandelbrot+Set+Fractal+Images&rlz=1C1CHBD_
enIN923IN923&sxsrf=ALiCzsZsWS5MB2LtK5rq7xD0YxbsmKYOJg:1663416600588&
source=lnms&tbm=isch&sa=X&ved=2ahUKEwjard3c5Zv6AhX_IrcAHT2yAU0Q_
AUoAXoECAEQAw&biw=1280&bih=657&dpr=1. (Accessed 11 May 2020).

[60] https://en.wikipedia.org/wiki/Mandelbrot_set. (Accessed 11 May 2020).

[61] https://www.sciencephoto.com/media/10373/view/fractal-image-of-the-mandelbrot-set. (Accessed 11 May 2020).

[62] https://www.pinterest.com/pin/354728908126735420/. (Accessed 11 May 2020).

[63] https://math.stackexchange.com/questions/2710/why-does-the-mandelbrot-set-contain-slightly-deformed-copies-of-itself. (Accessed 11 May 2020).

[64] https://www.dreamstime.com/illustration/mandelbrot-set.html. (Accessed 11 May 2020).

[65] https://www.nsf.gov/pubs/2002/nsf0120/nsf0120_man.html. (Accessed 11 May 2020).

[66] http://www.math.utah.edu/~alfeld/math/mandelbrot/mandelbrot.html. (Accessed 11 May 2020).

[67] http://math.bu.edu/DYSYS/FRACGEOM/node2.html. (Accessed 11 May 2020).

[68] http://themancave-rayc.blogspot.com/2011/12/mandelbrot-set-fractals.html. (Accessed 11 May 2020).

[69] https://wallpaperplay.com/board/mandelbrot-set-wallpapers. (Accessed 11 May 2020).

[70] http://www.math.utah.edu/~alfeld/math/mandelbrot/mandelbrot.html. (Accessed 11 May 2020).

[71] http://www.oceanlight.com/spotlight.php?img=10401. (Accessed 11 May 2020).

[72] https://www.pinterest.com/pin/576320083566847366/. (Accessed 11 May 2020).

[73] http://www.misterx.ca/Mandelbrot_Set/M_Set-IMAGES_&_WALLPAPER.html. (Accessed 11 May 2020).

[74] https://michaelheasell.com/blog/2014/10/14/functional-fractals/. (Accessed 11 May 2020).

[75] https://www.stratio.com/blog/sparkart-part-one/. (Accessed 11 May 2020).

[76] https://www.youtube.com/watch?v=TsJ3_PIBKto. (Accessed 11 May 2020).

[77] http://www.pygame.org/project-Mandelbrot+Set+Viewer-698-1238.html. (Accessed 11 May 2020).

[78] http://www.math.utah.edu/~alfeld/math/mandelbrot/mandelbrot.html. (Accessed 11 May 2020).

[79] http://www.fundamentalfinance.com/excel/projects/mandelbrot-set-in-excel.php. (Accessed 11 May 2020).

Chapter 3

Chaos-based image encryption

Sezgin Kaçar[a], Ünal Çavuşoğlu[b], and Hadi Jahanshahi[c]

[a]*Sakarya University of Applied Sciences, Electrical and Electronics Engineering Department, Sakarya, Türkiye,* [b]*Sakarya University, Software Engineering Department, Sakarya, Türkiye,* [c]*Department of Mechanical Engineering, University of Manitoba, Winnipeg, MB, Canada*

3.1 Chaos-based image encryption studies and encryption/decryption processes

3.1.1 Literature review

In this section, a summary of the chaos-based image encryption studies in the literature is presented. When the studies in the literature are examined, it is seen that there has been a significant increase in the number of chaos-based encryption studies, especially in recent years. In this section, examples from past to present studies are presented. In the studies presented, care was taken to select different techniques and systems. Wang et al. [1] proposed a new chaos-based image encryption algorithm. In the study, an architecture was presented that enabled all image pixels to be scanned and processed in one step by combining two separate steps that provide the two main requirements of encryption, mixing and spreading. Chen et al. [2] developed an asymmetric encryption algorithm using a 3D chaotic map for image encryption. It is stated that the designed 3D chaotic-based image encryption algorithm can be used in real-time applications because it provides better dispersion [3] having developed an image encryption algorithm using the logistic map. Hraoui et al. [4] designed a chaos-based algorithm using the logistic chaotic map to be used in encryption and performed comparative performance and security analyses with the AES algorithm. Wang et al. [5] proposed a novel method to design S-box based on chaotic map and genetic algorithm. In their study, Asim and Jeoti [6] designed a hybrid image encryption algorithm using S-Box produced with fragmented and logistic chaotic maps. S-Box was created by Farwa et al. [7] using chaotic systems, which is a common method for creating chaotic image encryption. They proposed an image cipher that combines a tent map-based chaotic substitution operation with the Arnold transform's scrambling effect. An S-box based on a one dimensional chaotic tent map is used in the proposed method. They employ this S-box to partly encrypt the image before applying the Arnold transform for a particular number of iterations to obtain the completely encrypted image. Hong et al.'s method [8] is

Copyright © 2024 Elsevier Inc. All rights are reserved, including those for text and data mining, AI training, and similar technologies.

also based on LEA that is a lightweight and quick block cipher. Modular addition, bitwise XOR and rotation are the sole operations used (ARX). In addition, the suggested approach divides the encryption operations into three stages. The rotation and XOR parts are used to provide the confusion requirement, while the round operation is utilized to provide the diffusion. Two key sequences used in this operation are generated using the two key pairs and two logistic maps during the confusion phase. For image encryption, Li et al. combined Tent and Lorenz chaotic systems [9]. They are attempting to build and normalize three Lorenz sequences. Thereafter, the Tent sequence was created by taking the average of these sequences. Finally, utilizing Tent and Lorenz sequences, an initial key was produced, and pixels were decoded using this key. Another common method of chaotic encryption is fractals. Fractals are used as a key in the encryption and decryption operations by Rozouvan [10]. Using parameters acquired from the encryption key, such as coordinates, iterations, zoom, and so on, a unique image is produced on a fractal set such as Mandelbrot or Julia. The encoded image is then modulated with the components chosen from this one-of-a-kind image. As just a few variables may represent a distinct fractal image, the fractal key takes up minimal memory. Sun et al. [11] made a Julia set, used the Hilbert curve to scramble it at the bit level, and then computed the modulo of the Julia set and plain pictures. The Julia set, forward and backward steps along the Hilbert curve, and diffusion keys all employed the secret key as a parameter. Wong et al. completed an encryption investigation based on simple insertion and replacement operations utilizing the chaotic Standard Map system. They attempted to reduce the encryption time by speeding up the chaos-based encryption algorithm they developed [12]. Wang et al., in their chaos-based image coding study, created different control parameters for each step, resulting in a strong. They aimed to obtain an algorithm. With this algorithm, the image entered the encryption process with a different key at each step, and successful results were achieved in terms of speed and security criteria in the encryption process [13]. Xiao et al. developed a grayscale image encryption algorithm by using Arnold Cat Map and Chen chaotic systems. As a result of their study, it is seen that successful results were obtained from the security tests [14]. Hongjun and Xingyuan developed a strengthened algorithm against noise in the image by using the Chebyshev Map chaotic system in their chaotic-based image encryption application. Despite the noise in the encoded image with the algorithm they created, they reached the original image with the least loss [15]. Liu and Wang performed encryption and decoding by combining different channels of three different images of the same size, respectively. They used the SHA-256 hash function for the key to be used in encryption and the phase output values of the Lorenz chaotic system as a random number generator [16]. Prusty et al. mixed the image using the Arnold Cat Map system in their image encryption study. They obtained successful results by generating keys and encrypting image data in different formats using the Henon Map system [17]. Liu et al. carried out studies on encrypting color images with a chaos-based block cipher system.

In the study, it was shown that a successful encryption process was performed with the results of histogram, correlation, differential attack, and entropy analyses [18]. Xian et al. introduced the fractal sorting matrix (FSM), a family of sorting matrices having fractal features, and its iterative computation technique. This paper described a new approach of pixel diffusion based on two chaotic arrays that provides strong security and great encryption efficiency. This study developed a more efficient and safe chaotic image encryption based on the FSM and global chaotic pixel diffusion than previous approaches [19]. Luo et al. proposed a unique image encryption technique based on double chaotic systems. As a single chaotic map's chaotic range and susceptibility are restricted, they employ a two-dimensional Baker chaotic map to manage the logistic chaotic map's system parameters and state variable. Based on the enhanced chaotic maps [20], a unique image encryption technique with shuffling and substitution processes is devised. Ye et al. introduced a novel and efficient pixel-level image encryption technique. In contrast to the classic permutation–diffusion architecture, the suggested technique strengthens the link between pixel location shuffling and grayness value change. The diffusion operation is done to the permuted image in a straightforward manner, which may be thought of as a workaround for the permutation's inability to modify the frequency of pixels [21]. The proposed algorithm in [22] converts a color image to 24 bit planes via RGB trying to bit-plane decomposition and split, then tends to perform 3D bit-plane possible combination on bit planes, obtaining position for permutation from a 3D Chen system, and finally obtaining the three confused elements. Secondly, three key matrixes are created using a 1D system and a multilayer discretization technique, and the color encrypted image is formed by diffusing the confused parts using a key matrix. Multiple chaotic iterative maps were used in this study to suggest a unique image encryption approach. The recommended encryption increased the amount of confusion and diffusion in the scheme, in which one of the most crucial characteristics related to encryption is confidentiality [23]. Liu et al. [24] presented a chaotic encryption approach based on simultaneous permutation–diffusion operations that is both secure and quick. The processes of diffusion and permutation are coupled. The secret keys and values of the previous encrypted pixels are connected to the beginning value of the current Sine map. The pixel values are also managed by row and column throughout the encryption process. The proposed approach offers a lower time complexity and faster operational performance than bit-pixel level image encryption solutions. Based on a chaos map and information entropy, this study [25] presents an image encryption technique. The suggested technique, unlike Fridrich's structure, includes permutation, modulation, and diffusion (PMD) operations. This solution addresses the flaw in existing systems that need pixel locations to be carefully shuffled before diffusion encryption. To impact the production of the keystream, entropy is used. The initial keys employed in the permutation and diffusion stages communicate with one another. Cheng et al. [26] presented a hyper-chaotic encryption approach using quantum genetic algorithms and compressive sensing, which is a hitherto unexplored encryption algorithm. To begin, QGA can use the quantum spin gate to refresh the population, increasing its unpredictability and preventing it from sliding into a local optimum. They used the SHA-512 hash of the image to identify the hyper-chaotic system's starting values. This study proposed a new chaotic image encryption approach based on an RNA process and extended Zigzag confusion [27]. The unique chaotic system was constructed that may be thought of as an upgrade on the standard Logistic and Sine map. It was intended to use an extended Zigzag confusion strategy. This approach can address the issue of some items remaining in the same place during each encryption cycle. The choice of operators in RNA computing is defined by the type of amino acids, and RNA matrix rules are totally controlled by chaotic outputs. Finally, this encryption algorithm was capable of encrypting color images.

Abbas et al. [28] presented a new image encryption system that uses pixel-level parallelism to speed up the process efficiency of creating chaotic outputs. This study used a group formed over elliptic points and the addition to make a discrete output. By making encryption and decryption methods very parallelizable, the suggested technique intended to make use of commonly available parallel processing platforms. Hua et al. [29] presented on the cosine transform using two chaotic maps based to encrypt images. They also described an encryption method based on one of the chaotic maps that resulted. The encryption approach employed high-efficiency scrambling to separate neighboring pixels and random substitution to send a small switch in the plaintext-image to all cipher-image pixels. According to the test results, the proposed systems generated chaotic maps with many more complex chaotic outputs. Liu et al. [30] proposed a new encryption technique based on the coupled piecewise sine map, a new chaotic system based on the sine map that employs a piecewise mechanism to obtain a more uniform probability density distribution of state values while also increasing the chaotic system's complexity through parameter coupling between sub-systems. Wu et al. [31] proposed a plaintext-coupled dynamic key encryption scheme with high security and minimal complexity. The RGB parts of the image were normalized and read to provide a key that was nearly connected to the plain image. The Arnold transform was then applied to the RGB parts of the image to stretch and fold them in order to modify the location of the pixel points in space, eliminating the connection between neighboring pixel points. The Arnold-transformed RGB matrix was then used to encrypt each of the resulting sequences independently. Liu et al. [32] presented a unique two-dimensional Hénon–Chebyshev transition unit by cascading the Chebyshev and Hénon chaotic systems. The dynamics of the suggested system were studied using a variety of objective evaluation methods. The proposed map, as well as the underlying ideas of genetic mutation and recombination, were used to present a novel method. Li et al. [33] presented a unique chaos image encryption approach that does not require any extra information. The link between the plaintext image and the private keys was built by a Henon map randomly and se-

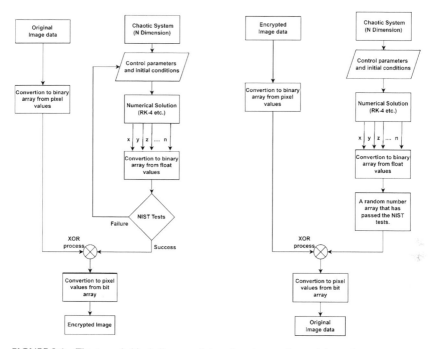

FIGURE 3.1 The example block diagram of chaos-based encryption and decryption.

curely picking certain pixels in the original data. The encryption procedure was split into two sections. First, the step was to encrypt the pixel while maintaining their hidden places untouched. The other step was to encrypt the remaining pixels using the generated keys and the hyperchaotic Lorenz system. Lyle et al. [34] described a two-dimensional chaotic Henon map and a quadratic map as an adaptive image encryption approach. The Henon map was used to produce the pseudo random number array for shuffling the pixels in the proposed encryption method. The pseudo random number sequence for the diffusion operation was generated using the standard quadratic map.

3.1.2 Explanation of the basic processes of chaos-based image encryption and decryption

An example block diagram of a chaos-based encryption and decryption process is presented in Fig. 3.1. In the figure, encryption was carried out with the XOR operation, which is used as the most basic encryption operator for the encryption process. However, in studies in the literature, safer algorithms were developed with many different components and operations instead of XOR operation. The commonly used methods can be exemplified as follows: pixel-based mixing and switching operations [1,20,24,34], S-box-based displacement operations [5], hybrid encryption methods [6,14], transform operations [7,22,32], fractal chaos-based encryption algorithms [10,19], and quantum and genetic methods [26,27]. The steps defined for encryption and decryption are explained below in order:

- First, the image file to be encrypted is converted to binary values.
- By determining the control parameters and initial conditions of the chaotic system, it is ensured that the system produces output.
- Float values produced by the chaotic system with numeric solution methods such as RK-4 are obtained. By displaying the produced float values in a binary base, random bit sequences are created by selecting a different number of bits.
- Whether the obtained bit sequences have sufficient randomness level is tested by NIST 800-22 tests.
- Random number sequences that have passed all the randomness tests are used in encryption processes.
- The bits in the sequence produced by resolving the chaotic system with the original image file converted on a bit basis are encrypted with the XOR operation.
- The XOR operation represents the most basic operation performed here. In the developed algorithms, encryption is done with many different arithmetic and numeric methods.
- After the encryption process, the binary bit sequence obtained is converted to integer values and a pixel-based encrypted image file is obtained.
- In the decryption process, after the bitwise conversion of the encrypted image file, the random bit sequence used in the encryption process is obtained by entering the same parameters and initial conditions of the chaotic system, and the original image file is obtained after the XOR process.

3.2 The evaluation methods

In this section, the methods frequently used in the literature for performance evaluation of chaos-based image encryption applications are discussed. These methods are examined in three parts. The first of these is the NIST test suite, which is used for random number analysis in chaos-based image encryption methods, which are generally developed based on random number generators. In the second part, statistical evaluation methods, which are mostly carried out on encrypted images, are explained. The most commonly used methods in the literature are; correlation, entropy, energy, contrast, homogeneity, MSE (Mean Square Error), MAD (Mean Absolute Deviation), MAE (Mean Absolute Error), Histogram, Encryption quality (Deviation), PSNR (Peak Signal to Noise Ratio), and Differential attack (NPCR, UACI) methods [35–42]. In the last part, keyspace and key sensitivity analyses [35] used for key evaluation are discussed.

3.2.1 Randomness tests

The main feature of chaotic systems is that although state variables exhibit random behavior, they have a regular and consistent structure within themselves. In other words, although the state variables of chaotic systems are non-periodic and have a random appearance, they have values that can be calculated with the mathematical models of chaotic systems. Thanks to these features, encryption applications can be performed with the obtained state variables. If the mathematical models of chaotic systems and the system parameters and initials used in these models are not known, it can be said that the obtained state variables change randomly. In cases where the mathematical model and other parameters are known, it is understood that the state variables can be calculated and actually do not show a random change. In this way, encryption applications are made based on random changes in state variables by using mathematical models, system parameters and initials as encryption keys. On the other hand, in order for the encryption to have sufficient randomness, the state variables of the chaotic system or the binary number sequences obtained from these systems must also have sufficient randomness. In this case, the state variables used or the number sequences obtained from chaotic systems should be tested to determine whether they meet this requirement. The most known and widely used test for this process is the internationally accepted NIST-SP 800-22 test suite.

The National Institute of Standards and Technology created NIST-SP 800-22, a statistical test suite for random and pseudorandom number generators for cryptographic purposes (NIST). As a result, it appears that this test packet is used in the vast majority of chaotic systems studies [43–45]. The NIST-SP 800-22 test packet includes 15 tests for determining the randomness of binary number sequences generated by true or pseudo random number generators. These tests look at a wide range of non-randomness that could occur in a sequence. These tests are listed below [46]:

1. The Frequency (Monobit) Test.
2. Frequency Test within a Block.
3. The Runs Test.
4. Tests for the Longest-Run-of-Ones in a Block.
5. The Binary Matrix Rank Test.
6. The Discrete Fourier Transform (Spectral) Test.
7. The Non-overlapping Template Matching Test.
8. The Overlapping Template Matching Test.
9. Maurer's "Universal Statistical" Test.
10. The Linear Complexity Test.
11. The Serial Test.
12. The Approximate Entropy Test.
13. The Cumulative Sums (Cusums) Test.
14. The Random Excursions Test.
15. The Random Excursions Variant Test.

In all these tests, P-values are calculated to determine the resulting sequences randomness [46]. If this P-value is calculated as P-value > 0.01, the resulting sequence is considered to provide sufficient randomness for the performed test. A random sequence consists of 1 000 000 bits and must satisfy the P-value > 0.01 requirement in all the above tests [46]. The random number generator that generated the sequence is regarded as having sufficient randomness if the resulting sequence passes all conditions.

3.2.2   Statistical evaluation methods

For a good encryption method or algorithm, it is expected to be resistant to cryptanalytic, statistical, and brute-force attacks of various types. In chaos-based image encryption methods, this resistance is commonly examined with statistical analysis methods [35–45]. In this section, statistical analysis methods commonly used in the literature are explained.

3.2.2.1   Correlation

The most basic method for detecting the similarity between two images, particularly in encryption applications, is correlation analysis [36–42]. Correlation can be explained as a metric of the relationship between two images. If two pixels in an image are neighboring pixels, there is a really close correlation between them; otherwise, they are considered to be less correlated. In an image, this is known as the adjacent pixel correlation. The following is a representation of the related correlation:

$$\text{corr} = \frac{\text{cov}(x, y)}{\sigma x \sigma y} \tag{3.1}$$

$$\sigma x = \sqrt{\text{var}(x)} \tag{3.2}$$

$$\sigma y = \sqrt{\text{var}(y)} \tag{3.3}$$

$$\text{var}(x) = \frac{1}{N} \sum_{i}^{N} (x_i - E(x))^2 \tag{3.4}$$

$$\text{cov}(x, y) = \frac{1}{N} \sum_{i}^{N} (x_i - E(x))(y_i - E(y)). \tag{3.5}$$

3.2.2.2   Entropy

Entropy is a measurable statistic of randomness that can be used to define image texture [36,39,40,42]. In cryptography, a higher increase in entropy is preferred because it indicates that the encrypted message is more complicated. The following is a definition of entropy:

$$\text{Entropy} = -\sum_{i}^{N} p(x_i) \log_b (p(x_i)),\qquad(3.6)$$

where $p(x_i)$ contains the histogram counts.

3.2.2.3 Histogram

Histogram is a frequently used metric for texture features of an n x m dimensional image [36,39–42]. The histogram can determine whether or not images are correctly exposed. In other terms, histogram graphs the number of pixels values to show how pixels in images are distributed. The statistical features of images are presented by histograms. The random numbers created from chaotic systems are uniformly spread out as white noise, according to the histogram of the encryption algorithm. In histograms, the distributions of strong encrypted images must be uniform.

3.2.2.4 Contrast

In principle, image contrast analysis allows the viewer to visually define the items in an image's texture [39]. Because of the high level of randomness produced by the encryption process, the encrypted image exhibits higher contrast levels so that the performance of an image encryption is evaluated using contrast. This analysis yields a metric of the intensity contrast of a pixel and its neighbor throughout the entire image. Contrast can be expressed mathematically as:

$$\text{Contrast} = \sum_{i,j} |i - j|^2 p(i, j),\qquad(3.7)$$

where $p(i, j)$ is the number of gray-level co-occurrence matrices.

3.2.2.5 Homogeneity

The homogeneity analysis determines how close the values with in a gray level co-occurrence matrix (GLCM), also well known as the gray tone spatial dependency matrix (GTSDM), are distributed to the GLCM diagonal [39]. In a tabular format, the GLCM displays the stats of combinations of pixel brightness or gray levels. The GLCM can be used to determine the frequency of gray level patterns. The homogeneity can be calculated as follows:

$$\text{Homogeneity} = \sum_{i,j} \frac{p(i, j)}{1 + |i - j|},\qquad(3.8)$$

where $p(i, j)$ represents the gray-level co-occurrence matrices.

3.2.2.6 Energy

The sum of squared components in the gray level co-occurrence matrix is calculated using this metric as follows:

$$\text{Energy} = \sum_{i,j} p(i, j)^2, \tag{3.9}$$

where the number of gray-level co-occurrence matrices is given by $p(i, j)$ [39]. Energy is projected to be greatly reduced with effective encryption.

3.2.2.7 Encryption quality

An effective encryption should uniformly randomize the input pixel values. This enables one to avoid situations where some pixels have a substantial change in their initial values while others have little change in their initial values [38,40,41]. The statistical distribution of the deviation tends to be uniform if the encryption algorithm handles the pixel values randomly. Although there are different encryption quality calculations in different sources for this and one of the most accepted calculations is the irregular deviation calculation. The irregular deviation is a statistic that indicates how close the statistical distribution of histogram deviation is to a uniform distribution [40]. The encryption procedure is regarded to be good if the irregular deviation is close to uniform distribution [40]. The ID (irregular deviation) is as follows:

$$D = |P - C|, \tag{3.10}$$

where D is the difference matrix of P (input image) and C (encrypted image):

$$h = \text{histogram}(D) \tag{3.11}$$

$$M_h = \frac{1}{256} \sum_{i=0}^{255} h_i \tag{3.12}$$

$$H_{D_i} = |h_i - M_h| \tag{3.13}$$

$$I_D = \sum_{i=0}^{255} H_{D_i}. \tag{3.14}$$

The encryption quality increases with the smaller the ID value. Using (3.14), the smaller value of ID implies that the histogram distribution belongs to the deviation between the input and encrypted images becomes closer to the uniform distribution [38,41,42].

3.2.2.8 MAD (mean absolute deviation) or mean absolute error (MAE)

The difference between the original image and the encrypted image is quantified in this analysis. To measure the difference of two images, the mean of the absolute deviation is calculated [39]. This analysis can be expressed numerically as:

$$MAD = \frac{1}{N \times N} \sum_{i=1}^{N} \sum_{j=1}^{N} |a_{ij} - b_{ij}|, \tag{3.15}$$

where a_{ij} are the plain image pixels, b_{ij} are the encrypted image pixels, and N indicates the sizes. The greater MAD value results from the higher the difference. This analysis is also known as the Mean Absolute Error (MAE) in some literature. The differences between the encrypted image (E_i) and the original image (P_i) are measured by MAE [36]. W and H are the width and height of the original image, respectively, and MAE is calculated as:

$$MAE = \frac{1}{W \times H} \sum_{i=1}^{H} \sum_{j=1}^{W} |P_i(i, j) - E_i(i, j)|. \tag{3.16}$$

3.2.2.9   Differential attack

A desirable aspect of any encryption technique is that a slight change in the plaintext image should result in a considerable change in the encrypted image. To determine the impact of a single pixel change on the whole image, two typical measures are utilized. Number of Pixel Change Rate (NPCR) and Unified Average Change Intensity (UACI) are the two measures [36,37,41]. The change in percent of different pixels between the original image (P) and the encrypted image (Q) is measured by NPCR. It is rated as follows:

$$D(i, j) = \begin{cases} 0 & P(i, j) = Q(i, j) \\ 1 & P(i, j) \neq Q(i, j) \end{cases} \tag{3.17}$$

$$NPCR = \frac{1}{W \times H} \sum_{i=1}^{H} \sum_{j=1}^{W} D(i, j) \times 100\%. \tag{3.18}$$

The average intensity change between P and Q is measured using UACI. It is rated as follows:

$$UACI = \frac{1}{W \times H} \sum_{i=1}^{H} \sum_{j=1}^{W} \frac{|P(i, j) - Q(i, j)|}{255} \times 100\%. \tag{3.19}$$

3.2.2.10   MSE (mean square error) and PSNR (peak signal to noise ratio)

An encryption algorithm's peak signal-to-noise ratio (PSNR) can be used to evaluate it. PSNR is a measure of encryption quality. It is a measurement that shows how the plaintext image and the encrypted image differ in pixel values [36,40].

PSNR is calculated as follows [40]:

$$PSNR = 10\log_{10}\left(\frac{L^2}{MSE}\right) \qquad (3.20)$$

$$MSE = \frac{1}{N \times N}\sum_{i=1}^{N}\sum_{j=1}^{N}(P(i, j) - Q(i, j))^2. \qquad (3.21)$$

In (3.20), the dynamic range of pixel values is denoted by L that can be 0 to 255 for gray-scale images. In MSE (3.21), $P(i, j)$, $Q(i, j)$ are the ith and jth pixels in the original and encrypted images, respectively. N is the width and height size of the images.

3.2.3 Key evaluation methods

3.2.3.1 Keyspace analysis

The encryption system's number of different key combinations is referred to as the keyspace size. A decent image encryption algorithm should have a large enough keyspace to make some attacks on encryption schemes, such as brute-force attacks, impossible [35,37,38,40,42]. The key search will require 2^k operations if an encryption technique utilizes a k-bit key. As an attacker must try all possible keys, this is a large number. For instance, if the key is 128 bits long, the attack will take 2^{128} operations to find the true key. This is an extremely long period of time, and it is practically impossible to achieve. In chaos-based image encryption algorithms, the keyspace depends on the number of parameters and initials in the chaotic system used. In addition, if the system used is a fractional-order system, different fractional-order numbers in the system can be included in the keyspace. Since the parameter, initials, and fractional orders used are generally floating-point numbers, a 32-bit sequence is added to the keyspace for each value. Accordingly, a larger keyspace is obtained when using a system with higher dimensions and more parameters. This makes the developed encryption algorithm more secure. For example, a keyspace value of $2^{(3+4)*32}$ can be predicted for an encryption algorithm using a 3-dimensional chaotic system with 4 system parameters.

3.2.3.2 Key sensitivity analysis

The key sensitivity analysis is the second test in terms of the key, and it determines how sensitive an encrypted image is to changes in the key. If there is a one bit change between the keys, the decryption algorithm cannot correctly decrypt the encrypted image for a secure cryptosystem [35,37,40,41]. For highly secure cryptosystems, this means that high key sensitivity is required. In the ideal situation, the cryptosystem must be sensitive to change of the key, with a single bit in the key resulting in an entirely different encrypted image [35,37,40,41]. For this analysis, the encryption process is run with the original input image for the first key and the first encrypted image is obtained. The process is run a second time with the same input image for a new (second) key, which is different from the first key by one bit and the second encrypted image is obtained. The difference image is obtained using these two encrypted images. If there is a homogeneous distribution in the difference image, it can be said that the key sensitivity is high. In addition, this analysis can be performed by the MSE method. If there is a high MSE value between two encrypted images, it can be said that the key sensitivity is high. Chaotic systems are extremely sensitive to parameter changes and initial values and chaos-based image encryption algorithms are highly key sensitive because when they use slightly different parameters or initial values, the encrypted images are vastly different. For chaos-based encryption algorithms, the key sensitivity analysis can be performed by changing the system parameters or initials in a very small range, for example 10^{-5}, and the resulting images should be very different from each other.

3.3 The security evaluation of chaos-based image encryption

In this section, the security evaluations of the studies on chaos-based cryptography are examined and some of the cryptanalysis studies in the literature are presented. Chaos-based cryptography is based on the nonlinear systems' complicated dynamics. Chaotic systems have features such as ergodicity, random behavior, sensitive dependence on control and initial conditions parameters, and long-period unstable trajectories [47,48]. The scrambling and propagation properties that are known as Shannon perfect secrecy principles [49] and that cryptological systems should have, can be achieved by ergodicity and the sensitive dependence of chaotic systems on initial conditions and control parameters [50]. The scrambling feature means that there should be no similarity or connection between plain text and ciphertext that can be resolved by any statistical analysis. In the propagation feature, the connection between the key and ciphertext must be as sophisticated as possible. Small changes to be made on the control and initial parameters of chaotic systems cause large changes in the values produced by the system and provide the propagation feature. The ergodicity characteristic of chaotic systems reveals that the behavior of the trajectory followed by the chaotic system in a long time period is sensitively dependent on control parameters and initial conditions [50]. This is because the value of the trajectory followed at time $t + 1$ in chaotic systems is completely dependent on the value at time t. Another evaluation criterion in cryptological designs is the concept of algorithm complexity. In chaos-based designs, this complexity is expressed as structural complexity. The rich dynamic features of the system are presented and used in encryption as a criterion similar to the complexity of the algorithm in modern cryptology. In conclusion, because of the features of systems, they may be employed in chaotic system encryption applications. The main features of chaotic systems can be listed as follows:

- The next state has non-periodic behavior in which it depends on the previous state.
- Being extremely sensitive to the initial conditions and system parameters of the system.
- Exhibiting complex behaviors with varying amplitude and frequency values within certain limits.
- Unlimited multitudes of different periodic orbits to follow.

When the literature is examined, it is seen that chaotic systems are used in many cryptology applications due to the above-mentioned features they have. Chaotic systems can generate high randomness bit sequences with simple iterations, significantly reducing the processing load and encryption time. However, it has been revealed that encryption studies carried out only with chaotic systems have disadvantages in terms of security, such as not analyzing the chaotic systems used well, errors in the decoding process, and inability to determine the keyspace correctly [51–56]. In the subsequent parts of this section, security evaluation of chaos-based picture encryption algorithms on some examples from cryptanalysis studies in the literature will be presented. The cryptanalysis of the schema is shown in [57] using three distinct images and a chosen plaintext attack. A realistic chosen plaintext attack is carried out in this paper, demonstrating that a cryptosystem relying only on multiplication and bitwise XOR may be easily cracked. The examination of the obtained image reveals the efficiency of the chosen plaintext attack, which successfully reproduces the original image. Alanazi et al. [58] examined two forms of cryptographic attacks against the diffusion based encryption technique in depth. The suggested attacks are successfully carried out using a single chosen image, demonstrating the susceptibility of multiple chaotic S-box-based cryptosystems to extract the key quickly. To ensure that the recovered data was accurate, it was subjected to statistical analysis. The cryptanalysis of a chaos-based encryption scheme developed by Pak et al. was carried out in [59]. In the permutation stage, the encryption technique constructed the new chaotic system structure, merged the two Sinemaps, and scrambled the image using the keystreams generated by the Sinemap. The encrypted image was then created using linear transformation. It has been demonstrated that the encrypted data can be entirely broken using the specified plaintext attack, indicating that this approach is not suitable for secure transmission. The article examined the original algorithm's weaknesses and proposed a Chosen Plaintext attack to break it. To eliminate the observed flaws, a new algorithm has been presented, and it has been shown that the suggested method is resistant to the weaknesses. Ma et al. [60] presented a complete security study of the chaos-based algorithm from the standpoint of current cryptography, as well as its flaws. The attacker can extract an equivalent secret key from five plain image graphs and their corresponding encrypted images in order to successfully decode further encrypted images encrypted with the identical private key. Furthermore, each security parameter used in the algorithm's security evaluation was questioned. The vulnerabilities discovered are usually relevant to a wide range of alternative image encryption techniques. The selected chaos encryption technique has been thoroughly examined and effectively attacked in [61,62]. Only an image is chosen as the diffusion step's selected plaintext. Only the permutation stage of the encrypted image may be recovered. The map matrix is discovered to be comparable to the permutation stage secret key by using the other selected images as the selected plaintexts of the permutation stage. Experiments and analysis show that the recommended assault technique is effective. An analysis of an encryption method chosen by the authors was carried out in [62]. The selected paper proposed a novel secure encryption system based on a new 2D sine-cosine cross chaotic map, which was claimed to be secure enough to withstand all known cryptanalytic techniques. However, it has been demonstrated that the flexible original encryption system with permutation-bit diffusion structure has three flaws and is comparable to the one-way encryption system with PBDS, allowing the original encryption system to be cracked successfully. The paper proposed two effective and feasible assault strategies. The efficiency and feasibility of the two assault concepts were validated by simulation and testing data. The cryptanalysis work was done on a novel encryption technique for embedded tools in [63]. The selected system was said to employ a lightweight implementation for usage in embedded tools, especially in an unmanned vehicle device. The architecture and construction of this cryptosystem were studied in this article, as well as its resistance to common attacks. The system was found to be vulnerable to differential attacks using only two selected plain images, as well as attacks requiring only selected plain text. Some recommendations were made to address these flaws and enhance the system. For cryptanalysis operations, [64] used a chaos based encryption technique based on information entropy. The article looked at the algorithm's security aspects and assessed the validity of the measured security measures implemented. When the number of rounds was simply one, a differential attack could recover the equivalent secret key of each fundamental information entropy operation independently. Some of the same security issues that exist in the field of chaotic image encryption can also be encountered in information entropy-based systems, according to the researchers. For example, the digital chaotic system's short paths and the incorrect sensitivity mechanism based on the flat image's information entropy. As a result, the article demonstrated that designs based on information entropy contain various security flaws, and it was suggested that they should not be used alone in designs. The security of hybrid encryption techniques integrated with DNA encryption techniques was investigated in [65]. According to several research articles, security vulnerabilities develop as a result of ignoring the security analysis of the proposed algorithm under the plaintext attack. This paper did a security study belonging to recent years published hyper chaotic encryption system employing DNA techniques. According to the test results using the suggested attack method, it was shown that it can fully recover plaintext data without knowing anything about the security key and solutions for closing the holes and making the algorithm more resistant.

The security vulnerabilities of a permutation–diffusion structure (PDS) based encryption method were explored and cryptanalysis was carried out in [66]. The work introduced a new attack technique that fully reveals any chaotic mappings or parameters that are functionally identical to the keys used in the permutation and diffusion phases of the original cryptosystem. The mathematical and experimental results revealed that encrypted images may be acquired entirely without knowing the secret keys. The design issues of chaos-based encryption algorithms were highlighted in [67], and the weaknesses of two distinct chaos-based image encryption techniques were examined. The first study's security concern was addressed and improved in the second study, and a new version with security was offered. Furthermore, a road map based on a proven secure driven design method was offered in the report for future investigations to prevent similar concerns. Zhu et al. [68] employed chaos-based S-box cryptanalysis to conduct cryptanalysis of an encryption algorithm. The encryption algorithm's merits include its basic structure, good encryption performance and high efficiency. However, the suggested system has been demonstrated to be totally breakable by a specific plaintext attack. If you do not know what the S-box, the encrypted image is decoded. A feedback mechanism and a bidirectional propagation method were also created, and an advanced new encryption strategy was provided, in which more parameters are handled in each ciphertext propagation. According to the findings of the security study, the newly suggested system has a greater level of security. As a result, it has been seen that chaos-based image encryption studies are used for many different applications in the literature. In the studies, it has been determined that similar methods are used on image files, such as changing the positions of image pixels, masking the generated key value and pixel values, using hybrid systems by using more than one chaotic system together. When the cryptanalysis studies of these studies were examined, it was concluded that some of the developed algorithms could be easily broken and there were vulnerabilities as a result of the cryptanalysis processes performed as a result of the errors made in the system design. Some studies have been found to be more resistant and have good security levels at the point of security. In the designs of chaos-based encryption algorithms, these points should be taken into account, and the systems to be used as a result of more detailed analyses should be selected and the designs should be made by considering the weaknesses in the studies. Another important issue is that the developed algorithms are not presented under a security framework, but are presented individually. In order for chaos-based encryption algorithms to become more secure and efficient, cryptanalysis studies in the literature should be increased and general attack scenarios that can be applied to the designs to be developed should be developed and standardized.

3.4 A case study: chaos-based medical image encryption

In this section, encryptions are applied for two different medical images based on the four-dimensional chaotic system [69] seen in Eq. (3.22):

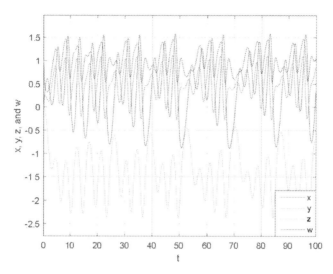

FIGURE 3.2   Time series of the chaotic system.

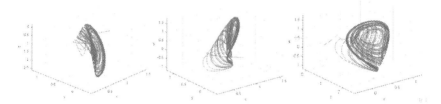

FIGURE 3.3   Phase portraits of the obtained state variables.

$$\begin{cases} x' = a(y - x) - z \\ y' = b(x - y) - w^2 \\ z' = c(z - w) - xz \\ w' = d(x + y) + z + e \end{cases} \tag{3.22}$$

When the initial value of $x_0 = y_0 = z_0 = w_0 = 0$ and the system parameters $a = 2.75$, $b = 1.25$, $c = 0.75$, $d = 0.95$, and $e = 0.02$ are given to the system in Eq. (3.22), the obtained time series and phase portraits with the Runge–Kutta 4 (RK4) algorithm are shown in Figs. 3.2 and 3.3. If Fig. 3.2 is examined, it is seen that the time series obtained from the chaotic system have a non-periodic random structure within a certain value range. In addition, when the phase portraits in Fig. 3.3 are examined, it is understood that the orbits are in a certain order and continue without ending. Accordingly, it can be said that the system shows chaotic behavior.

While calculating the time series of the state variables, the values obtained in each iteration were converted from floating to 32 bit binary arrays and the 10

TABLE 3.1 NIST-800-22 tests results.

Statistical tests	X		Y		Z		W	
	P-value	Results	P-value	Results	P-value	Results	P-value	Results
Frequency (Monobit) Test	0.3461	Successful	0.7964	Successful	0.2955	Successful	0.6906	Successful
Block-Frequency Test	0.1519	Successful	0.7775	Successful	0.8904	Successful	0.1856	Successful
Runs Test	0.5008	Successful	0.6470	Successful	0.0487	Successful	0.6702	Successful
Longest-Run Test	0.0201	Successful	0.0532	Successful	0.3829	Successful	0.1397	Successful
Binary Matrix Rank Test	0.5884	Successful	0.3530	Successful	0.1591	Successful	0.0850	Successful
Discrete Fourier Transform Test	0.1522	Successful	0.1658	Successful	0.4855	Successful	0.1231	Successful
Non-Overlapping Templates Test	<0.01	Failure	0.0171	Successful	<0.01	Failure	0.0116	Successful
Overlapping Templates Test	0.4904	Successful	0.5743	Successful	0.6417	Successful	0.8685	Successful
Maurer's Universal Statistical Test	0.3462	Successful	0.9738	Successful	0.6688	Successful	0.3363	Successful
Linear-Complexity Test	0.5958	Successful	0.9777	Successful	0.5952	Successful	0.7390	Successful
Serial Test-1	0.7506	Successful	0.7664	Successful	0.7457	Successful	0.0925	Successful
Serial Test-2	0.7722	Successful	0.8567	Successful	0.7458	Successful	0.1392	Successful
Approximate Entropy Test	0.9394	Successful	0.4690	Successful	0.1813	Successful	0.0893	Successful
Cumulative-Sums Test	0.4799	Successful	0.7026	Successful	0.1418	Successful	0.6951	Successful
Random-Excursions Test ($x = -4$)	0	Failure	0.1013	Successful	0.2142	Successful	0.7849	Successful
Random-Excursions Variant Test ($x = -9$)	0	Failure	0.8547	Successful	0.4616	Successful	0.1704	Successful

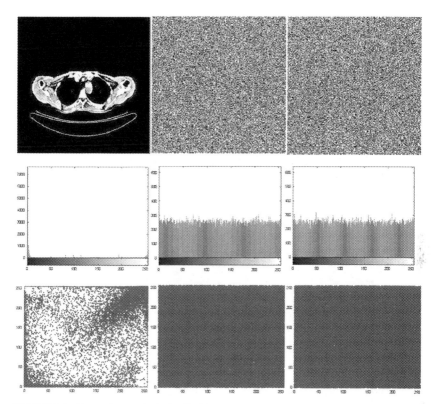

FIGURE 3.4   Encryption results for medical image 1.

LSB bits of each binary array were taken. In this way, the NIST-800-22 test was applied by obtaining bit sequences consisting of 1 000 000 bits of each state variable. The test results are shown in Table 3.1. It can be seen in Table 3.1 that there are failure results in the results of the x and z state variables. All of the results of y and w are successful. Accordingly, it is understood that the x and z state variables, which do not have complete randomness, are not suitable enough for encryption applications, and y and w state variables can be used safely in encryption applications.

After the NIST test, encryption is performed using the appropriate bit sequences. During this process, the random bit sequences obtained for y and w state variables and the bit arrays taken from the pixel values of the medical images are subjected to XOR operation and the encryption process is performed. Original and encrypted images, histograms and correlation distributions of the encryption operations performed for sample medical images are shown in Figs. 3.3 and 3.4. In Fig. 3.4, in the first row, the original medical image 1, encrypted image with y and the encrypted image with w are seen, respectively. In the second row of Fig. 3.4, histograms of the images in the first row are seen. In

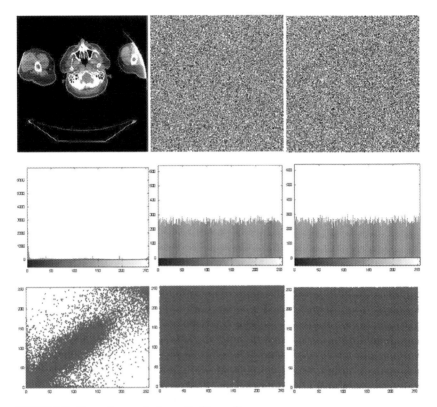

FIGURE 3.5 Encryption results for medical image 2.

the last row of Fig. 3.4, correlation distributions of the images are placed. It is seen that the histograms and the correlation distributions belonging to encrypted images are completely different from the histogram and the correlation distribution of medical image 1. Also, they are entirely homogeneous distributions. This is also true for medical image 2, as seen in Fig. 3.5. According to the results seen in Figs. 3.4 and 3.5, encryptions realized with bit sequences obtained from the y and w state variables passed the NIST tests by performing quite well.

Although the results seen in Figs. 3.4 and 3.5 are satisfactory in terms of visual aspects, they are not sufficient to demonstrate the encryption performance. Therefore statistical analysis results are given in Table 3.2. The explanations of all the analysis methods used here are explained above under the heading Evaluation methods. When the results in Table 3.2 are examined, it is understood that the results obtained using the y and w state variables for medical image 1 and 2 are very close to each other and the encryptions performed quite well.

After the statistical results, finally, keyspace and key sensitivity analyses were carried out, which are very effective on the security of encryption. In this application, considering 5 system parameters, 4 initial values, and one calcula-

TABLE 3.2 Statistical analysis results.

Tests	Medical Image 1			Medical Image 2		
	Original	Encrypted (y)	Encrypted (w)	Original	Encrypted (y)	Encrypted (w)
Correlation	0.8836	−0.0001	−0.0018	0.9347	0.0027	−0.0016
Entropy	2.4863	7.9973	7.9968	3.2073	7.9974	7.9966
Contrast	0.4982	0.4121	0.3798	0.5352	0.3813	0.4038
Homogeneity	0.9440	0.9926	0.9932	0.9337	0.9932	0.9928
Energy	0.6885	0.9833	0.9845	0.5923	0.9845	0.9835
MAD or MAE	—	121.2121	121.2589	—	116.7514	116.7403
Enc. Quality	—	5618	5752	—	9070	9030
NPCR	—	99.5819	99.6033	—	99.6185	99.5850
UACI	—	47.5341	47.5525	—	45.7849	45.7805
MSE	—	2.0131e+04	2.0150e+04	—	1.8982e+04	1.8973e+04
PSNR	—	5.0921	5.0880	—	5.3474	5.3495

TABLE 3.3 Key sensitivity results.

	Medical Image 1		Medical Image 2	
	Encrypted (y)	Encrypted (w)	Encrypted (y)	Encrypted (w)
MSE	6186.2	6174.4	6747.9	6782.6

tion step interval, it is seen that 10 different float parameters includes 32 bits are used. Accordingly, the keyspace is calculated as $2^{(10x32)} = 2320$. This value is a very good keyspace value and provides sufficient security. The $x0$ initial value for key sensitivity analysis was changed to $x0 = 0.000001$ only by the amount of 10^{-6}. The MSE values between encrypted images obtained for $x_0 = 0$ and $x_0 = 0.000001$ are presented in Table 3.3. According to these results, quite large MSE values were obtained for both y and w for both medical images. With this result, it can be said that there is sufficient key sensitivity for encryption.

References

[1] Y. Wang, K.-W. Wong, X. Liao, G. Chen, A new chaos-based fast image encryption algorithm, Applied Soft Computing 11 (1) (2011) 514–522.

[2] G. Chen, Y. Mao, C.K. Chui, A symmetric image encryption scheme based on 3D chaotic cat maps, Chaos, Solitons and Fractals 21 (3) (2004) 749–761.

[3] Z. Zhang, S. Sun, Image encryption algorithm based on logistic chaotic system and s-box scrambling, in: Proceedings - 4th International Congress on Image and Signal Processing, CISP 2011, vol. 1, 2011, pp. 177–181.

[4] I.P. Faculty, S. Dhar-mahraz, Benchmarking AES and chaos based logistic map for image encryption, in: Computer Systems and Applications (AICCSA), 2013 ACS International Conference on, IEEE, 2013.

[5] Y. Wang, K. Wong, C. Li, Y. Li, A novel method to design S-Box based on chaotic map and genetic algorithm, Physics Letters A 376 (2012) 827–833.

[6] M. Asim, V. Jeoti, Hybrid chaotic image encryption scheme based on S-box and ciphertext feedback, in: 2007 International Conference on Intelligent and Advanced Systems, ICIAS 2007, 2007, pp. 736–741.

[7] S. Farwa, T. Shah, N. Muhammad, N. Bibi, A. Jahangir, S. Arshad, An image encryption technique based on chaotic s-box and Arnold transform, International Journal of Advanced Computer Science and Applications 8 (6) (2017) 360–364.

[8] D. Hong, J.-K. Lee, D.-C. Kim, D. Kwon, K.H. Ryu, D.-G. Lee, Lea: a 128-bit block cipher for fast encryption on common processors, in: International Workshop on Information Security Applications, Springer, 2013, pp. 3–27.

[9] J. Li, Y. Xing, C. Qu, J. Zhang, An image encryption method based on tent and Lorenz chaotic systems, in: 2015 6th IEEE International Conference on Software Engineering and Service Science (ICSESS), IEEE, 2015, pp. 582–586.

[10] V. Rozouvan, Modulo image encryption with fractal keys, Optics and Lasers in Engineering 47 (1) (2009) 1–6.

[11] Y. Sun, L. Chen, R. Xu, R. Kong, An image encryption algorithm utilizing Julia sets and Hilbert curves, PLoS ONE 9 (1) (2014) e84655.

[12] K.W. Wong, B.S.H. Kwok, W.S. Law, A fast image encryption scheme based on chaotic standard map, Physics Letters A 372 (15) (2008) 2645–2652.

[13] Y. Wang, K.W. Wong, X. Liao, T. Xiang, G. Chen, A chaos-based image encryption algorithm with variable control parameters, Chaos, Solitons and Fractals 41 (4) (2009) 1773–1783.

[14] D. Xiao, X. Liao, P. Wei, Analysis and improvement of a chaos-based image encryption algorithm, Chaos, Solitons and Fractals 40 (5) (2009) 2191–2199.

[15] H. Liu, X. Wang, Color image encryption based on one-time keys and robust chaotic maps, Computers & Mathematics with Applications 59 (10) (2010) 3320–3327.

[16] H. Liu, X. Wang, Triple-image encryption scheme based on one-time key stream generated by chaos and plain images, The Journal of Systems and Software 86 (3) (2013) 826–834.

[17] A.K. Prusty, A. Pattanaik, S. Mishra, An image encryption decryption approach based on pixel shuffling using Arnold Cat Map Henon Map, in: 2013 International Conference on Advanced Computing and Communication Systems, IEEE, December 2013, pp. 1–6.

[18] H. Liu, A. Kadir, Y. Niu, Chaos-based color image block encryption scheme using S-box, AEÜ. International Journal of Electronics and Communications 68 (7) (2014) 676–686.

[19] Y. Xian, X. Wang, Fractal sorting matrix and its application on chaotic image encryption, Information Sciences 547 (2021) 1154–1169.

[20] Y. Luo, J. Yu, W. Lai, L. Liu, A novel chaotic image encryption algorithm based on improved baker map and logistic map, Multimedia Tools and Applications 78 (15) (2019) 22023–22043.

[21] G. Ye, C. Pan, X. Huang, Q. Mei, An efficient pixel-level chaotic image encryption algorithm, Nonlinear Dynamics 94 (1) (2018) 745–756.

[22] Z.H. Gan, X.L. Chai, D.J. Han, Y.R. Chen, A chaotic image encryption algorithm based on 3-D bit-plane permutation, Neural Computing & Applications 31 (11) (2019) 7111–7130.

[23] M. Khan, F. Masood, A novel chaotic image encryption technique based on multiple discrete dynamical maps, Multimedia Tools and Applications 78 (18) (2019) 26203–26222.

[24] L. Liu, Y. Lei, D. Wang, A fast chaotic image encryption scheme with simultaneous permutation-diffusion operation, IEEE Access 8 (2020) 27361–27374.

[25] G. Ye, C. Pan, X. Huang, Z. Zhao, J. He, A chaotic image encryption algorithm based on information entropy, International Journal of Bifurcation and Chaos 28 (01) (2018) 1850010.

[26] G. Cheng, C. Wang, C. Xu, A novel hyper-chaotic image encryption scheme based on quantum genetic algorithm and compressive sensing, Multimedia Tools and Applications 79 (39) (2020) 29243–29263.

[27] X. Wang, N. Guan, A novel chaotic image encryption algorithm based on extended zigzag confusion and RNA operation, Optics and Laser Technology 131 (2020) 106366.

[28] A.M. Abbas, A.A. Alharbi, S. Ibrahim, A novel parallelizable chaotic image encryption scheme based on elliptic curves, IEEE Access 9 (2021) 54978–54991.

[29] Z. Hua, Y. Zhou, H. Huang, Cosine-transform-based chaotic system for image encryption, Information Sciences 480 (2019) 403–419.

[30] J. Liu, Y. Wang, Z. Liu, H. Zhu, A chaotic image encryption algorithm based on coupled piecewise sine map and sensitive diffusion structure, Nonlinear Dynamics 104 (4) (2021) 4615–4633.

[31] Z. Wu, P. Pan, C. Sun, B. Zhao, Plaintext-related dynamic key chaotic image encryption algorithm, Entropy 23 (9) (2021) 1159.

[32] Y. Liu, Z. Qin, X. Liao, J. Wu, A chaotic image encryption scheme based on Hénon–Chebyshev modulation map and genetic operations, International Journal of Bifurcation and Chaos 30 (06) (2020) 2050090.

[33] M. Li, M. Wang, H. Fan, K. An, G. Liu, A novel plaintext-related chaotic image encryption scheme with no additional plaintext information, Chaos, Solitons and Fractals 158 (2022) 111989.

[34] M. Lyle, P. Sarosh, S.A. Parah, Adaptive image encryption based on twin chaotic maps, Multimedia Tools and Applications 81 (6) (2022) 8179–8198.

[35] Shiguo Lian, Jinsheng Sun, Zhiquan Wang, Security analysis of a chaos-based image encryption algorithm, Physica A. Statistical Mechanics and Its Applications (ISSN 0378-4371) 351 (2–4) (2005) 645–661, https://doi.org/10.1016/j.physa.2005.01.001.

[36] S. Yadav, N. Tiwari, Recent advancements in chaos-based image encryption techniques: a review, in: R. Shukla, J. Agrawal, S. Sharma, N. Chaudhari, K. Shukla (Eds.), Social Networking and Computational Intelligence, in: Lecture Notes in Networks and Systems, vol. 100, Springer, Singapore, 2020.

[37] Y. Mao, G. Chen, Chaos-based image encryption, in: Handbook of Geometric Computing, Springer, Berlin, Heidelberg, 2005.

[38] Didier Lopez-Mancilla, Juan H. Garcia-Lopez, Rider Jaimes-Reategui, Roger Chiu, Edgar Villafaña-Rauda, Carlos E. Castañeda-Hernandez, Guillermo Huerta-Cuellar, Statistical analysis of imaging encryption using chaos, in: Latest Trends in Circuits, Systems, Signal Processing and Automatic Control, vol. 86, 2014.

[39] Tariq Shah, Iqtadar Hussain, Muhammad Asif Gondal, Hasan Mahmood, Statistical analysis of S-box in image encryption applications based on majority logic criterion, International Journal of Physical Sciences 6 (16) (18 August, 2011) 4110–4127.

[40] Kalyani Mali, Shouvik Chakraborty, Mousomi Roy, Study on statistical analysis and security evaluation parameters in image encryption, IJSRD - International Journal for Scientific Research Development (ISSN 2321-0613) 3 (08) (2015).

[41] Shrija Somaraj, Mohammed Ali Hussain, Performance and security analysis for image encryption using key image, Indian Journal of Science and Technology 8 (35) (December 2015), https://doi.org/10.17485/ijst/2015/v8i35/73141.

[42] L. Roohi, S. Ibrahim, R. Moieni, Analysis of statistical properties of chaos based image encryption by different mappings, International Journal of Computer Applications 62 (20) (2013).

[43] Jiancheng Liu, Karthikeyan Rajagopal, Tengfei Lei, Sezgin Kaçar, Burak Arıcıoğlu, Ünal Çavuşoğlu, Abdullah Hulusi Kökçam, Anitha Karthikeyan, A novel hypogenetic chaotic jerk system: modeling, circuit implementation, and its application, Mathematical Problems in Engineering 2020 (2020) 8083509, https://doi.org/10.1155/2020/8083509.

[44] A. Akgül, M.Z. Yıldız, Ö.F. Boyraz, E. Güleryüz, S. Kaçar, B. Gürevin, Doğrusal olmayan yeni bir sistem ile damar görüntülerinin mikrobilgisayar tabanlı olarak şifrelenmesi, Gazi Üniversitesi Mühendislik Mimarlık Fakültesi Dergisi 35 (3) (2020) 1369–1386, https://doi.org/10.17341/gazimmfd.558379.

[45] Ü. Çavuşoğlu, S. Kaçar, A novel parallel image encryption algorithm based on chaos, Cluster Computing 22 (2019) 1211–1223, https://doi.org/10.1007/s10586-018-02895-w.

[46] Andrew Rukhin, Juan Soto, James Nechvatal, Miles Smid, Elaine Barker, Stefan Leigh, Mark Levenson, Mark Vangel, David Banks, Alan Heckert, James Dray, San Vo, Lawrence E. Bassham III, A statistical test suite for random and pseudorandom number generators for cryptographic applications Sp 800-22 rev. 1a., available at: https://nvlpubs.nist.gov/nistpubs/Legacy/SP/nistspecialpublication800-22r1a.pdf, 2010.

[47] R.C. Hilborn, Chaos and Nonlinear Dynamics: An Introduction for Scientists and Engineers, Oxford University Press, 2003.

[48] T.S. Parker, L.O. Chua, Practical Numerical Algorithms for Chaotic Systems, Springer-Verlag, 1989.

[49] C.E. Shannon, Communication theory of secrecy system, The Bell System Technical Journal 28 (1949) 656–715.

[50] G. Alvarez, S. Li, Some basic cryptographic requirements for chaos-based cryptosystems, International Journal of Bifurcation and Chaos 16 (8) (2006) 2129–2151.

[51] D. Arroyo, C. Li, S. Li, G. Alvarez, W.A. Halang, Cryptanalysis of an image encryption scheme based on a new total shuffling algorithm, Chaos, Solitons and Fractals 41 (5) (2009) 2613–2616.

[52] G. Alvarez, S. Li, Cryptanalyzing a nonlinear chaotic algorithm (NCA) for image encryption, Communications in Nonlinear Science and Numerical Simulation 14 (11) (2009) 3743–3749.

[53] D. Arroyo, G. Alvarez, J.M. Amigó, S. Li, Cryptanalysis of a family of self-synchronizing chaotic stream ciphers, Communications in Nonlinear Science and Numerical Simulation 16 (2) (2011) 805–813.

[54] E. Solak, R. Rhouma, S. Belghith, Cryptanalysis of a multi-chaotic systems based image cryptosystem, Optics Communications 283 (2) (2010) 232–236.

[55] C. Li, S. Li, K.T. Lo, Breaking a modified substitution-diffusion image cipher based on chaotic standard and logistic maps, Communications in Nonlinear Science and Numerical Simulation 16 (2) (2011) 837–843.

[56] F. Özkaynak, A. Bedri, Cryptanalysis of a new image encryption algorithm based on chaos, Optik - International Journal for Light and Electron Optics 127 (13) (2016) 5190–5192.

[57] N. Munir, M. Khan, S.S. Jamal, M.M. Hazzazi, I. Hussain, Cryptanalysis of hybrid secure image encryption based on Julia set fractals and three-dimensional Lorenz chaotic map, Mathematics and Computers in Simulation 190 (2021) 826–836.

[58] A.S. Alanazi, N. Munir, M. Khan, M. Asif, I. Hussain, Cryptanalysis of novel image encryption scheme based on multiple chaotic substitution boxes, IEEE Access 9 (2021) 93795–93802.

[59] H. Wang, D. Xiao, X. Chen, H. Huang, Cryptanalysis and enhancements of image encryption using combination of the 1D chaotic map, Signal Processing 144 (2018) 444–452.

[60] Y. Ma, C. Li, B. Ou, Cryptanalysis of an image block encryption algorithm based on chaotic maps, Journal of Information Security and Applications 54 (2020) 102566.

[61] M. Li, K. Zhou, H. Ren, H. Fan, Cryptanalysis of permutation–diffusion-based lightweight chaotic image encryption scheme using CPA, Applied Sciences 9 (3) (2019) 494.

[62] M. Li, P. Wang, Y. Yue, Y. Liu, Cryptanalysis of a secure image encryption scheme based on a novel 2D sine–cosine cross chaotic map, Journal of Real-Time Image Processing 18 (6) (2021) 2135–2149.

[63] I. El Hanouti, H. El Fadili, K. Zenkouar, Cryptanalysis of an embedded systems' image encryption, Multimedia Tools and Applications 80 (9) (2021) 13801–13820.

[64] C. Li, D. Lin, B. Feng, J. Lü, F. Hao, Cryptanalysis of a chaotic image encryption algorithm based on information entropy, IEEE Access 6 (2018) 75834–75842.

[65] W. Feng, Y.G. He, Cryptanalysis and improvement of the hyper-chaotic image encryption scheme based on DNA encoding and scrambling, IEEE İmagenics Journal 10 (6) (2018) 1–15.

[66] M. Li, Y. Guo, J. Huang, Y. Li, Cryptanalysis of a chaotic image encryption scheme based on permutation-diffusion structure, Signal Processing. Image Communication 62 (2018) 164–172.

[67] Z.M.Z. Muhammad, F. Özkaynak, Security problems of chaotic image encryption algorithms based on cryptanalysis driven design technique, IEEE Access 7 (2019) 99945–99953.

[68] C. Zhu, G. Wang, K. Sun, Cryptanalysis and improvement on an image encryption algorithm design using a novel chaos based S-box, Symmetry 10 (9) (2018) 399.

[69] S. Kaçar, F. Yalçin, B. Aricioğlu, A. Akgül, A pseudo random number generator design based on a four dimension chaotic system, in: International Conference on Advanced Technologies, Computer Engineering and Science (ICATCES'18), Safranbolu, Turkey, May 11-13, 2018.

Chapter 4

Fractal feature based image classification

Soumya Ranjan Nayak[a] and Utkarsh Sinha[b]

[a]*School of Computer Engineering, KIIT Deemed to be University, Bhubaneswar, Odisha, India,*
[b]*Amity School of Engineering and Technology, Amity University, Noida, Uttar Pradesh, India*

4.1 Introduction

Fractal geometry is being used extensively in the field of image processing for estimating the roughness of the surface of complex objects. Initially, fractals were presented by Mandelbrot to express the characteristics of different shapes and surfaces, Fractal Dimension (FD) has been used widely in recent years as an important feature extractor for analysis of complex problems such as analysis of texture, shape management, and texture segmentation and classification have been simplified by the use of fractal dimension [1]. Texture analysis is a very effective method to quantify the qualities of an image such as roughness/smoothness by evaluation of pixel intensities. It is a necessary process for robust and accurate classification of images and involves techniques such as feature selection [2]. The motive of feature extraction is to recognize those features that can represent the image in the best way possible for object recognition. In this study, we focus on the use of FD for Brain MRI image classification. Brain MRI images contain vital information about the brain structure and function, and their accurate classification is crucial for proper diagnosis and treatment. Moreover, various Gray-Level-Co-Occurrence Matrix (GLCM) algorithms [3] have been considered along with seven different FD based features extracted from 1000 Brain MRI Images. The features examined are Contrast, Correlation, Homogeneity, Energy, Entropy, Smoothness, and Hausdorff Dimension. These features can represent the image meaningfully with fewer parameters, as a result of which the computation time for classification is less. The dataset originally contained 253 images, of which 98 were of Normal patients and 155 were of patients with Brain Tumor. This dataset was augmented to increase the number of images to 1000 without any class imbalance. The images have been resized to 224×224 and normalized and the aforementioned features have been aggregated. The interdependence of Fractal features with GLCM features has been analyzed by exploratory data analysis and classification has been performed utilizing various Machine Learning classifiers such as SVM, KNN, Random

Intelligent Fractal-Based Image Analysis. https://doi.org/10.1016/B978-0-44-318468-0.00010-6
Copyright © 2024 Elsevier Inc. All rights reserved, including those for text and data mining, AI training, and similar technologies.

Forest, and Artificial Neural Networks. The results have been scrutinized by various evaluation metrics and the impact of Fractal Dimension on the accuracy of classification has been critically analyzed. The main purpose of this study is to determine the effects of the use of the Fractal Dimension in image classification. As a result, various feature extraction algorithms and classification techniques have been evaluated and the results are presented.

The rest of the chapter is organized as follows: related studies and the corresponding literature are presented in Section 4.2. Section 4.3 represents the materials and methods used in the proposed study. Section 4.4 represents the complete experimental setup adopted in this chapter. Concluding remarks are presented in Section 4.5.

4.2 Related works

The initial development of fractal geometry was pioneered by Mandelbrot. According to his theory, a fractal means irregular fragments and many natural objects having self-similarity features exhibit the fractal properties [4]. Self-similarity properties can be further observed in mathematical fractal structures such as Koch's curve, the Sierpinski triangle, and a Cantor set and a huge number of natural objects such as coastlines and snowflakes [5]. In his elementary research, Mandelbrot successfully calculated the fractal dimension of Britain's coastline by the slope of a log-log curve [6]. This opened up a lot of scope for the use of fractal geometry for the analysis of many natural surfaces. Moreover, the utilization of the Hausdorff dimension also led researchers to calculate the fractal dimension of complex surfaces [7]. Over the years, many novel approaches have been proposed for calculating the fractal dimension of digital images. Pentland [8] proposed the calculation of Fractal Dimension by considering image surface intensity as a Fractal Brownian function. A box-counting approach was proposed by Gangepain and Roques-Carmes [9]. Differential box-counting methodology was proposed by Sarkar and Chaudhuri [10]. This method was extended by Nayak and Mishra [11] who calculated the fractal dimension of color images [12]. They also worked on a modified triangle box-counting approach for a precise calculation of fractal dimension solving the problem of over-counting and under-counting simultaneously [13]. In this study, the Hausdorff dimension and smoothness of Brain MRI images have been calculated on an augmented dataset of 1000 images. The interdependence of these features with GLCM [14] features has been examined. Further, binary classification of images has been demonstrated. The major contributing factors of this study are as follows:

- Use of fractal dimension based features of images on a complex problem such as classification.
- Demonstration of the correlation between fractal dimension and GLCM features by various Exploratory Data Analytics techniques.
- Accurate classification despite using fewer but meaningful features to represent the image.

It is important to note that while each of the previously mentioned methods has its own advantages and disadvantages, this study aims to further elaborate on the results obtained by these methods and evaluate their effectiveness in the context of Brain MRI image classification.

4.3   Materials and method

In this study, experimental analysis of binary classification on Brain MRI Images has been performed. The methodology adopted is based on fractal feature extraction of images. Gray-Level Co-Occurrence Matrix (**GLCM**) features of the images are also explored and their relationship with fractal dimension has been determined. In the subsequent section, the dataset used, methods implemented for data augmentation, and the proposed methodology to classify the images are mentioned.

4.3.1   Proposed methodology

The proposed model from extracting features to classification of images is illustrated in the Work-Flow diagram in Fig. 4.1. The diagram clearly represents the experimental analysis that has been implemented in this study. This section gives the detailed description of all the methods that have been used to procure the results.

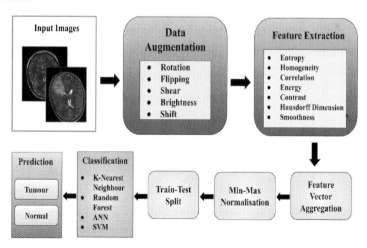

FIGURE 4.1   Work flow diagram of the proposed study.

4.3.2   Dataset used

For the study, a distinctive and publicly available dataset of Brain MRI images for detecting Tumor has been selected [15]. The obtained dataset consists of two classes namely Normal and Tumor. It contains 98 normal brain MRI images and 155 MRI images of patients with tumors. The sample brain MRI images of both normal and tumor classes are depicted in Fig. 4.2. To deal with the class imbalance, data augmentation has been performed on the dataset. This increased the number of images to 1000 (500 Normal and 500 Brain tumor images). The main purpose of the selection of this dataset is that it is publicly available, hence it is accessible for researchers. Therefore further studies based on this database may be more helpful in the diagnosis and treatment of brain tumor.

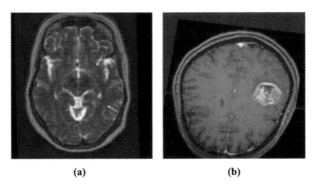

(a)   (b)

FIGURE 4.2   The dataset used for the study. (a) Normal, (b) Tumor.

4.3.3   Data augmentation

There has been a major concern of class imbalance while performing analysis on medical image datasets. The dataset used in the study had the same issue as well. Therefore to tackle this issue an offline Data Augmentation technique is used for expanding the dataset using a set of transformations. This set of transformations are: (1) the images are rotated between a range from 0 to 15 degree clockwise; (2) width and height shift is defined for a range of 0 to 0.1; (3) brightness of the image is increased in a range from 0 to 30%; (4) images were also flipped vertically. After all these transformations the dataset size increased from a total of 253 to 1000 images (500 normal and 500 brain tumor images). Sample images after performing data augmentation are illustrated in Fig. 4.3.

4.3.4   Feature extraction

In this section, all the features extracted from the images are discussed. These features have been further aggregated together in order to perform classification and make predictions.

4.3.4.1   Hausdorff dimension

There are several approaches to obtain the FD of an image, in this study we have used the Box counting method to extract the feature vector of fractal dimension. Fractal Dimension is estimated using the concept of self-similarity.

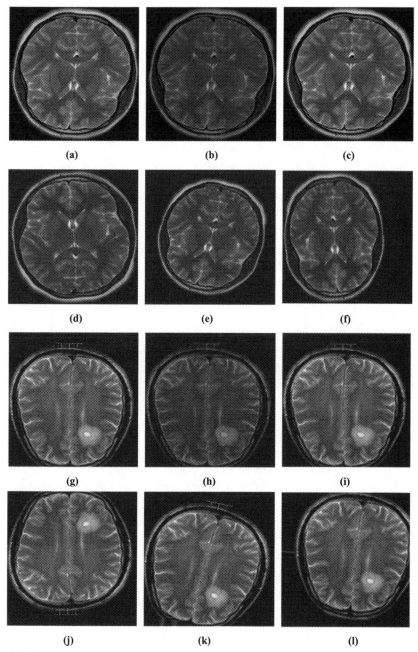

FIGURE 4.3 Sample results of augmentation: (a) normal image, (b, h) brightness, (c, i) shear, (d, j) vertical flip, (e, k) rotated, (f, l) shift, (g) tumor image.

FD can be evaluated by means of a least square regression line of $\log Nr(A)$ versus $\log(1/r)$. In general, fractal dimensions are represented in terms of D, which can be formulated as Eq. (4.1):

$$D = \frac{\log(N)}{\log(1/r)}.$$ (4.1)

4.3.4.2 Smoothness

Smoothness is regarded as a function of color gradient. To calculate smoothness a 2d gradient on three color channels has been taken. The magnitude is calculated by $\sqrt{dx^2 + dy^2}$ and averaging this over 3 channels. However, there are also cases when the linear changes in color are also smooth. In such cases a second-order differential equation is suitable and the Laplacian operator gives suitable results.

4.3.4.3 GLCM

A Gray Level Co-Occurrence Matrix (GLCM) is a commonly used statistical technique for characterizing the texture of an image. The GLCM method considers the relationship between the intensity values of a pixel and its neighbor pixels, and it describes the texture of the image based on the resulting co-occurrence matrix.

There are a total of fourteen texture features that can be calculated using the GLCM method, and the study focuses on five of these features, which are contrast, correlation, energy, homogeneity, and entropy.

The five GLCM features that are used in the study are as follows:

- **Contrast**
  This measures the intensity of contrast between a pixel and its neighbor throughout the image. This is given by Eq. (4.2), the value of contrast is zero for a constant image:

$$\sum_{i,j=0}^{N-1} P_{ij}(i-j)^2.$$ (4.2)

- **Correlation**
  Correlation measures how the pixels are correlated to its neighbor throughout the image. This is given by Eq. (4.3). The value of correlation lies between 1 and -1. 1 signifies perfect positive correlation and -1 signifies perfect negative correlation. Correlation is NAN for constant image:

$$\sum_{i,j=0}^{N-1} P_{ij}\frac{(i-\mu)(j-\mu)}{\sigma^2}.$$ (4.3)

- **Energy**
  This measures the sum of squared elements in the GLCM, energy is 1 for a constant image, as represented in Eq. (4.4):

$$\sum_{i,j=0}^{N-1} (P_{ij})^2. \tag{4.4}$$

- **Homogeneity**
  This measures how close the elements are distributed in the GLCM with respect to the GLCM diagonal. This is given by Eq. (4.5), homogeneity is 1 for a diagonal GLCM:

$$\sum_{i,j=0}^{N-1} \frac{P_{ij}}{(i-j)^2+1}. \tag{4.5}$$

- **Entropy**
  This measures the amount of information that is needed for the image compression. Entropy actually measures the loss of information or message in a transmitted signal. This is given by Eq. (4.6):

$$\sum_{i=0}^{Ng-1} \sum_{j=0}^{Ng-1} -P_{ij} * \log(P_{ij}), \tag{4.6}$$

where,
P_{ij} = Elements i, j of the normalized symmetrical GLCM.
N = Number of gray levels in the image, as specified by the number of levels under quantization on the GLCM.
μ = The GLCM mean.

The elements in the normalized symmetrical GLCM, the number of gray levels in the image, and the mean of the GLCM are all considered when calculating these texture features.

4.4 Experimental analysis

This section is based on the complete experimental study that has been performed after extracting the features of the augmented Brain MRI images. There are a total of 7 features extracted from 1000 images (500 Normal and 500 Tumor) to perform classification using various classification algorithms.

4.4.1 Data aggregation

The previous section briefly described the transformations used for increasing the number of images. The increase in images is summarized in Table 4.1.

TABLE 4.1 Details of dataset with and without augmentation.

Class	Original Dataset Images	Augmented Dataset Images
Normal	98	500
Tumor	155	500

All seven features (Hausdorff Dimension, Smoothness, Entropy, Energy, Homogeneity, Contrast, and Correlation) of the final image set were explored. This resulted in a vector of 1000 elements for each of these features that contains values for both normal and brain tumor MRI images. Further, these seven feature vectors were used to create a dataframe containing 8 attributes. Out of these 8 attributes 7 are feature vectors and 1 is a Target column, as illustrated in Table 4.2.

4.4.2 Exploratory data analysis and normalization

Exploratory Data Analysis (EDA) is performed to inspect the relationship that exists among the columns of the data frame. To attain insights from the data frame, various visualization techniques were applied. A pair plot is illustrated in Fig. 4.4. This plot is a collection of scatter plots between all the columns. The left diagonal of the plot gives the distribution of Targets (Normal or Tumor) for that specific feature.

Observing the left-diagonal of the pair plot, it is evident that there is overlapping between Orange (light gray in print version) (Tumor images) and Blue (dark gray in print version) (Normal images). This overlapping means that no feature alone is enough to classify images with higher accuracy. The graph shows a negative relation between energy and entropy. It also shows a positive relation between entropy and smoothness. In order to confirm the assumption drawn from the pair plot, a correlation plot was created between the key features, as shown in Fig. 4.5.

By observing the Correlation plot, it is evident that the Hausdorff dimension has a slight positive relation with Entropy and smoothness. On the other hand, Hausdorff dimension is negatively correlated to Energy. Homogeneity and Contrast show a very slight negative correlation to Hausdorff dimension.

The classification algorithms may lead to substandard results if columns of the data frame are on a different scale. The reason behind this is that the effectiveness of a feature may become diluted if the values in that column are on a lower scale. To tackle this issue, min–max normalization is used that scales values stored in each column between 0 and 1. This scaling process is formulated in Eq. (4.7):

$$v' = \frac{v - \min(A)}{\max(A) - \min(A)}, \tag{4.7}$$

TABLE 4.2 Aggregated table after feature extraction.

S. No.	Contrast	Correlation	Energy	Homogeneity	Smoothness	Hausdorff Dimension	Entropy	Targets
1	1.016543	0.949751	0.310925	0.801436	0.377213	1.996968	6.758202	Tumor(1)
2	1.782887	0.887716	0.291443	0.73472	0.438868	1.985406	6.722942	Tumor(1)
3	1.196655	0.941792	0.345015	0.780175	0.381853	1.992981	7.05706	Tumor(1)
4	0.91244	0.955165	0.405508	0.854584	0.341603	1.999639	5.962271	Tumor(1)
5	1.225797	0.941186	0.353812	0.785529	0.386948	1.992516	7.054832	Tumor(1)
⋮	⋮	⋮	⋮	⋮	⋮	⋮	⋮	
996	3.651607	0.777192	0.28612	0.678527	0.340451	1.985634	5.968029	Normal(0)
997	2.196811	0.784942	0.429963	0.766403	0.320496	1.935729	5.303099	Normal(0)
998	2.878766	0.828484	0.344209	0.714429	0.314314	1.9921	5.817743	Normal(0)
999	2.734177	0.718091	0.315889	0.708743	0.377344	1.970022	6.17599	Normal(0)
1000	1.622848	0.843332	0.644444	0.862529	0.203673	1.913061	3.887738	Normal(0)

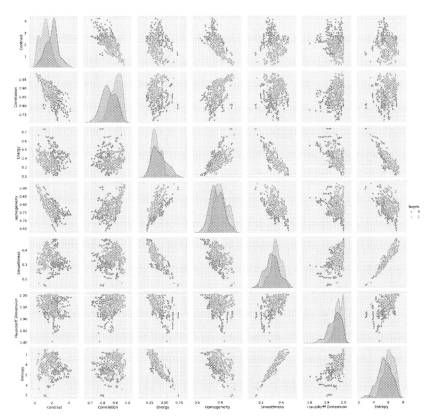

FIGURE 4.4   Pair plot of the extracted dataset.

	Contrast	Correlation	Energy	Homogeneity	Smoothness	Hausdorff Dimension	Entropy
Contrast	1	-0.78	-0.19	-0.78	0.083	-0.27	0.038
Correlation	-0.78	1	-0.056	0.51	0.043	0.37	0.18
Energy	-0.19	-0.056	1	0.65	-0.9	-0.61	-0.92
Homogeneity	-0.78	0.51	0.65	1	-0.56	-0.12	-0.51
Smoothness	0.083	0.043	-0.9	-0.56	1	0.58	0.95
Hausdorff Dimension	-0.27	0.37	-0.61	-0.12	0.58	1	0.61
Entropy	0.038	0.18	-0.92	-0.51	0.95	0.61	1

FIGURE 4.5   Correlation plot of the entire feature vector.

where v is the original value, v' is the normalized value, and A is the column that has to be normalized.

4.4.3 Classification methods and performance metrics

This section deals with a comprehensive analysis of the experiment performed for classifying tumor and normal images. After obtaining the scaled Data frame, the dataset was split into 70% train and 30% test set. Subsequently, four classification algorithms were evaluated. These 4 algorithms are SVM, Random Forest, ANN, and K-Nearest Neighbor.

4.4.3.1 Classification methods

a. **Random Forest**

Random Forest is a machine learning algorithm that uses an ensemble of decision trees for classification and regression. The algorithm builds multiple decision trees, each trained on a random subset of the data, and the prediction is made by taking a majority vote of all the trees. In the current problem statement, Random Forest can be used to classify the brain MRI images based on their fractal dimension and GLCM features. The algorithm can provide a robust and accurate classification performance by combining the results of multiple decision trees.

b. **Support Vector Machine (SVM)**

Support Vector Machine (SVM) is a supervised learning algorithm that can be used for both classification and regression tasks. SVM works by finding the optimal boundary between classes that maximizes the margin between them. In the current problem statement, SVM can be used to classify the brain MRI images based on their fractal dimension and GLCM features. The algorithm can provide a high level of accuracy in classifying the images, especially for complex and non-linear data distributions.

c. **K-Nearest Neighbors (KNN)**

K-Nearest Neighbors (KNN) is a non-parametric and instance-based machine learning algorithm that can be used for classification and regression tasks. The algorithm works by assigning the class label to a new data point based on the majority class of its k nearest neighbors in the training data. In the current problem statement, KNN can be used to classify the brain MRI images based on their fractal dimension and GLCM features. The algorithm can provide a simple and effective classification performance, especially for small datasets.

4.4.3.2 Classification techniques

The first classification technique used in this study is SVM. Each data item in SVM is plotted in n-dimensional space (here, n is the number of columns) and the classification is performed by segregating two classes using hyperplane. The linear kernel of SVM has been used that achieved an accuracy of 87.67%.

The Random Forest classifier is the next classification algorithm used in this study. It uses a set of decision trees that is formed by randomly selected subsets of training data. The votes of different decision trees are aggregated to predict the final class on the test data. In this study a set of 300 decision trees are used.

The next classifier used in the study is an Artificial Neural Network (ANN). This network is based on the neural structure of the brain. It consists of various parameters that allows the machine to learn and fine-tune itself analyzing the new data. Every parameter, also known as a neuron, is a function that produces an output after receiving a single or multiple inputs. In the current study, the Artificial Neural Network consists of 1 input layer taking the 7 features of images as input, 2 hidden layers, and 1 output layer.

A feature vector can be created from this information by representing the 7 image features as a single array or column. This array can then be used as input to the Artificial Neural Network (ANN) in the study. The input layer of the ANN takes this feature vector as input and processes it through 2 hidden layers before producing the final output. The resulting output represents the prediction made by the ANN based on the input feature vector, as depicted in Fig. 4.6.

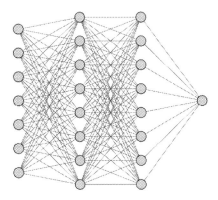

FIGURE 4.6 ANN architecture created for classification.

The fourth classification technique used in the study is the method of K-Nearest Neighbors. KNN is based on the selection of K values. The K value is the decision making factor for classifying the coordinate into different classes. If the K value is 1 then the classifier is bound to over fit as it will make decisions on the basis of only one neighboring coordinate. Hence, the K value used in this study is chosen experimentally to achieve the best results. The K value of 5 gave the optimum result and the classification accuracy of 94.67% was achieved.

The performance of each classifier was evaluated on the basis of different metrics, like $F1$-score, specificity, sensitivity (or recall), precision, and accuracy, which are defined as follows:

Accuracy is the ratio correctly predicted to the total observations. It is given by Eq. (4.8):

$$Accuracy = \frac{TP + TN}{TP + TN + FP + FN}. \tag{4.8}$$

Precision is the ratio of correctly predicted positive observations to the total predicted positive observations. It is given by Eq. (4.9):

$$Precision = \frac{TP}{TP + TN}. \tag{4.9}$$

Recall is the ratio of correctly predicted observation to all observation in the same class. It is given by Eq. (4.10):

$$Recall = \frac{TP}{TP + FN}. \tag{4.10}$$

The $F1$ score is the weighted average of Precision and Recall. It is demonstrated by the formula given in Eq. (4.11):

$$F1\text{-}score = 2 * (Recall * Precision)/(Recall + Precision). \tag{4.11}$$

4.4.4 Results

The accuracy metrics that are described in the previous section can only be calculated by using a confusion matrix. The confusion matrices of all the classifiers for the test set are shown in Fig. 4.7.

The results of all the metrics are computed on the test set. The detailed classification results obtained from all the Classifiers are compared in terms of aforementioned metrics and are shown in Table 4.3.

TABLE 4.3 Classification result comparison.

Classifier	Precision	Sensitivity	Specificity	$F1$-Score	Accuracy
SVM	83.23	93.06	82.69	87.87	87.67
Random Forest	91.93	92.5	90.71	92.21	91.67
ANN	93.17	92.02	91.97	92.59	92
K-Nearest Neighbor	95.03	95.03	94.24	95.03	94.67

4.4.4.1 Accuracy loss curve for ANN

In this section, the accuracy loss curve against the number of iterations has been presented. The aim of this analysis is to demonstrate how the accuracy of the model changes as the number of iterations increases. This information is critical in determining the optimal number of iterations for the model to achieve maximum accuracy. The accuracy loss curve for ANN used in the study has been

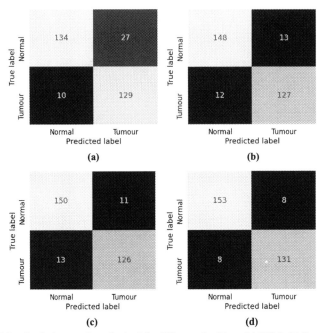

FIGURE 4.7 Confusion matrices obtained for different classifiers: (a) SVM, (b) Random Forest, (c) ANN, (d) KNN.

plotted and analyzed by the behavior of the curve over the range of iterations. The analysis of accuracy loss curves helps to identify the overfitting and underfitting of the model and make necessary adjustments to improve the performance of the model, as demonstrated in Fig. 4.8.

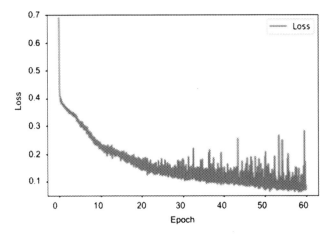

FIGURE 4.8 Accuracy loss curve for ANN.

The obtained experimental results in this study demonstrate a significant improvement compared to the existing literature. By utilizing fractal dimension and GLCM features, the proposed method was able to accurately classify the brain MRI images with a high level of precision. This is a clear indication that the combination of these two features provides a powerful representation of the images and is highly effective for the classification task.

This highlights the superiority of the proposed approach in comparison to the state-of-the-art techniques, and opens up new avenues for further research in this field. Overall, the experimental results indicate that the proposed method is highly effective for the classification of brain MRI images and provides a promising solution for real-world applications.

4.5 Conclusions

In this chapter, the effects of using Fractal features for image classification have been demonstrated by using various Machine Learning classifiers. This chapter also presents the interdependence between GLCM and Fractal features of images and how closely they are related to each other. Various pre-processing techniques, namely offline data augmentation and data normalization, have been applied for meaningful feature extraction and better analysis of the dataset. The features to be extracted have been empirically selected based on previous research. Various classification algorithms have been evaluated and it has been determined that the computation time for the analysis is less. Moreover, it has been deduced that using fractal dimension as a feature extractor is an efficient method to perform classification analysis of digital images. Future scope of this work is the use of various other methodologies of Fractal Geometry. In addition, efficient classification can be done by using a modeled Deep Learning approach. Overall, we hope that this approach will be utilized more often by the community.

References

[1] S.R. Nayak, J. Mishra, A. Khandula, G. Palai, Fractal dimension of RGB color images, Optik 162 (2018) 196–205.

[2] Ş. Öztürk, B. Akdemir, Application of feature extraction and classification methods for histopathological image using GLCM, LBP, LBGLCM, GLRLM and SFTA, Procedia Computer Science 132 (2018) 40–46.

[3] A.L.V. Coelho, C.A.M. Lima, Assessing fractal dimension methods as feature extractors for EMG signal classification, Engineering Applications of Artificial Intelligence 36 (2013) 81–98.

[4] B.B. Mandelbrot, Fractal Geometry of Nature, Freeman, San Francisco, CA, 1982.

[5] S.R. Nayak, J. Mishra, A modified triangle box-counting with precision in error fit, Journal of Information & Optimization Sciences 39 (2018) 113–128.

[6] B.B. Mandelbrot, How long is the coast of Britain? Statistical self-similarity and fractional dimension, Science 156 (1967) 636–638.

[7] K.J. Falconer, The Hausdorff dimension of self-affine fractals, Mathematical Proceedings of the Cambridge Philosophical Society 103 (1988) 339–350.

[8] A.P. Pentland, Fractal based description of natural scenes, IEEE Transactions on Pattern Analysis and Machine Intelligence 6 (Nov. 1984) 661–674.

[9] J. Keller, R. Crownover, S. Chen, Texture description and segmentation through fractal geometry, Computer Vision, Graphics, and Image Processing 45 (1989) 150–160.

[10] N. Sarkar, B.B. Chaudhuri, An efficient differential box-counting approach to compute fractal dimension of image, IEEE Transactions on Systems, Man and Cybernetics 24 (1) (Jan. 1994) 115–120.

[11] S.R. Nayak, A. Khandual, J. Mishra, Ground truth study on fractal dimension of color images of similar texture, Journal of the Textile Institute 109 (2018) 1159–1167.

[12] S.R. Nayak, J. Mishra, G. Palai, Analysing roughness of surface through fractal dimension: a review, Image and Vision Computing 89 (2019) 21–34.

[13] S.R. Nayak, J. Mishra, G. Palai, A modified approach to estimate fractal dimension of gray scale images, Optik 161 (2018) 136–145.

[14] S.R. Nayak, J. Mishra, G. Palai, An extended DBC approach by using maximum Euclidean distance for fractal dimension of coloured images, Optik 166 (2018) 110–115.

[15] Kaggle, Brain MRI images for brain tumour detection, https://www.kaggle.com/navoneel/brain-mri-images-for-brain-tumor-detection.

Part II

Recognition model using fractal features

Chapter 5

The study of source image and its futuristic quantum applications: an insight from fractal analysis

Ghulam Bary[a,b] and Riaz Ahmad[a]

[a]Faculty of Science, Yibin University, Yibin, Sichuan, China, [b]Key Laboratory of Computational Physics of Sichuan Province, Yibin University, Yibin, China

5.1 Introduction

Intensity interferometry is a tool widely used to examine the geometry and the peculiarities of particles emanating from sources. The characterization of such sources can provide new perspectives on how partons evolve toward chemical and dynamical freeze-out. To investigate the structure of the pion-particle emissions zone produced by collisions with unprecedented energy, two-particle interferometry is frequently applied in the image analysis and pattern recognition of the source [1–3]. These techniques are investigated for a crucial change in the pattern recognition from the hot, dense hadronic matter to the unconfined plasma phase that contains the basic building blocks of matter in a particular volume. Assuming that the particles are generated from somewhat stochastic sources and the widely exercised parameter lambda can be retrieved and used to examine the coherence as well as the geometry of the source with two-particle Bose–Einstein correlations. For partially coherent particle-producing sources, this value ranges from 0 to 1, but for fully chaotic emissions, it attains its maximum value of one [2–4]. More frequently, the coherent proportion and the particular details about the cause of the chaotic image aspect that is observed during particle interactions are investigated directly using two-particle interferences. It has been explored that if the coherent proportion is as high as 50 percent but the reduction in the intercept of the correlations is just 25 percent. Due to this, employing just this two-particle interference methodology to find the given coherent percentages is not achievable practically. Specifically, the relatively high correlations contain additional facts about the sources that are not visible from the simpler relatively low correlations phenomenon [5]. According to experimental data collected at the largest colliders in the world, the expected

Copyright © 2024 Elsevier Inc. All rights reserved, including those for text and data mining, AI training, and similar technologies.

reduction of the chaotic limit from the particles coherent ratio significantly rises for quantum interferences at higher orders [6,7].

The source condensate and chaotic proportion are also determined using multiparticle pion interferometry that also has several recent innovative improvements due to the elimination of the presence of the FT phase along with the impact of resonances from resonances of long-lived particles [8–12]. Practical measurements show that the three-bosonic particle intensity interferences in the image analysis led to the conclusion that the normalized correlations showed a considerable curtailment from the thermal chaotic limit. The suppression could be caused by the coherent proportion in the source image at smaller transverse momenta for the specific freeze-out during the formation of new particles data [11,12]. Additionally, several other types of nonchaotic emissions prevent the chaotic limit from being reached, such as coherence resulting from color gluonic or pionic Bose–Einstein processes [13,14]. It is found that pulse emissions also create coherent sources in the form of multiple components. It is important to note that the huge colliders at extraordinary energy are accessible for new tasks to explore the potential development of a distributed assemblage of fluid droplets in the undergone significant materials that comprise the extent of partonic freedoms that support to explore the image analysis [15,16].

Various speculations suggested that such a special state of matter is broadly contemplated to evolve with the generation of a hybrid phase of new high-energy plasma and dilute hadron gas, respectively [17,18]. The secrets of the probed hybrid phase for the observations of the particle debris have been widely debated, particularly concerning the significance of the mixed phase that evolves gradually from the hydrodynamics of the system analysis studied by [19]. A more appropriate approach explained the granular nature of the mixed phase that occurs due to the velocity distribution fluctuations of the final detected hadrons [20].

Recently, numerous investigations have been conducted and after extensive thought, it has been determined that the particle radiating sources images produced by collisions do not follow the simple distribution function. Due to the probability distribution fit, the traditional Bose interferences approach is inapplicable and therefore needs some improvements for the granular droplets to examine the source images and pattern recognition [21,22]. To assess the properties of the particle-emission sources, it is therefore essential to explore the evaluation of model independence. It should be mentioned that the most practical and adequate concept not only motivates the computation of various correlations with the assistance of the granular framework of the ionization beads but also characterizes the granular features in order to probe the source peculiarities [23–25]. The main point of this study is that we compute the multiparticle correlations to distinguish the image analysis and pattern recognition by using the concept of machine learning along with the simulation results from the data collected [11,26]. Such a comparison inspires the calculation of the cohesive portion for the considered source data analysis.

This chapter is divided into six sections. In Section 5.2, we describe the multiparticle correlations with source peculiarities for analyzing the source. We investigate the correlations within the completely and partially chaotic emission sources in Section 5.3 and Section 5.4, respectively. We discuss the model image results of our study in Section 5.5. Finally, in Section 5.6, we demonstrate our findings.

5.2 Source description and correlation functions

The data of source production from extraordinary energy systems have been widely used to seek deconfined states of particles generated at quantum-mechanical energies. These side impact energies shed new light on the formation of highly energetic plasma droplets of matter which are comprised of partonic degree and interferometry is a good tool to analyze the image formation of the sources for the pattern recognition [16]. During the majority of its entire lifespan, the matter mixture phase may comprise plasma droplets. We investigate particle expulsion from distributed droplets to examine the unique characteristics of matter created and use them to explore the source peculiarities within the image analysis. It is essential to emphasize that the two-particle quantum interferences and coherent portion are the fundamental components of correlation coefficients in order to measure the creation of coherence and symmetrical source structures [21,22]. Therefore the primary interference for two particles can be written as:

$$I_2(p_1, p_2) = 1 + w\Gamma^s(1, 2) + (1 - w)\Gamma^d(1, 2), \tag{5.1}$$

where $\Gamma^s(1, 2) = \exp(-q_{12}^2 r_s^2)$ and $\Gamma^d = \exp[-q_{12}^2(r_s^2 + R_m^2)]$ shows the correlators of two particles when they originate from one and several droplets, respectively. Here, $w = 1/D$ determines the inverse relation with the source droplets and the radii of the droplets as well as the whole source are represented by the symbols r_s, R_m, respectively. In Eq. (5.1), the subcomponents of the 2nd and 3rd terms occur due to the ejection of pions from alike and several droplets.

The correlations at higher levels comprised extra particulars than the primary intensity interferences and thus the investigation for the source geometry as well as the chaotic behaviors take place with the intensity interferometry by using the three particles that possess the boson nature:

$$I_3(p_1, p_2, p_3) = 1 + w\Omega_{2l}(\mu, \nu) + (1 - w)R_{2g}(\mu, \nu) + 2w^2\Omega_{3l}(\mu, \nu, \omega)$$
$$+ 2(1 - w)(1 - 2w)\Omega_{3d}(\mu, \nu, \omega)$$
$$+ 2w(1 - w)\Omega_{3d1}(\mu, \nu, \omega), \tag{5.2}$$

where p_μ shows the momenta of the particles that are ejected from the droplets D. The 2nd and 3rd terms in Eq. (5.2) indicate the participation of two particles when they are emitted from the one and various droplets, respectively. The contribution of three particles is presented in the 4th and 5th terms when they are emitted from the same and several droplets. In particular, the demonstration of three-particle emission in which two of them come from the same droplet and one from a different composition can be mentioned in the last term. It is quite interesting to explain the mathematical formulas in order to explore their characteristics and contributions.

Furthermore, it is more interesting and meaningful to illustrate the higher-level correlations in the form of the corresponding correlators in order to explore the internal degree of associations that helps us to investigate the source image and peculiarities:

$$
\begin{aligned}
I_4(p_1, p_2, p_3, p_4) = {} & 1 + w\Omega_{2l}(\mu, v) + (1-w)\Omega_{2D}(\mu, v) + w^2\Omega_{22l}(\mu v, \omega\lambda) \\
& + (1-w)^2\Omega_{22d}(\mu v, \omega\lambda) + 2w^2\Omega_{3L}(\mu, v, \omega) \\
& + 2(1-w)(1-2w)\Omega_{3D}(\mu, v, \omega) \\
& + 2w(1-w)\Omega_{26D}(\mu, v, \omega) + 2w^3\Omega_{4l}(\mu, v, \omega, \lambda\lambda) \\
& + 2(1-w)(1-2w)(1-3w)\Omega_{4d}(\mu, v, \omega, \lambda) \\
& + 2w^2(1-w)\Omega_{41d}(\mu, v, \omega, \lambda) \\
& + 2w(1-w)(1-2w)R_{42d}(\mu, v, \omega, \lambda).
\end{aligned}
\tag{5.3}
$$

The 2nd and 3rd expressions in Eq. (5.3) illustrate the participation when single pairs of bosons are ejected from the same and several droplets. The 4th and 5th terms indicate the interferences of boson particles when double pairs are ejected from the same and various droplets to show the quantum interferences. Moreover, the interference of three bosonic particles when they are ejected from the same and distinct droplets are shown in the 6th and 7th terms, respectively. However, the emission of two bosons from the same and one from the various droplets is shown in the 8th term. In particular, the interferences of quadruplets when all particles are ejected from the same and distinct droplets are shown in the 9th and 10th components of the above equation. The second to last term occurs due to the ejection of three bosons from the same and one from the various droplets. The contribution of the last terms appears when two boson particles come from the same and the remaining two bosons are emitted from different droplets.

In addition, the correlation that is extracted after the removal of single-pair interferences from the full four-boson correlations is known as the partial cumulant correlation I_a and can be expressed as:

$$
\begin{aligned}
I_a(p_1, p_2, p_3, p_4) = {} & 1 + w^2\Omega_{22l}(\mu v, \omega\lambda) + (1-w)^2\Omega_{22d}(\mu v, \omega\lambda) \\
& + 2w^2\Omega_{3L}(\mu, v, \omega) + 2(1-w)(1-2w)(R_{3G}(\mu, v, \omega) \\
& + 2w(1-w)\Omega_{31D}(\mu, v, \omega) + 2w^3\Omega_{4l}(\mu, v, \omega, \lambda) \\
& + 2(1-w)(1-2w)(1-3w)\Omega_{4d}(\mu, v, \omega, \lambda)
\end{aligned}
$$

$$+ 2w^2(1 - w)\Omega_{41d}(\mu, \nu, \omega, \lambda)$$
$$+ 2w(1 - w)(1 - 2w)\Omega_{42d}(\mu, \nu, \omega, \lambda). \tag{5.4}$$

Similarly, the second type of the partial cumulant interference I_b obtained in the absence of single- and double-pair interferences can be illustrated in the compact form as:

$$I_b(p_1, p_2, p_3, p_4) = 1 + 2w^2\Omega_{3L}(\mu, \nu, \omega) + 2(1 - w)(1 - 2w)(\Omega_{3D}(\mu, \nu, \omega)$$
$$+ 2w(1 - w)\Omega_{31D}(\mu, \nu, \omega) + 2w^3\Omega_{4l}(\mu, \nu, \omega, \lambda)$$
$$+ 2(1 - w)(1 - 2w)(1 - 3w)\Omega_{4d}(\mu, \nu, \omega, \lambda)$$
$$+ 2w^2(1 - w)\Omega_{41d}(\mu, \nu, \omega, \lambda)$$
$$+ 2w(1 - w)(1 - 2w)\Omega_{42d}(\mu, \nu, \omega, \lambda). \tag{5.5}$$

Furthermore, the particular correlation that is known as the cumulant or genuine correlation I_c for four-particle quantum interferences deals only with the quadruplet interferences and it is more meaningful than those of the full three- and four-particle interferences. Mathematically, it can be expressed in terms of the correlators as:

$$I_c(p_1, p_2, p_3, p_4) = 1 + 2w^3\Omega_{4l}(\mu, \nu, \omega, \lambda)$$
$$+ 2(1 - w)(1 - 2w)(1 - 3w)\Omega_{4d}(\mu, \nu, \omega, \lambda)$$
$$+ 2w^2(1 - w)\Omega_{41d}(\mu, \nu, \omega, \lambda)$$
$$+ 2w(1 - w)(1 - 2w)\Omega_{42d}(\mu, \nu, \omega, \lambda). \tag{5.6}$$

It is important to note that the distributions of the freeze-out zone are still uncertain, which causes the momentum-dependent interferences in the space of momentum to be muddled as well. This drives us to establish a solid method for carefully examining the particle-emission region and leads to exploring image analysis and pattern recognition with machine learning precisely. Investigating the characteristics of the particle-expelling sources, while taking into account granular and partially chaotic gaussian origins, is the main goal of this study. The related functions of the correlation for both scenarios can be stated independently for an attractive and insightful source presentation.

5.3 Methods explanation for coherent droplets

The small radius of a little droplet makes the production of particles from it substantially coherent and it is also considered that the generation of such particles from small solitary droplets seems to be coherent. There exist a substantial number of coherent particle radiation to affect the multiparticle couplings that have been significantly suppressed and such suppression explores the image analysis of the sources. We investigate the multipion BEC processes in a granular reference system with coherence emanations droplets to understand the exhibited destructions in the intercepts. We discover that the quantity of droplets in the particulate source has an impact on how the detections of multiparticle correlations behave to recognize the source pattern through machine vision. As a result, the quantum interferences of two particles for a particulate source can be demonstrated as:

$$I_2(p_1, p_2) = 1 + (1 - w)\Gamma^d(1, 2). \tag{5.7}$$

It is observed that the droplet count in the particulate source affects the multiparticle correlation equations intercepts at low relative momentum. As the quantity of droplets declines then the correlation starts to decrease gradually and they do as well. The triple-particle correlation values within the coherent emission of the stationary granular emitters are in general expressed as [27–29]:

$$I_3(p_1, p_2, p_3) = 1 + (1 - w)\Omega_{2d}(\mu, \nu) + 2(1 - w)(1 - 2w)\Omega_{3d}(\mu, \nu, \omega), \tag{5.8}$$

$$\begin{aligned} I_4(p_1, p_2, p_3, p_4) = {} & 1 + (1 - w)\Omega_{2D}(\mu, \nu) + (1 - w)^2\Omega_{22d}(\mu\nu, \omega\lambda) \\ & + 2(1 - w)(1 - 2w)\Omega_{3D}(\mu, \nu, \omega) \\ & + 2(1 - w)(1 - 2w)(1 - 3w)\Omega_{4d}(\mu, \nu, \omega, \lambda). \end{aligned} \tag{5.9}$$

Rather than estimating the coherent proportion at the correlation function's unquantified intercept, one might instead employ the idea of built or constructed response curves. Thus the elimination of interferences due to two particles and the two-particle plus pair-particle symmetrizations can be represented by the two partial cumulants I_a and I_b, respectively [30]. Particularly noteworthy are the partial cumulant Ia correlations that achieved after removing the single-couple interference and thus the partial cumulant correlations I_b, on the contrary measures a kind of interference in which only one- as well as double-sequence interferences are excluded and only the triplet and quadruplet play a significant role in such constructive interference:

$$\begin{aligned} I_a(p_1, p_2, p_3, p_4) = {} & 1 + (1 - w)^2\Omega_{22d}(\mu\nu, \omega\lambda) \\ & + 2(1 - w)(1 - 2w)\Omega_{3D}(\mu, \nu, \omega) \\ & + 2(1 - w)(1 - 2w)(1 - 3w)\Omega_{4d}(\mu, \nu, \omega, \lambda), \end{aligned} \tag{5.10}$$

$$\begin{aligned} I_b(p_1, p_2, p_3, p_4) = {} & 1 + 2(1 - w)(1 - 2w)\Omega_{3D}(\mu, \nu, \omega) \\ & + 2(1 - w)(1 - 2w)(1 - 3w)\Omega_{4d}(\mu, \nu, \omega, \lambda). \end{aligned} \tag{5.11}$$

Most saliently, the genuine correlations for the quantum interference of four bosons can be achieved by eliminating the consequences of the single pairs of bosons, double pairs, and triplet quantum interferences, respectively. Therefore the cumulant (genuine) correlation would appear only after all four particles take part in the quantum entanglements, as shown below [30,31]:

$$I_c(p_1, p_2, p_3, p_4) = 1 + 2(1 - w)(1 - 2w)(1 - 3w)\Omega_{4d}(\mu, \nu, \omega, \lambda). \tag{5.12}$$

Furthermore, when the emanations of bosonic particles occur due to the various droplets then the chaotic coefficients measure the degree of chaotic coherence for four, three, and two particles, respectively.

5.4 Formulation within partially chaos emission

In this section, we compute the quantum correlations of the multipion system and the system–environment entanglement with the partially coherence peculiarities. In these studies, we can procure information about the source's initial conditions and it is assertable to harvest meaningful correlations from the hybrid systems. More surprisingly, under such a particular condition when $r_s^2 + R_m^2 = R_{ga}^2$, the chaotic coefficient for two bosons that is more meaningful procures the unique feature and we can demonstrate this as $\Pi^{2g} = \Pi^{2d}$ [32]. Correspondingly, if the examined source seems to as only one drop to like $D = 1$ and the radius of the considered small source achieves the radius of a partly chaotic Gaussian source then symbolically it can be written as $r_s = R_{ga}$. Therefore the combined interpretation needs to behave as a single system coherent existence that curtails the partially chaotic systems specific behavior. The related relationship between the correlator and correlation can be manipulated as:

$$I_2(p_1, p_2) = 1 + \Omega(1, 2), \tag{5.13}$$

where $\Omega(1, 2)$ presents the two-boson correlator for the partial coherence image-analysis system of boson particles. Such a correlator exhibits the numerical values 1 and 0 for the chaotic and coherence peculiarities, respectively. Therefore it can manifest within the density matrix that contains the whole information about the source and its geometry [27,28]:

$$\Omega(v, \mu) = \frac{|\rho^{(1)}(p_v, p_\mu)|^2 - n_c^2 |u_c(p_v)|^2 |u_c(p_\mu)|^2}{\rho^{(1)}(p_v, p_v)\, \rho^1(p_\mu, p_\mu)}. \tag{5.14}$$

It is notable that the correlations of quantum-statistical interferences of extremely similar bosons are affected by the spatial degree and complexities of the particle-emissive source. Whereas two particle interferences are commonly evaluated in experimental studies, computational correlations of 3 bosons or above are underexplored. Here, is described a set of methodologies for isolating and analyzing 3- and 4-particle quantum numerical correlations. The method of distinctive designed correlation functions makes it easier to investigate the effects of coherence at limited comparative momenta rather than at the unquantified intercepts of the considered correlations. Therefore the three-boson correlation due to quantum interference acquired a mathematical form as:

$$I_3(p_1, p_2, p_3) = 1 + \Omega_{2l}(\mu, v) + \Omega_{3l}(\mu, v, \omega). \tag{5.15}$$

One can note that the correlators of two and three bosons contribute to the interference phenomenon in order to build the correlation of three particles. Specifically, the correlator $R_{3l}(\mu, \nu, \omega)$ due to the interference of three bosons that possess the numerical value 2 for an ideal chaos emission and the suppression in the measured values occur for the partially nonchaotic sources. According to the quantum-statistical principle and Bose statistics, the three-boson correlation exhibits the numerical value 6 at the intercept provided that the source has completely chaotic peculiarities. However, any reduction from the thermal limit presents an indication of an ideally coherent or partially coherence system.

In addition, the expression for the interferences of four bosons that are composed of correlators of quadruplets, triplets contributions, double-pair, and single-pair interferences can be expressed as:

$$I_4(p_1, p_2, p_3, p_4) = 1 + \Omega_{2L}(\mu, \nu) + \Omega_{22l}(\mu\nu, \omega\lambda) + \Omega_{3L}(\mu, \nu, \omega)$$
$$+ \Omega_{4l}(\mu, \nu, \omega, \lambda). \tag{5.16}$$

From Eq. (5.16) it is obvious that the considered correlation shows the value of 24 in the case of chaotic sources but it deviates from this chaos value due to the tiny source that behaves partially coherent. Moreover, there are two types of partial cumulant correlation functions while studying the correlation about four pions. One of them is represented by I_a, as shown in Eq. (5.17) that contains only double pairs, triplets, and four-pion interference and removed only the single-pair interference. It possesses the value of 18 for chaotic sources and any suppression indicates the appearance may be due to the coherent components.

According to the aforementioned equation, the evaluated correlation has a real worth of 24 in the particular instance of turbulent sources, but that diverged from such a chaotic system valuation because of the small source that performs slightly coherently. Furthermore, when studying the significant relation between four particles, there are two kinds of partial cumulant scaling functions. The first of these is signified by I_a in Eq. (5.17), which includes double combinations, triplets, and four-particle intervention and excludes only one pair interruption. It has a real numerical worth of 18 for chaos sources as well as any deprivation demonstrates that the presence could be due to the constituents of coherent components:

$$I_a(p_1, p_2, p_3, p_4) = 1 + \Omega_{22l}(\mu\nu, \omega\lambda) + \Omega_{3L}(\mu, \nu, \omega) + \Omega_{4l}(\mu, \nu, \omega, \lambda).$$
$$\tag{5.17}$$

Moreover, Eq. (5.18) presents the cumulant correlations I_b that are composed of the quadruplets and triplets interferences and it isolates the double pairs, as well as single boson pair quantum interferences. Such a correlation exhibits a thermal limit for the chaotic source of the order of 15. However, the coherence peculiarities suppress the thermal limit, which shows the characteristics of the coherent source. Mathematically, it can be illustrated as:

$$I_b(p_1, p_2, p_3, p_4) = 1 + \Omega_{3l}(\mu, \nu, \omega) + \Omega_{4l}(\mu, \nu, \omega, \lambda). \tag{5.18}$$

There is a specific kind of correlation that deals with pure quadruplets interference and it is obtained by the elimination of triplet-, single-pair, and double-pair interferences. Such special correlation illustrates the influence of coherence and source geometry significantly more than those of the full and cumulant correlations. It possesses the numerical value of seven at the intercept where all particles acquire identical momenta and measured the zero relative momenta. It explores the image analysis and pattern recognition of the source with the help of machine learning. The mathematical expression for such a correlation can be expressed for Gaussian sources as:

$$I_c(p_1, p_2, p_3, p_4) = 1 + \Omega_{4l}(\mu, \nu, \omega, \lambda). \tag{5.19}$$

In particular, the more captivated parameters that measure the internal peculiarities of the source are called chaotic parameters or coefficients. Specifically, various other considerations also influence the correlations that imply that the hybrid emanated source is composed of granular characteristics. We know that if the particles are disseminated after their emanation then some specific pieces of information can be lost. The measurements with long path differences and the ejection of copious particles can help to save all the information carried by particles from the source to the detectors.

However, the elucidation of the quantum correlations undergoes convolution if the mechanism of pion emanation proceeds via the familiar effect of resonances [21,25]. The elimination of these considerations can be achieved with the normalization of higher-level correlations

5.5 Model results and discussion

5.5.1 Source performance with three-particle correlations

In this section, we discuss how the granular structure model affects the observed correlation functions when used to investigate condensate and the geometry of the source. It is due to the purpose of this study drives the peculiarities of quantum statistics approaches to look for both the coherence and chaotic fraction for the image analysis. With the use of machine-learning data, we can arrive at our goal of quantifying the outcomes of our computational measurements and determining the coherence percentage during collisions at exceptional energies.

We first examine the correlation functions I_3 against invariant comparative momenta $q_3 = \sqrt{q_{12} + q_{13} + q_{23}}$ for origins with varying numbers of small sources [32–34]. According to quantum statistics, for an ideal chaos source at the interception ($q_3 = 0$) all particles possess the identical momenta and I_3 must preserve its maximum values of six. The I_3 increases as the number of small droplets grow significantly and conversely, as shown in Fig. 5.1. The explanation about the source peculiarities and why a source behaves in a random manner steadily rises with droplet count is patently clear: When there are plenty of droplets present there is a high likelihood that particles may erupt from different drops and seem to be the origin of the chaotic system. On the contrary, fewer

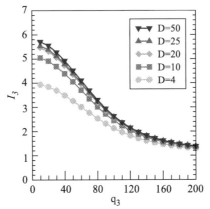

FIGURE 5.1 Quantum correlations I_3 vs q_3 for three particles within the granular sources that possessed the droplets D at $R_m = 6.0$ fm and $r_s = 1.5$ fm.

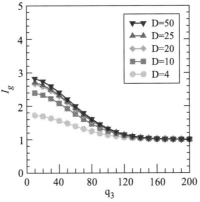

FIGURE 5.2 Genuine quantum correlations I_g vs q_3 for three particles within the granular sources that possessed the droplets D at $R_m = 6.0$ fm and $r_s = 1.5$ fm.

droplets match the high coherence percentage, which results in rapid suppression of the correlations. Therefore the high coherence proportion in the image analysis is caused by the emission of particles from the smallest droplets. We also observe that in the scenario of the tiny proportion of droplet $D = 4$ the corresponding correlations for three particles are dampened, which also explores the image analysis and pattern recognition very well.

Additionally, Fig. 5.2 displays the genuine correlations I_g for three bosons against q_3 within the source that composed the droplets. It is obvious from the spectrum that the presented correlation approaches unity slightly faster than the full correlations. This is because the genuine correlations only comprise the interference of triplets and the quantum interferences effect shows its absence if any pair emanate from the same droplets. It seems strange behavior in the availability of strong coherence portions that would reduce such connections significantly. However, in the circumstances of genuine correlations, the intercept decreases substantially more than that of full interferences for three bosons due to the occurrence of the condensation phenomenon, as shown in the figures, which are substantially more obvious in true correlations than in full quantum interferences for three pions.

The stochastic maximum limit at zero relative momentum seems to be almost three with the numerous droplets, $D = 50$, as can be seen from the figure that explores the consequences of the source characteristics. The correlations are severely decreased by the magnitude of small droplets ($D = 4$) associated with condensation production that causes the emanation of pions from those identical coherent droplets to fluctuate greatly at large q_3. The correlations I_g exhibit an odd reduction even at low q_3 when the concentration of droplets is relatively small, which illustrates the condensation aspects and analyzes the source image analysis.

Furthermore, the average transverse momenta with the parameters $R_g = 6$ fm, $r_s = 1.5$ fm have been extricated corresponding to the droplets of the particle-emission region that provides extensive information about the hadronic zone consistent with the measured practical and our simulated data for image analysis. The reason for this is quite obvious; due to the presence of boost velocity of expansion particles in the sense of minute droplets for a larger momentum in the transverse direction, which also contribute to lower temperature at the freeze-out region.

Moreover, in Ref. [24], the mean momentum of transverse with said specifications $R_g = 6$ fm, $r_s = 1.5$ fm about the measured temperature of the particle emanated area has been extracted, which provides comprehensive information about the particle zone sustained with the evaluated functional and our computational results. The possible explanation for this is indeed very noticeable about the source because of the existence of enhanced kinetic energy of exploration particles in the context of a small temperature for a wider momentum in the direction are perpendicular, which further contributes to a lower temperature at the region of freeze-out.

5.5.2 Source analysis with four-particle correlations

Particulate interferometry is a vital tool for investigating the space–time configuration of the emanated sources formed by collisions of heavy particles. Experimental results not only allow us to acquire a high level of resolution in all three aspects for the four-particle correlation functions but they also offer additional sufficient sequence data to probe numerous correlation coefficients for exploring the structure of matter. The fundamental question that three- or more particle femtoscopy seeks to address and answer is whether there are extra significant relationships much further than the Bose interference of equivalent particles from the chaotic source. On these suppositions, we presume the four

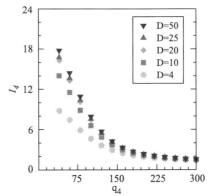

FIGURE 5.3 Quantum correlations I_4 vs q_4 for four particles within the granular sources that possessed the droplets D at $R_m = 6.0$ fm and $r_s = 1.5$ fm.

and many more particulate correlation functions in this paper and investigate if the two-boson correlation is sufficient to explain the multiparticle system.

We make a comparison of recent statistics on multiparticle interferences specifically to probe the image analysis and pattern recognition for the emanated sources. Any discrepancy from this predictive model in the computation indicates fundamental science, such as cohesive emissions or true relatively high coherence due to an identical image pattern. Using the evaluated primary correlation as the contribution we also acquire a parametric prognostication of significant correlation features.

Higher-level correlations are identified as four-boson correlations and it is denoted by I_4 that have been projected versus variational relative momenta $q_4 = \sqrt{q_{12}^2 + q_{13}^2 + q_{14}^2 + q_{23}^2 + q_{24}^2 + q_{34}^2}$, as shown in Fig. 5.3. The quantum statistic I_4 should have a numeric value of 24 at the intercept $q_4 = 0$, where all bosons acquire identical momenta for the chaotic particles expelling sources. This is also known as the limit of the chaotic or thermal system. As the number of tiny small droplets intensifies then the relevant particles possess the quantum interferences that show their effectiveness on the correlation images. Also, since sources with a significant number of droplets have a strong chance of emitting bosons from independent droplets they obey Bose statistics. Such meaningful interferences, in contrast, are markedly decreased due to a lack of tiny droplets due to the coherence aspects of the source [35,36].

The obvious reason for this is that fewer droplets directly correlate to more condensation configuration that incentivizes reducing the strength of the correlations. It is more important and meaningful to discuss that the stability of the correlation at higher numerical order is significantly useful, particularly in comparison to the three correlations or primary correlation, even though the quantity of drops is much greater. In actuality, within the higher-level interfer-

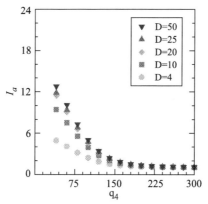

FIGURE 5.4 Partial cumulant quantum correlations I_a vs q_4 for four particles within the granular sources that possessed the droplets D at $R_m = 6.0$ fm and $r_s = 1.5$ fm.

ence scenario, the condensate composition possesses the peculiarity to continue its domination significantly.

Moreover, the kind of higher-level correlation in which single-pair interferences are eliminated is known as the partial cumulant and it is represented by I_a. Such a correlation at the intercept where all the bosons possess alike momenta to acquire the thermal limit 18, as shown in Fig. 5.4. One can see that the correlation increases with rising number of droplets and decreases its numerical value below a certain number of droplets. The results show considerable suppression at the droplet quantity $D = 4$ due to the influence of coherence components. The reduction of droplets invites the Bose condensation that affects the correlation significantly at the intercept and the results further start to decrease with the wide regime of the q_3. These results indicate that the interference effect disappears at wide relative momentum and such meaningful expressions are useful for exploring the source geometry as well as the peculiarities of the image analysis.

The pillars of the theories of quantum computation and quantum information, as well as having basic implications for quantum physics, are quantum correlations and coherence. The field of quantum data processing has long been interested in finding physically acceptable and mathematically sound quantum-mechanical notation for them. Previously, various steps have been implemented. In this research, we investigate the numerous quantum interference concepts that multiparticle phenomena contain as well as the number of suggestions about coherence metrics for a unique quantum system [37,38]. We make an effort to offer comprehensive details concerning quantum coherent development such as its applicability throughout many systems.

Furthermore, the correlation that deals only the triplets and quadruplets is known as the partial cumulant correlation and it is expressed by the symbol I_b. Such a correlation is obtained by eliminating the single and double pairs interferences from the full four-boson interferences. Fig. 5.5 versus q_4 demonstrates

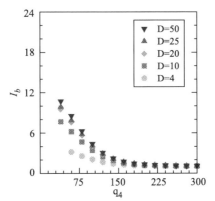

FIGURE 5.5 Partial cumulant quantum correlations I_b vs q_4 for four particles within the granular sources that possessed the droplets D at $R_m = 6.0$ fm and $r_s = 1.5$ fm.

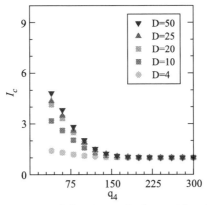

FIGURE 5.6 Genuine quantum correlations I_c vs q_4 for four particles within the granular sources that possessed the droplets D at $R_m = 6.0$ fm and $r_s = 1.5$ fm.

the partial cumulant boson correlation and it is obvious that the I_b examines the influences of coherence on the correlations. It shows the enhancement with the increasing magnitude of the droplets and decreases with the deficiency of sub-sources. The thermal limit of such a correlation possesses the numerical value of 15 at the intercept. The deviations for the spherical source from the thermal limit represent the indication of the coherence and the asymmetric characteristics. One can see an obvious suppression with the droplets $D = 4$ due to the coherence emissions.

In particular, the correlation for four bosons is known as the genuine correlation and is represented by I_c. This particular type of correlation possesses pure quadruplet interferences and the triplet-, double-, and single-pair interference are eliminated. Fig. 5.6 against relative momentum q_4 demonstrates the pure correlations in order to analyze the source geometry and condensations, respectively. The thermal limit of genuine correlation at q_4 is seven for symmetric and chaotic sources. The enhancement of results increases with the droplets and starts to suppress the source exhibiting the limited droplets [39–41]. The appearance of such a reduction phenomenon is caused by the fact that the existence of three- or two-particle emissions from the coherence component contribute zero interference with the real full correlations, since the production of any three or two bosons from identical droplets has such a substantial impact. This specific correlation shows a larger suppression than those of the three- as well as fourth-order correlations, which suggests that the genuine correlation is more sensitive to exploring the source peculiarities in wide regimes of relative momentum. It should be obvious that even though there are many droplets worth 50 the results show an obvious decline well below the catastrophic limit of 7. Analyzing the aforementioned figures in particular reveals that the I_c values are substantially more susceptible to suppression than the I_g, I_a, and I_b responses. The interesting explanation is that the higher-order interferences contained significant aspects of additional signals that are not present in primary-level interferences.

In this chapter, we concentrate on quantum interferences by using granular and nongranular model correlation equations. We compute the quantum correlations of the multipion system and the system–environment entanglement with coherence peculiarities in order to investigate the image analysis for deep learning. In these models, we can procure information about the source's initial conditions and it is assertable to harvest meaningful correlations from the hybrid systems for pattern recognition. We explore how the quantum interferences oscillate with the droplet's magnitude and relative momentum at constant radii. We noted that the correlations due to quantum interferences in particle emanated system declined first with the deficiency of the droplets due to coherence pattern and then start to increase monotonically to a persistent as the droplets increased from a value close to the chaotic system.

5.6 Summary and conclusions

We have explored several quantum correlations for the sources image analysis that emanated particles with traditional as well as granular peculiarities and perceived the true three- and four-particle interferences that are noticeably suppressed. The specific connections among the correlations and the source droplet magnitude are probed to examine the coherent components as well as pattern recognition in wide momenta regimes. It is ascertained that the particular correlations are delicated to the quantity and size of the droplets during the evolved source characteristics. Such meaningful correlations dominated an obvious suppression by decreasing the droplets magnitude due to the occurrence of high condensation constituents at low droplet quantity, but induce obvious clear chaotic image peculiarities with the substantially huge numbers of emanated sources. The partial coherence of the pion production within the collisions is one tenable and practical explanation for these novel results. Therefore harnessing particle femtoscopy in this analysis, we analyzed various special correlations in this chapter to investigate the information about chaotic-coherent dynamics and the intrinsic morphology of the particle-generating sources for pattern recognition.

We analyzed the source-probing concept with the granular tiny subsources configuration of increased energy extracellular fluid that can identify the information of multiparticle correlations at relativistic energies using a futuristic and innovative method of data analysis. We conclude that pions exhaled from relatively energetic smashing demonstrated the specificity of their source materials and also contribute substantiation for phase transformation inside of the considered drops in particular regimes by measuring distinguishable correlations. As a result, in the final phases of the evaluated sources, the Bose interferences reveal explicit chaos principles about multiple sizes measurements in order to analyze the conclusive measurements about pattern recognition to support the machine-vision concept. We also realized that as the number of droplets intensifies then the corresponding correlation effectiveness of the scaling function increases exponentially, resulting in large transverse momenta, which is consistent with the innovative notions.

As a consequence, we evaluated by comparing the four- and three-boson interferences along with the genuine correlations in order to explore the image analysis. These findings investigated the evaluated substantial gorgeous tentatives in these computations with the obtained results that are susceptible to the geometrical and consistency particularities for four-particle comparison to third-order interference to demonstrate the presence of coherent constituents during the image formation for machine vision. Such marginalization is continuous with the source coherence configuration that emanates particles cohesively at low momentum and is persistent to support the experimental measurement statistics at pattern recognition. It can also be corroborated that such an innovative scientific method applies to a wide range of engineering applications. In the future, we should indeed focus on fractional coherence-chaotic formation within the applications of the different equations about the source image peculiarities in oceanography.

Acknowledgments

The authors would like to thank Prof. Wei-Ning Zhang for invigorating discussions about this chapter.

References

[1] J.P. Blaizot, F. Gelis, J.F. Liao, Nucl. Phys. A 873 (2012) 68.
[2] M. Gyulassy, S.K. Kauffmann, L.W. Wilson, Phys. Rev. C 20 (1979) 2267.
[3] U.A. Wiedemann, U.W. Heinz, Phys. Rep. 319 (1999) 145.
[4] M.M. Aggarwal, et al., Phys. Rev. C 67 (2003) 014906.
[5] G. Bary, Chaos Solitons Fractals 152 (2021) 111414.
[6] G. Goldhaber, S. Goldhaber, W.Y. Lee, A. Pais, Phys. Rev. 120 (1960) 300.

[7] I.V. Andreev, M. Plumer, R.M. Weiner, Int. J. Mod. Phys. A 8 (1993) 4577–4626.
[8] H. Boggild, et al., Phys. Lett. B 455 (1999) 77.
[9] J. Adams, et al., Phys. Rev. Lett. 91 (2003) 262301.
[10] G. Bary, et al., Fractals 31 (2023) 2340161.
[11] B.B. Abelev, et al., Phys. Rev. C 89 (2014) 024911.
[12] U.W. Heinz, Q.H. Zhang, Phys. Rev. C 56 (1997) 426.
[13] U. Ornik, M. Plumer, D. Strottmann, Phys. Lett. B 314 (1993) 401–407.
[14] J.P. Blaizot, F. Gelis, J. Liao, et al., Nucl. Phys. A 904–905 (2013) 829c–832c.
[15] E. Ikonen, Phys. Rev. C 78 (2008) 051901.
[16] P. Ru, G. Bary, W.N. Zhang, Phys. Lett. B 777 (2018) 79–85.
[17] T. Csorgo, Acta Phys. Hung. A 15 (2002) 203–257.
[18] B.L. Friman, K. Kajantie, P.V. Ruuskanen, Nucl. Phys. B 266 (1986) 468–486.
[19] G. Bary, P. Ru, W.N. Zhang, J. Phys. G 45 (2018) 065102.
[20] D. Seibert, Phys. Rev. D 41 (1990) 3381.
[21] S. Pratt, P.J. Siemens, A.P. Vischer, Phys. Rev. Lett. 68 (1992) 1109.
[22] W.N. Zhang, Y.M. Liu, L. Huo, et al., Phys. Rev. C 51 (1995) 922.
[23] W.N. Zhang, Y.Y. Ren, C.Y. Wong, Phys. Rev. C 74 (2006) 1832–1838.
[24] W.N. Zhang, M.J. Efaaf, C.Y. Wong, Phys. Rev. C 70 (2004) 024903.
[25] W.N. Zhang, Z.T. Yang, Y.Y. Ren, et al., Phys. Rev. C 80 (2009) 044908.
[26] J. Adam, et al., Phys. Rev. C 93 (2016) 054908.
[27] G. Bary, W. Ahmed, R. Ahmad, Eur. Phys. J. Plus 138 (2023) 771.
[28] G. Bary, P. Ru, W.N. Zhang, J. Phys. G 46 (2019) 115107.
[29] G. Bary, Chaos Solitons Fractals 164 (2022) 112572.
[30] G. Bary, W. Ahmed, R. Ahmad, Results Phys. 32 (2022) 105075.
[31] G. Bary, W. Ahmed, R. Ahmad, Fractals 30 (2022) 2240186.
[32] Y. Karaca, Y.D. Zhang, M. Khan, Expert Syst. Appl. 144 (2020) 113098.
[33] Y. Karaca, C. Cattani, Adv. Math. Mod. Appl. 4 (2019) 5–14.
[34] Y. Karaca, C. Cattani, M. Moonis, Complexity 2018 (2018).
[35] R. Ahmad, A. Farooqi, G. Bary, et al., Fractals 30 (2022) 2240171.
[36] R. Ahmad, A. Farooqi, J. Zhang, et al., IEEE Access 8 (2020) 141057–141065.
[37] Y. Karaca, Z. Aslan, C. Cattani, et al., J. Med. Syst. 41 (2017) 1–10.
[38] G. Bary, W. Ahmed, Riaz Ahmad, Fractals 30 (2022) 2240125.
[39] R. Ahmad, A. Farooqi, R. Farooqi, Complexity 2021 (2021).
[40] A. Sohail, Y. Zhang, G. Bary, Int. J. Theor. Phys. 57 (2018) 2814–2827.
[41] Y. Wang, G. Bary, R. Ahmad, Math. Probl. Eng. 2021 (2021).

Chapter 6

Deep CNNS and fractal-based sequence learning for violence detection in surveillance videos

Fath U Min Ullah[a], Khan Muhammad[b], and Sung Wook Baik[c]

[a]School of Engineering and Computing, University of Central Lancashire (UCLan), Preston, United Kingdom, [b]Department of Applied Artificial Intelligence, Sungkyunkwan University, Seoul, South Korea, [c]Sejong University, Seoul, South Korea

6.1 Introduction

Video-based violence detection (VD) has attracted a large amount of attention for its wide range of applications such as security, safety, etc. Regarding this, wide usage of surveillance cameras, smart phone cameras, and body-worn cameras have played a vital role in the development of automatic VD. Real-time monitoring of crowded and abnormally susceptible areas requires a system that supports the law-enforcement and security agencies. As action recognition has widely focused on detection of simple activities, such as running, walking, etc., the VD or detection of abnormal events have been less explored and studied. Such a capability and involvement will be highly convenient in surveillance-based scenarios held in markets, organizations, institutes, or subways. Different VD methods with support of machine learning, deep learning, and traditional learning have been described for their efficiency and accuracy improvement.

6.1.1 Related work

Violence is an aggressive or abnormal action that intends to affect or harm the state of an object, pet, or human through a physical force. Detection of this activity is mandatory to ensure the safety and security of certain areas under surveillance. Numerous techniques have been proposed by researchers dealing with the detection of violent scenes in surveillance videos using conventional and DL-based approaches to analyze the abnormal activities. We categorize the VD methods into two major sub-sections and their details are given below.

Intelligent Fractal-Based Image Analysis. https://doi.org/10.1016/B978-0-44-318468-0.00013-1
Copyright © 2024 Elsevier Inc. All rights reserved, including those for text and data mining, AI training, and similar technologies.

6.1.1.1 Conventional learning-based VD methods

Over the last years, a great amount of research has been done in the VD regarding conventional learning approaches. For instance, Deepak et al. [1] extracted the motion cues via a spatiotemporal features descriptor for the detection of unusual scenes in surveillance videos. They applied a support vector machine (SVM) to distinguish the violent class from non-violent cases. Next, Senst et al. [2] focused on the crowd violence through Lagrangian direction fields and considered the long-term motion information, background motion, appearance, and spatio-temporal model for the detailed investigation. They applied an extended bag-of-words (BoW) procedure via late fusion and demonstrated the vitality of the Lagrangian integration for capturing the temporal scale. Similarly, Zhang et al. [3] enhanced the motion weber local descriptor for VD and added the temporal component to collect motion information and an improved weber local descriptor to depict the lower image appearance. They developed a classification model based on sparse representation to control the reconstruction error. Considering this model, a dictionary having atoms corresponding to labels is learned to detect the final violent scenes. Furthermore, Cosar et al. [4] fused the output object trajectory to integrate the pipeline and apply pixel-based analysis to identify the abnormal behavior in the surveillance videos. The trajectories helped them in computing the speed and direction of the objects movement as well as the behavior related to the object's finer motion. Ullah at el. [5] shed light on the internal mechanism of VD considering both the neural networks and traditional featuring engineering methods. They provided the emerging trends that existed in VD and the working mechanism applied for surveillance scenarios.

6.1.1.2 VD methods using deep learning

DL has already played its significant role in a different research community since it emerged. Several ConvNets have been practiced dealing with distant tasks on account of their better performance for activity analysis, edge vision [6], time series analysis [7,8], etc. Recently, Kang et al. [9] proposed a pipeline combined with a conventional 2D convolutional neural network (CNN) where a frame-grouping strategy is applied to make 2D CNN capable of learning the spatio-temporal representations. They averaged the input frames channels and made their group an input to the 2D CNN and developed a lightweight spatial and temporal attention module to improve the performance of VD. Next, Accattoli et al. [10] combined the 3D CNN and SVM to detect the individual's aggressiveness that lead to violent activity. Similarly, Samuel et al. [11] proposed a spark framework where the features from individual frames are extracted using a histogram of oriented gradients (HOG) after the frames are separated. They labeled the frames based on the features as violence model, negative model, and human part model that are applied for training the bi-directional long short-term memory (B-LSTM) to detect the violent scenes at the end. Further, Ullah et al. [12] investigated 3D ConvNet for VD by incorporating object detection prior to the final VD. Furthermore, Serrano et al. [13] proposed a fusion approach where the binary robust invariant scalable keypoints (BRISK) are extracted to capture motion and appearance from the sequence, which are then fed to a Hough–Forest classifier. Wang et al. [14] investigated the brute force detection and deep learning-based facial recognition and proposed a method by combining CNN and the trajectory to detect the violent scene.

6.1.2 Violence detection working flow

Detection of violent scenes using DL is generally based on CNN, which has achieved a remarkable performance regarding the object detection and classification task. The working flow of VD consists of data acquisition, its preprocessing, features extraction, features learning, and classification. Initially, the data collection step held where the data is acquired from a surveillance camera installed in indoor and outdoor environments. With the increase in demand for security and safety, surveillance-based video analysis has become an important arena in the research community [15]. Therefore we strongly emphasize surveillance-based setups. First, the video data obtained from the acquisition step is converted into frames/sequences of frames. Subsequently, the data is refined using diverse cleaning filters to enrich the quality of the images that assists the network in effective processing. These enhancement filters contain smoothing filters and contrast adjustment that affects the image corners and edges. Sometimes, the Gaussian filters and ideal low-pass ones are also applicable to improve the data. Once the data are cleaned, the data are ready for processing and training. Another step is the features extraction that is one of the basic steps for meaningful information collection from the huge amount of data for processing. Various features are extracted from video frames such as appearance with motion, motion space and time, speed direction and centroid, optical flow and motion region, motion vector, acceleration and movement, speed direction and density, motion and postures, motion speed and direction, movement direction and speed, spatiotemporal and motion streams, and corners with motion blobs [16]. VD methods based on these features were explained in a previous section. After the feature's extraction, the features are learned and the scene is classified as violent/non-violent in the last step. The overall steps held in the process are given in Fig. 6.1.

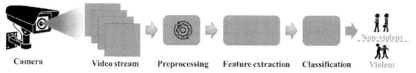

FIGURE 6.1 Generic overview of the VD system where the video data is acquired from the surveillance camera and fed into preprocessing step prior to features extraction. After the features extraction, the label is assigned as violent or non-violent based on the probability.

6.1.3 Recurrent neural network

Recurrent neural networks (RNN) are the class of deep neural networks assisting in modeling the sequence data. RNN exhibits the same behavior as the human brain works as they are derived from a feedforward neural network. RNN are a type of artificial neural network that became more popular in the recent decade due its powerfulness for learning the time series data. RNN acts as a special network, unlike conventional feedforward networks, they have recurrent connections that made the network able to process the arbitrary sequential input data. RNN investigate the hidden sequential patterns in both the spatial and temporal sequential data [17]. RNN process the data via considering the input at different time steps over time. It has two inputs, where the first input is obtained from the previous RNN hidden state, while the second input is acquired from the series data. A standard form RNN network is given in Fig. 6.2(a), where the data is processed via taking the input at distinct time steps with time. RNN obtains the input at each time stamp and provides the output. At each time stamp, taking the input disturbs the sequence patterns in series data because of weight multiplication with data and addition with bias. Therefore data multiplication at several times with hidden states forget the impact of early sequential series. This problem is knowns as the vanishing gradient problem, as illustrated in Fig. 6.2(b).

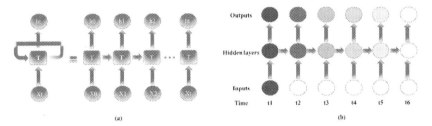

(a) (b)

FIGURE 6.2 (a) Standard RNN network (b) Vanishing gradient problem in RNN.

6.1.4 Long short-term memory

Compared to RNN, long short-term memory (LSTM) handles the data for an extensive time period in memory. LSTM solves the vanishing gradient problem and possesses the ability of learning long-term sequential dependencies. It has a special structure containing input, output, and forget gates controlling the long-term sequential pattern recognition in series data. These gates are adjusted via a sigmoid function that is connected to each unit in a training process that determines what state it requires to be open or close. Eqs. (6.1)–(6.7) describe the operations held in the LSTM unit, where x_t acts as the input at time t, where the input is given in small pieces from a long sequence. f_t represents forget gates that clear the information from a memory cell when required and retain the information record of the previous frame where information needs to be cleared from memory. The output gate o_t retains information regarding the forthcoming step where the recurrent unit g has the activation function "tanh" that is computed from the current frame input and the previous state frame s_{t-1}. Through the memory cell c_t and tanh activation, the RNN hidden state is calculated. Each state output is used in different tasks, although VD does not need the intermediate LSTM output. Therefore the final decision is accomplished through applying the softmax classifier at the final LSTM state that gives the label and probabilities of the violent and non-violent class:

$$i_t = \sigma((x_t + s_{t-1})W^i + b_i) \tag{6.1}$$

$$f_t = \sigma((x_t + s_{t-1})W^f + b_f) \tag{6.2}$$

$$o_t = \sigma((x_t + s_{t-1})W^o + b_o) \tag{6.3}$$

$$g = tanh((x_t + s_{t-1})W^g + b_g) \tag{6.4}$$

$$c_t = c_{t-1}.f_t + g.i_t \tag{6.5}$$

$$s_t = tanh(c_t).o_t \tag{6.6}$$

$$Predictions = soft\,max(V s_t). \tag{6.7}$$

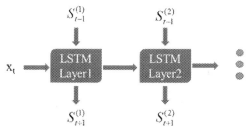

FIGURE 6.3 LSTM network sample layers. The input at the first layer is managed at time 't', while the latter is fed into the 2nd layer to be processed in a detailed manner.

Recent studies reveal that increasing the amount of data boosts the accuracy of any machine learning task. To this purpose, learning the complex video data patterns is a challenging task and, in such cases, a single LSTM is not effective. Therefore researchers have introduced a multi-layer LSTM network that is formed via stacking various LSTM cells for learning the long-term dependencies. See Fig. 6.3.

6.1.5 Gated recurrent unit (GRU)

A GRU network has similarities to LSTM, however, it is easier and simpler to implement where the gating signal is reduced to two from the LSTM. It has two gates, such as an update gate and a reset gate. Its internal structure of working gates is given in Fig. 6.4(a) and Fig. 6.4(b). The functionality of reset gates is the same as a forget gate of LSTM, i.e., the GRU possesses various similarities to LSTM. Hence, the Adam optimizer is applied for back propagation through time (BPTT) that minimizes the training error by avoiding a local minimal point.

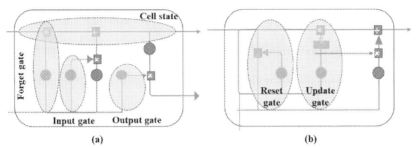

(a)   (b)

FIGURE 6.4 Internal structure of each sequential network, where (a) LSTM has three main gates such as input, output, and forget gates, while (b) GRU has two gates including a reset gate and an update gate.

6.2 Violence detection methods

Over the decades, several deep-nets and hand-crafted-based VD approaches have been presented by researchers. Most of the earlier work was based on hand-crafted features where an action has to be taken in a simple background. These systems extract the low-level features from video data and feed them into a classifier such as KNN, decision tree, and SVM to detect the violent scenes, although, current methods utilize the most realistic video data to the violent scenes. These methods widely rely on large-scale video data and are based on CNN. For instance, a VD method is presented by Ullah et al. [12] where they developed a three-staged DL-based framework. Initially, they detected the persons in the surveillance video streams using a light-weight CNN model to overcome and reduce the processing of useless frames. Similarly, they passed the sequence of 16 frames having the detected persons into a 3D CNN to extract the spatiotemporal features and fed them into a softmax classifier giving the final prediction. An alert is generated in the case when a violent scene is detected in the sequence for immediate action to be taken. Their framework is presented in Fig. 6.5 for VD.

Another VD method was proposed by Ullah et al. [15] where industrial surveillance scenarios are considered for their detail investigation. First, they processed the incoming surveillance frames by a CNN model to obtain the important shots by detecting the persons and vehicles using Mask R-CNN. This procedure helped them in reducing the processing time and computational costs. The sequence of frames with the detected objects (Humans and Vehicles) are conceded into the feature's extraction stage. Optical flow is used to extract the temporal features from these frames and the extracted features are fused with the features extracted via the CNN model. Finally, they added an LSTM network for final feature map generation to learn the violence patterns. Their proposed framework based on CNN-LSTM is illustrated in Fig. 6.6.

Another method proposed by Ullah et al. [18] introduced an AI-assisted industrial internet of things (IIoT)-based VD framework. First, they passed the input frames into a lightweight CNN model to extract the meaningful frames

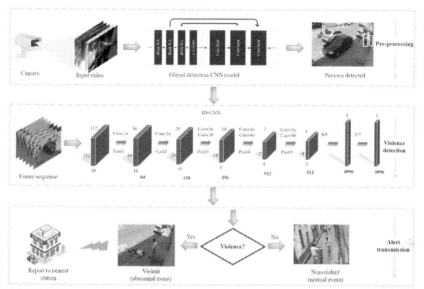

FIGURE 6.5 Framework of the VD method in videos using 3D CNN. Data acquisition and its preprocessing for object detection is held in the first step, while the next step extracts deep features from the refined sequence via a 3D CNN model for final violent activity detection. In the last step, if an activity is detected as violent, then the information is reported to the nearest station [12].

containing object information such as a human or any suspicious object such as a knife or a gun. If any suspicious object is detected, an alert is produced over the IoT network as an earlier VD. Only the frames having the object are forwarded into the cloud to extract the features using ConvLSTM. The latter obtained by the ConvLSTM is given to a GRU for final detection of the violent scene. The framework is given in Fig. 6.7.

Seerano et al. [13] proposed a method that aimed to classify the fight and non-fight sequences. First, they assumed that a single image can summarize a sequence. The frames from the sequence are accumulated to build a single representative image. Next, they proposed to leverage the zones in the frames that may be important to describe the motion in sequence. In this way, the significant motion parts are weighted while the rest of the regions receive less attention, the later part usually corresponds to static background and noise. The features collection phase intends to attain a descriptive image from a sequence of video, while 2D CNN is applied for final classification of the image and obtain the final decision. This is well illustrated in Fig. 6.8.

Roman et al. [19] introduced a method for violent action detection and localization via temporal video annotations. They followed a two-stage approach: first, classifying video as violent or non-violent, secondly, localization of the violent region. Their method is based on convolutional networks and dynamic images, and achieved results closer to the state-of-the-art. They analyzed the worth of dynamic images for violent motion representation in videos. Instead of

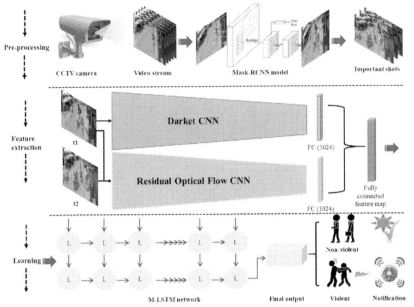

FIGURE 6.6 Proposed framework for VD in video streams for secure surveillance [15]. First, the video frames are preprocessed and scrutinized for object detection using a CNN model. After object detection, the frames are fed into the features extraction phase. Similarly, the temporal optical flow features from the pair of two consecutive frames are extracted that are concatenated with the features extracted from the Darknet CNN (Darket). Finally, the final features map is forward propagated into an LSTM network for final VD.

using an optical flow, the dynamic images allowed them to analyze the temporal information and avoid the computational cost of applying optical flow. Their proposed framework is given in Fig. 6.9.

The region proposal is generated after the saliency mask is computed. Adaptive thresholding is applied over the mask to detect the moving regions with various motion contrast [20]. They applied morphological transformations for filtering the small regions and obtaining the complete region proposals. This procedure alleviated the problem of incomplete regions. Next, the salient regions in dynamic images move in a sequence of video frames. To produce the saliency mask, [19] trained the supervised model. Fig. 6.10 depicts the architecture of the masking model.

Ghosh et al. [21] presented a VD architecture named a dual spatiotemporal convolutional network (DSTCN) that extracts the spatial and temporal features via 1D-, 2D-, and 3D-CNN from the video frames. This network combined multi-dimensional CNNs and remained lightweight. They claimed the lower channel capacity for their model, therefore the model learned to obtain significant spatial and temporal information. The DSTC block receives the input from the previous layer, extracting the temporal and spatial features by passing to

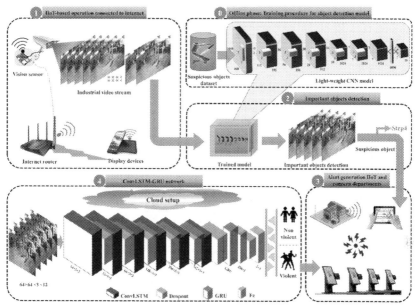

FIGURE 6.7   Basic four steps are held in an online fashion in the IoT-based VD framework. In step 0, a procedure for training a light-weight ConvNet model for detection of suspicious object is provided, while in step 1, the vision sensor on the resource-constrained captures the scenes where the frames are passed into step 2 for screening the important information collection including suspicious objects or humans via a trained CNN model. Once the suspicious object appears in step 2, the frames are forwarded into step 4 for detailed analysis. Step 4 comprises of VD-Net to perform features extraction, and learning procedure, to distinguish the violent and non-violent scenes [18].

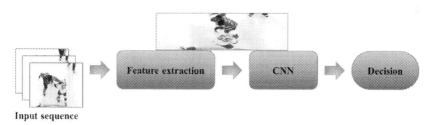

FIGURE 6.8   Overview of the general flow of the VD method in [13].

the next layer. The input shape is $(C \times D \times H \times W)$ where 'C' indicates the channel numbers, while 'D' shows the input frames number. H and W represent the height and width, respectively. The spatial features from an individual frame are present in $(H \times W, D)$ and show the temporal details for a pixel. The single block includes three convolutional layers for temporal and spatial information collection. The main pipeline of the method is given in Fig. 6.11. A 1D convolutional layer is applied for temporal information collection having the kernel size $(Kt \times 1 \times 1)$. A 1D convolutional layer convolves on a single pixel and extracts the temporal information from a specific pixel. Similarly, a

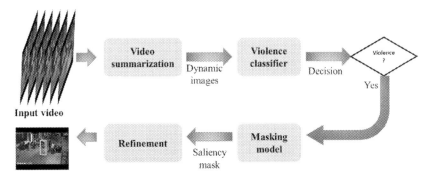

FIGURE 6.9 Different stages of the method in [19].

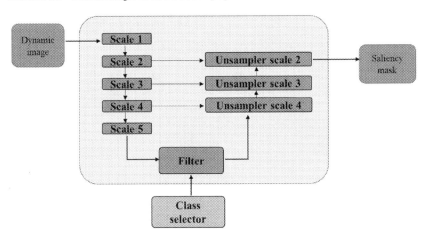

FIGURE 6.10 Masking model of the method in [19].

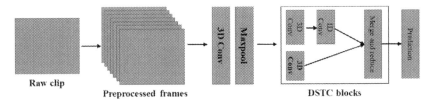

FIGURE 6.11 Pipe line of end-to-end DSTCN [21].

2D convolutional layer collects the spatial information from a single frame via 2D-CNN on a specific frame and has kernel size $(1 \times Ks \times Ks)$. Next, a 3D convolutional layer extracts both the temporal and spatial features from the input with kernel size $(Kt \times Ks \times Ks)$. Furthermore, a 3D convolutional layer extracts both the temporal and spatial information from the given input with a kernel having size $(Kt \times Ks \times Ks)$. The 3D convolution is performed on both the spatial and temporal dimensions.

TABLE 6.1 Detail explanation of VD methods based on LSTM, domain, and deep features.

Method	Sequence		Sequential learning			Domain	
	Deep-CNN	Conventional-based	GRU	LSTM	Bi-LSTM	Surveillance	Non-surveillance
[23]	–	✓	–	–	–	✓	–
[24]	✓	–	–	✓	✓	–	✓
[25]	✓	–	✓	–	–	✓	–
[26]	✓	–	–	✓	–	–	✓
[27]	✓	–	–	–	–	✓	–
[13]	✓	✓	–	–	–	–	–
[28]	–	✓	–	–	–	–	✓
[29]	–	✓	–	–	–	–	✓
[22]	✓	✓	–	✓	–	–	–
[30]	–	–	–	✓	–	–	✓
[31]	–	✓	–	–	–	–	✓
[32]	–	✓	–	–	–	✓	–
[3]	–	✓	–	–	–	✓	–
[33]	✓	✓	✓	–	–	✓	–
[18]	✓	–	–	–	–	✓	–
[21]	✓	–	–	–	–	✓	–
[34]	✓	–	–	–	–	✓	–
[35]	–	✓	–	✓	–	✓	–
[36]	✓	–	–	✓	✓	✓	–
[15]	✓	–	–	✓	–	✓	–
[37]	✓	–	–	–	–	✓	–
[38]	✓	–	–	✓	–	✓	–

Recently, Liang [22] proposed a CNN-based recurrent network with an attention mechanism. Several attention mechanisms are introduced in processing video. GhostNet architecture and convLSTM are chosen in the feature extraction stage. The combined approach classified the behaviors. The statistics of the exiting VD methods based on LSTM, domain, and deep features analysis are shown in Table 6.1.

6.3 Discussion on experimental results

This section discusses and compares the results achieved by different deep CNN and LSTM-based VD methods on different benchmarks. The comprehensive summary of the VD datasets with their statistics are given in Table 6.2, while the detailed comparison of VD methods is given in Table 6.3. Similarly, the graphs that represent the accuracy for Surveillance fight, Violent Flow, and Hockey fight datasets are given in Fig. 6.12 and the samples of these datasets are given in Fig. 6.13.

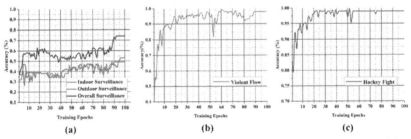

FIGURE 6.12   Accuracy details of the VD datasets. (a) Surveillance Fight Dataset, (b) Violent Flow Dataset, (c) Hockey Fight Dataset [15].

Most recently, Islam et al. [27] proposed a two-stream approach comprising a pre-trained MobileNet and Separable Convolutional LSTM (SepConvLSTM). One of the streams obtained the suppressed frames, while the other stream processes the difference in the adjacent frames. They applied a pre-processing technique highlighting the moving objects in frames through suppressing any non-moving background to capture the motion between the frames. These inputs assist the method in producing discriminative features as the violent actions widely rely on characterization by the body movement. The convolution operation in each gate of ConvLSTM is replaced by depthwise separable convolution forming SepConvLSTM. This method obtained 99%, 89.75%, and 100% accuracy on Hockey Fight, RWF-2000, and Movies dataset, respectively. Next, Liang et al. [22] investigated the network using Hockey Fight and the RWF-2000 dataset and obtained 97.5% and 87.5%, respectively. Another method [18], as discussed in a previous section, investigated surveillance-based scenarios for violent scene description. To develop their network, they performed experiments on RWF-2000, Surveillance Fight, and Hockey Fight dataset by obtaining 88.2%, 75.9%, and 98.5%, respectively.

TABLE 6.2 VD datasets with their detailed statistics including the accuracy obtained on their base method.

Reference	Fps	Frame resolution	Accuracy (%)	Total Video Samples	Surveillance?
Hockey Fight [39]	25	360 × 288	91.7	1000	–
Behave [40]	25	640 × 480	93.67	4	✓
Violence in Movies [39]	25	360 × 250	89.5	200	–
Web Abnormality [41]	–	–	–	20	✓
RWF-2000 [34]	–	Diverse	86.75	2000	✓
CAVIAR [42]	25	384 × 288	–	–	–
SDHA 2000 [43]	30	720 × 480	–	–	✓
USD [44]	30	–	–	–	–
Real-world Fight [45]	–	Diverse	–	1000	✓
Violent Flows [28]	25	320 × 240	81.30	246	–
UMN [41]	–	320 × 240	–	–	–
Surveillance Camera Fight [36]	25	480 × 360	72	300	✓

TABLE 6.3 Results achieved for Violent Flow, Hockey Fight, and Surveillance Fight dataset in [15] in terms of precision, F1-score, specificity, and recall.

Dataset	Values				Precision	F1 score	Specificity	Recall
	True Positive	True Negative	False Positive	False Negative				
Violent Flow [28]	24	24	0	0	1	1	1.0	1
Indoor Surveillance	10	10	10	10	0.5	0.5	0.5	0.5
Hockey Fight [39]	98	98	2	2	0.9810	0.9810	0.98	0.9810
Outdoor Surveillance	5	5	5	4	0.5555	0.5263	0.5	0.5556
Surveillance Fight [36]	22	22	8	8	0.7333	0.733	0.733	0.730

FIGURE 6.13 Samples considered from VD dataset. (a) Surveillance Fight frames, (b) Hockey Fight Frames, (c) RWF-2000 samples [18].

Similarly, Kang et al. [38] proposed a pipeline combined with 2D CNN. In particular, they proposed frame grouping to make 2D CNN capable of learning the spatiotemporal representation in videos. Their proposed pipeline brought significant improvement compared to 2D CNN followed by LSTM. They verified their network performance using Violent Flow, Hockey Fight, Surveillance Fight, RWF-2000, and Movies dataset by achieving 98%, 99.6%, 92%, and 100% accuracy, respectively.

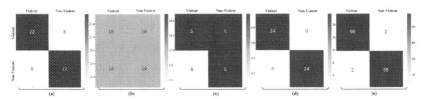

FIGURE 6.14 Confusion matrix of proposed VD method. (a) Confusion matrix obtained for Surveillance Fight Dataset, while (b) and (c) shows confusion matrix for indoor and outdoor scenes in Surveillance Fight Dataset, while (d) and (e) show the confusion matrix of Violent Flow and Hockey Fight dataset, respectively [15].

Furthermore, Ullah et al. [15] used a hybrid connection of a DL-based network where its details are covered in a previous section, used Hockey Fight, Violent Flow, and Surveillance Fight datasets and attained 98.21%, 98%, and 74% accuracy, respectively. Moreover, Traore et al. [26] proposed an architecture that combined bidirectional GRU and 2D CNN for VD in video frames. In this method, the CNN extracts the spatial information, while the bidirectional GRU collects the temporal and local motion information via CNN features from numerous frames. They performed the experiments on Hockey Fight and Violent Flow achieving 98% and 95.5% accuracy, respectively. The confusion matrix of this method is given in Fig. 6.14.

Traore et al. [46] proposed a deep VD architecture that combines RNN and 2D CNN. They also used optical flow that is computed via captured sequences to encode the motion in the scenes. Their method obtained 99% and 93.75% accuracy on Hockey Fight and Violent Flow dataset, respectively. Finally, Akti et al. [36] exploited different LSTM-based approaches by utilizing the attention

TABLE 6.4 Comparison of different CNN and LSTM-based VD methods results obtained on different challenging datasets. Yearly-based comparison held where the bold shows the highest accuracies, while the underlined shows the runner up.

Method	Dataset accuracy (%)				
	Violent flow	Hockey fight	Surveillance fight	RWF-2000	Movies
SepConvLSTM, 2 stream [27]	-	<u>99</u>	-	<u>89.75</u>	100
GhostNet and ConvLSTM [22]	-	97.5	-	87.5%	-
ConvLSTM, GRU [18]	-	98.5	75.9-	88.2	-
DSTCN [21]	-	<u>99</u>	-	-	100
MSM, T-SE block, EfficientNet-B0 [38]	98.0	**99.6**	**<u>92.0</u>**	**92.0**	100
CNN-LSTM [15]	**98.21**	98	74	-	-
2D-bidirectional GRU [26]	95.5	98	-	-	-
Skeleton points interaction learning (SPIL) [35]	94.5	96.8	-	89.3	-
2D CNN, RNN [46]	93.75	99	-	-	-
BrutNet [47]	83.19	-	-	-	88.74
Fight-CNN, BD-LSTM [36]	-	96	72	-	100

layer. They achieved 96%, 72%, and 100% accuracy on Hockey Fight, Surveillance Fight Dataset, and Movies dataset, respectively. See Table 6.4.

6.4   Challenges and future research guidelines

Over the decade, several researchers have focused on the problem of VD by applying distant strategies in sequential learning based on the need of the domain. The current literature reveals that several researchers used a variety of features to represent a sequence followed by LSTM or RNN. There also exist VD methods inspired by deep features, although, most of these methods face the problem of time complexity. Therefore there is an intense need of state-of-the-art procedures to establish a tradeoff between time complexity and time.

First, the datasets utilized by the researchers have limited amounts of data and remain insufficient to train a DL or AI model from the scratch. However, some of the video-based VD datasets can be utilized for effective detection of violent scenes. Table 6.2 shows a few of the most popular VD datasets. Some of these datasets are based on surveillance scenarios, while the rest of them are recorded with a camera or smart phone. RWF-2000 [34] is one of the largest amongst these datasets with 2000 surveillance-based videos. Nevertheless, this dataset is not able to train a DL model from scratch. Therefore large-scale datasets are highly demanded for their involvement in complex networks.

Next, detection of violent scenes over time by the researchers, has faced the challenges of complex networks where the data is passed over different stages in the framework for effective detection. However, the detection process over these networks makes the process more computational in terms of generating a large number of parameters. Therefore it is highly recommended to develop end-to-end models and explore their working mechanism. Image classification achieved greater results using CNN that beat the human error rate in deep models [48]. Such models are able to perform frame level representation in the video data. Several CNN models such as ShuffleNet [49], SqueezeNet [50], and MobileNet [49] can assist the VD for real-time processing. Regarding the VD literature, we have seen that many methods are based on single scene analysis from a single camera in a video that decides whether a whole scene is Violent or Non-violent. However, if a person passes from one camera, he/she appears on another camera from a different angle. Therefore it is highly recommended to consider multi-views for the VD in surveillance scenes.

6.5   Conclusion

Over the decades, crime has become a serious threat to our society and environment and there has emerged an increase of surveillance camera installation in markets, streets, schools, institutions, etc. The increase in the number of these cameras made the monitoring process difficult due to human involvement and manual setup. This problem leads in the need for automatic VD development for secure surveillance. In this chapter, we discussed the concept of learning the sequence for VD using CNN, LSTM and its variants including multi-layer, GRU, etc. Similarly, the CNN and LSTM-based methods are surveyed with their main contributions and drawbacks. Next, we explained their working flow along with their frameworks, experimental performance, and their visual results. We also shed light on the working of RNN and their limitations, explaining why LSTM produces a better performer than RNN. Finally, we concluded the chapter with discussion and the recommendations in terms of future research directions that may assist the forthcoming challenges of VD. In the near future, we aim to investigate VD methods in terms of IoT in consideration with multi-view surveillance scenarios.

References

[1] K. Deepak, L. Vignesh, G. Srivathsan, S. Roshan, S. Chandrakala, Statistical features-based violence detection in surveillance videos, in: Cognitive Informatics and Soft Computing, Springer, 2020, pp. 197–203.

[2] T. Senst, V. Eiselein, A. Kuhn, T. Sikora, Crowd violence detection using global motion-compensated Lagrangian features and scale-sensitive video-level representation, IEEE Transactions on Information Forensics and Security 12 (2017) 2945–2956.

[3] T. Zhang, W. Jia, X. He, J. Yang, Discriminative dictionary learning with motion Weber local descriptor for violence detection, IEEE Transactions on Circuits and Systems for Video Technology 27 (2016) 696–709.

[4] S. Coşar, G. Donatiello, V. Bogorny, C. Garate, L.O. Alvares, F. Brémond, Toward abnormal trajectory and event detection in video surveillance, IEEE Transactions on Circuits and Systems for Video Technology 27 (2016) 683–695.

[5] F.U.M. Ullah, M.S. Obaidat, A. Ullah, K. Muhammad, M. Hijji, S.W. Baik, A comprehensive review on vision-based violence detection in surveillance videos, ACM Computing Surveys (2022).

[6] K. Muhammad, R. Hamza, J. Ahmad, J. Lloret, H. Wang, S.W. Baik, Secure surveillance framework for IoT systems using probabilistic image encryption, IEEE Transactions on Industrial Informatics 14 (2018) 3679–3689.

[7] F.U.M. Ullah, A. Ullah, I.U. Haq, S. Rho, S.W. Baik, Short-term prediction of residential power energy consumption via CNN and multilayer bi-directional LSTM networks, IEEE Access (2019).

[8] F.U.M. Ullah, N. Khan, T. Hussain, M.Y. Lee, S.W. Baik, Diving deep into short-term electricity load forecasting: comparative analysis and a novel framework, Mathematics 9 (2021) 611.

[9] M.-s. Kang, R.-H. Park, H.-M. Park, Efficient spatio-temporal modeling methods for real-time violence recognition, IEEE Access (2021).

[10] S. Accattoli, P. Sernani, N. Falcionelli, D.N. Mekuria, A.F. Dragoni, Violence detection in videos by combining 3D convolutional neural networks and support vector machines, Applied Artificial Intelligence (2020) 1–16.

[11] E. Fenil, G. Manogaran, G. Vivekananda, T. Thanjaivadivel, S. Jeeva, A. Ahilan, Real time violence detection framework for football stadium comprising of big data analysis and deep learning through bidirectional LSTM, Computer Networks 151 (2019) 191–200.

[12] F.U.M. Ullah, A. Ullah, K. Muhammad, I.U. Haq, S.W. Baik, Violence detection using spatiotemporal features with 3D convolutional neural network, Sensors 19 (2019) 2472.

[13] I. Serrano, O. Deniz, J.L. Espinosa-Aranda, G. Bueno, Fight recognition in video using hough forests and 2D convolutional neural network, IEEE Transactions on Image Processing 27 (2018) 4787–4797.

[14] P. Wang, P. Wang, E. Fan, Violence detection and face recognition based on deep learning, Pattern Recognition Letters 142 (2021) 20–24.

[15] F.U.M. Ullah, M.S. Obaidat, K. Muhammad, A. Ullah, S.W. Baik, F. Cuzzolin, et al., An intelligent system for complex violence pattern analysis and detection, International Journal of Intelligent Systems (2021).

[16] W. Lejmi, A.B. Khalifa, M.A. Mahjoub, Fusion strategies for recognition of violence actions, in: 2017 IEEE/ACS 14th International Conference on Computer Systems and Applications (AICCSA), 2017, pp. 178–183.

[17] K.-i. Funahashi, Y. Nakamura, Approximation of dynamical systems by continuous time recurrent neural networks, Neural Networks 6 (1993) 801–806.

[18] F.U.M. Ullah, K. Muhammad, I.U. Haq, N. Khan, A.A. Heidari, S.W. Baik, et al., AI-assisted edge vision for violence detection in IoT-based industrial surveillance networks, IEEE Transactions on Industrial Informatics 18 (2021) 5359–5370.

[19] D.G.C. Roman, G.C. Chávez, Violence detection and localization in surveillance video, in: 2020 33rd SIBGRAPI Conference on Graphics, Patterns and Images (SIBGRAPI), 2020, pp. 248–255.

[20] N. Otsu, A threshold selection method from gray-level histograms, IEEE Transactions on Systems, Man and Cybernetics 9 (1979) 62–66.

[21] D.K. Ghosh, A. Chakrabarty, N. Mansoor, D.Y. Suh, M.J. Piran, Learning-driven spatiotemporal feature extraction for violence detection in IoT environments, in: 2021 International Conference on Information and Communication Technology Convergence (ICTC), 2021, pp. 1807–1812.

[22] Q. Liang, Y. Li, K. Yang, X. Wang, Z. Li, Long-term recurrent convolutional network violent behaviour recognition with attention mechanism, in: MATEC Web of Conferences, 2021, p. 05013.

[23] I. Serrano, O. Deniz, G. Bueno, G. Garcia-Hernando, T.-K. Kim, Spatio-temporal elastic cuboid trajectories for efficient fight recognition using Hough forests, Machine Vision and Applications 29 (2018) 207–217.

[24] A. Hanson, K. Pnvr, S. Krishnagopal, L. Davis, Bidirectional convolutional lstm for the detection of violence in videos, in: Proceedings of the European Conference on Computer Vision (ECCV) Workshops, 2018, pp. 280–295.

[25] N. Singh, O. Prasad, T. Sujithra, Deep learning-based violence detection from videos, in: Intelligent Data Engineering and Analytics, Springer, 2022, pp. 323–332.

[26] A. Traoré, M.A. Akhloufi, 2D bidirectional gated recurrent unit convolutional neural networks for end-to-end violence detection in videos, in: International Conference on Image Analysis and Recognition, 2020, pp. 152–160.

[27] Z. Islam, M. Rukonuzzaman, R. Ahmed, M.H. Kabir, M. Farazi, Efficient two-stream network for violence detection using separable convolutional lstm, in: 2021 International Joint Conference on Neural Networks (IJCNN), 2021, pp. 1–8.

[28] T. Hassner, Y. Itcher, O. Kliper-Gross, Violent flows: real-time detection of violent crowd behavior, in: 2012 IEEE Computer Society Conference on Computer Vision and Pattern Recognition Workshops, 2012, pp. 1–6.

[29] C. Gong, H. Shi, J. Yang, J. Yang, Multi-manifold positive and unlabeled learning for visual analysis, IEEE Transactions on Circuits and Systems for Video Technology 30 (2019) 1396–1409.

[30] J. Mahmoodi, A. Salajeghe, A classification method based on optical flow for violence detection, Expert Systems with Applications 127 (2019) 121–127.

[31] J. Yu, W. Song, G. Zhou, J.-j. Hou, Violent scene detection algorithm based on kernel extreme learning machine and three-dimensional histograms of gradient orientation, Multimedia Tools and Applications 78 (2019) 8497–8512.

[32] I. Febin, K. Jayasree, P.T. Joy, Violence detection in videos for an intelligent surveillance system using MoBSIFT and movement filtering algorithm, Pattern Analysis & Applications 23 (2020) 611–623.

[33] S. Chandrakala, L. Vignesh, V2AnomalyVec: deep discriminative embeddings for detecting anomalous activities in surveillance videos, IEEE Transactions on Computational Social Systems (2021).

[34] M. Cheng, K. Cai, M. Li, RWF-2000: an open large scale video database for violence detection, arXiv preprint, arXiv:1911.05913, 2019.

[35] Y. Su, G. Lin, J. Zhu, Q. Wu, Human interaction learning on 3D skeleton point clouds for video violence recognition, in: European Conference on Computer Vision, 2020, pp. 74–90.

[36] Ş. Aktı, G.A. Tataroğlu, H.K. Ekenel, Vision-based fight detection from surveillance cameras, in: 2019 Ninth International Conference on Image Processing Theory, Tools and Applications (IPTA), 2019, pp. 1–6.

[37] Ş. Aktı, F. Ofli, M. Imran, H.K. Ekenel, Fight detection from still images in the wild, in: Proceedings of the IEEE/CVF Winter Conference on Applications of Computer Vision, 2022, pp. 550–559.

[38] Waseem Ullah, Fath U. Min Ullah, Zulfiqar Ahmad Khan, Sung Wook Baik, Sequential attention mechanism for weakly supervised video anomaly detection, Expert Systems with Applications (2023) 120599.

[39] E.B. Nievas, O.D. Suarez, G.B. García, R. Sukthankar, Violence detection in video using computer vision techniques, in: International Conference on Computer Analysis of Images and Patterns, 2011, pp. 332–339.

[40] S. Blunsden, R. Fisher, The BEHAVE video dataset: ground truthed video for multi-person behavior classification, Annals of the BMVA 4 (2010) 4.

[41] R. Mehran, A. Oyama, M. Shah, Abnormal crowd behavior detection using social force model, in: 2009 IEEE Conference on Computer Vision and Pattern Recognition, 2009, pp. 935–942.

[42] Robert Fisher, Jose Santos-Victor, James Crowley, CAVIAR: context aware vision using image-based active recognition, http://homepages.inf.ed.ac.uk/rbf/CAVIAR/, 2004.

[43] P.D. Garje, M. Nagmode, K.C. Davakhar, Optical flow based violence detection in video surveillance, in: 2018 International Conference on Advances in Communication and Computing Technology (ICACCT), 2018, pp. 208–212.

[44] C.-H. Demarty, C. Penet, M. Soleymani, G. Gravier, VSD, a public dataset for the detection of violent scenes in movies: design, annotation, analysis and evaluation, Multimedia Tools and Applications 74 (2015) 7379–7404.

[45] M. Perez, A.C. Kot, A. Rocha, Detection of real-world fights in surveillance videos, in: ICASSP 2019-2019 IEEE International Conference on Acoustics, Speech and Signal Processing (ICASSP), 2019, pp. 2662–2666.

[46] A. Traoré, M.A. Akhloufi, Violence detection in videos using deep recurrent and convolutional neural networks, in: 2020 IEEE International Conference on Systems, Man, and Cybernetics (SMC), 2020, pp. 154–159.

[47] M. Haque, S. Afsha, H. Nyeem, Developing BrutNet: a new deep CNN model with GRU for realtime violence detection, in: 2022 International Conference on Innovations in Science, Engineering and Technology (ICISET), IEEE, 2022 Feb 26, pp. 390–395.

[48] S. Khan, K. Muhammad, S. Mumtaz, S.W. Baik, V.H.C. de Albuquerque, Energy-efficient deep CNN for smoke detection in foggy IoT environment, IEEE Internet of Things Journal (2019).

[49] A.G. Howard, M. Zhu, B. Chen, D. Kalenichenko, W. Wang, T. Weyand, et al., Mobilenets: efficient convolutional neural networks for mobile vision applications, arXiv preprint, arXiv:1704.04861, 2017.

[50] F.N. Iandola, S. Han, M.W. Moskewicz, K. Ashraf, W.J. Dally, K. Keutzer, Squeezenet: Alexnet-level accuracy with 50x fewer parameters and <0.5 mb model size, arXiv preprint, arXiv:1602.07360, 2016.

Chapter 7

Wavelets for anisotropic oscillations in nanomaterials

Wavelets for anisotropic oscillations

Anouar Ben Mabrouk[a,b,c], Mourad Ben Slimane[d],
Belkacem-Toufik Badeche[e], Carlo Cattani[f], and Yeliz Karaca[g]

[a]*Laboratory of Algebra, Number Theory and Nonlinear Analysis Lab UR11ES50, Department of Mathematics, Faculty of Sciences, University of Monastir, Monastir, Tunisia,* [b]*Department of Mathematics, Higher Institute of Applied Mathematics and Computer Science, University of Kairouan, Kairouan, Tunisia,* [c]*Department of Mathematics, Faculty of Sciences, University of Tabuk, Tabuk, Saudi Arabia,* [d]*Department of Mathematics, College of Sciences, King Saud University, Riyadh, Saudi Arabia,* [e]*Collaborating Academics - University of Montpellier, Montpellier, France,* [f]*Department of Economics, Engineering, Society and Business Organization - DEIM, Tuscia University, Viterbo, Italy,* [g]*University of Massachusetts Chan Medical School, Worcester, MA, United States*

7.1 Introduction

In nature, the discontinuity and/or the irregularity and also the anisotropy structures, such as trees, rocks, clouds, the human body, and so forth, are observed everywhere. In sciences, especially in academia, the concept of regularity and/or singularity is met everywhere in mathematical as well as physical contexts such as singularities of solutions of PDEs especially in the case of fractional calculus, which have generated in the last years a great interest. They are met also in many mono-dimensional and multidimensional signals and images ([56,57,77,85,95,96,99]).

Nowadays, with the technological developments, and the discovery of the nano-microscopes, new horizons of image processing are being pointed out, accompanied as usual with the necessary mathematical tools. The present chapter aims to highlight the utility of wavelets, the discovery of oscillations, especially singular trigonometric anisotropic ones in nanomaterials, and the eventual link between the two concepts; wavelets as a tool and nanomaterials as analyzed objects.

This chapter aims to provide a detailed analysis of developing the wavelet-based study to understand the anisotropic image behavior, such as nano-images and oscillations in nanomaterials. Various methods have been applied generally

Intelligent Fractal-Based Image Analysis. https://doi.org/10.1016/B978-0-44-318468-0.00014-3

Copyright © 2024 Elsevier Inc. All rights reserved, including those for text and data mining, AI training, and similar technologies.

to the recognition problems such as neural networks, fuzzy logic, morphological methods, genetic algorithms, etc. Nevertheless, more development is still necessary for mathematical models such as wavelets for Artificial Intelligence (AI) object recognition. Among application areas, nano-images still need major development because of their link and application in many fields nowadays.

Recall also that machine learning itself necessitates AI in developing computational methods to improve performance or make accurate predictions by developing advanced algorithms that go through the data and learn patterns, rules and also hidden structures such as fractality, chaotic behavior, oscillations, anisotropy, etc., see [37,101].

In its simple definition, pattern recognition may be described as a tool permitting some systems to observe their environment, distinguishing the main patterns from their proper background, to make final decisions on classification and/or categorization. One of the main methods used is the artificial neural networks, which are well-known tools in AI and machine learning. Neural networks have been already generalized, extended, and improved by the involvement of wavelets to obtain the wavelet neural networks.

Another concept that is worth noting in nano-structures is the temporal dynamic texture that is strongly related to the concept of self-similarity and thus to fractals. Before the discovery of fractals and nano-structures, scientists used, for example, the particle image velocimetry for making a flow visible and measurable by injecting many small particles that scatter light and show the fluid motion. Now, with the discovery of nano-structures, and the comprehension of the nano-fluids, more sophisticated ways may be applied to facilitate the task [42].

In [70], fractal dimension has been proved to be a good predictor for surface roughness description. The method is applied on implant materials from scanning electron microscopy images. We noted that the authors there applied a pre-processing step in which the image is transformed into 1D signals. An analysis step is next adopted for such 1D signals. This may induce, in our opinion, some loss of information due to the anisotropy and/or the slicing method according to the directions. In [71], Kockentiedt and collaborators proposed an automatic detection method of scanning electron microscopy images to select engineered nanoparticles from others in the same ambient air based on fractal dimensions and wavelet filtering.

In [16] an exploitation of fractals in nano-material understanding is pointed out, by investigating the formation of fractal nano-structures on gold-like films, via the analysis of the thin film growth process. Anisotropic structures, and the possibility of applying fractals for nano-material surfaces are pointed out. Volchuk et al. [97] proposed a one-to-one correspondence relating mechanical properties of materials to fractal dimensions. The method has been proved to be useful for the prediction of mechanical properties of the material based on its fractal structure. In [11], the use of fractals in image recognition has been investigated based on special types of fractals such as Cantor or generally iterated function systems encoded fractals. It is shown that fractal encoding may be exploited in information extraction for object recognition tasks by introducing a new type of feature, called fractal features. Such features are jointly combined with machine learning techniques in 2D image recognition tasks, see also [38,48,68,74,98].

A first kind of fractal analysis of images is the detection of singularities or irregularities. A first type of regularity is the so-called Hölder regularity that permits one to study the global behavior of functions at irregular points. The most important disadvantage in this type of regularity is the fact that it did not reflect the microscopic behavior of the function near the singular point and did not describe the oscillatory behavior of the function near such a point either. This leads to introducing a second kind of regularity and/or singularity called the oscillating singularity that reflects both the Hölder regularity and the oscillation of the function. Our aim in the present chapter is to extend the notion of oscillating singularity that is investigated only in mono-dimensional cases to some higher dimensional ones taking into account the different oscillating behaviors of the function according to the direction.

Oscillating singularities and spaces were first introduced in [9,10]. In such a reference, the obtained results can be understood as mono-dimensional and global formulations. Indeed, let F be a real valued function defined on $\mathbb{R}^d$ and satisfying in a neighborhood of $x \in \mathbb{R}^d$

$$F(x + h) - F(x) \sim |h|^\alpha \sin \frac{1}{|h|^\beta}. \tag{7.1}$$

According to [9,10], F has an oscillating singularity (α, β) at the point x. Note that these formulations may be understood as mono-directional and/or global relatively to the directions or coordinates of the system. As it appears in (7.1), one cannot control well the directional behavior of f near the point x. All that we can know is the global behavior and it remains questionable to know about the macroscopic behavior of F. See also [80,81,84,85].

In the present chapter, we will focus on such a problem and we will try to give some "similar" but large definition of oscillating singularities taking into account the direction-wise behavior of F.

There are several ideas behind our work. The first one is to consider on $\mathbb{R}^d$ some oscillating but separated singularities like

$$F(x + h) - F(x) \sim \sum_{i=1}^{d} |h_i|^{\alpha_i} \sin \frac{1}{|h_i|^{\beta_i}}, \tag{7.2}$$

where $x = (x_1, x_2, \ldots, x_d)$ and $h = (h_1, h_2, \ldots, h_d)$ in $\mathbb{R}^d$. Secondly, is to consider some oscillations like

$$F(x + h) - F(x) \sim \sum_{i,j=1}^{d} |h_i|^{\alpha_i} \sin \frac{1}{|h_j|^{\beta_j}}. \tag{7.3}$$

Another interesting case consists of controlling the following oscillating singularities:

$$F(x+h) - F(x) \sim \sum_{i=1}^{d} |h_i|^{\alpha_i} \sin \frac{1}{\displaystyle\sum_{i=1}^{d} |h_i|^{\beta_i}}. \qquad (7.4)$$

Finally, another example of singular oscillations may be described by the equation

$$F(x+h) - F(x) \sim \prod_{i=1}^{d} |h_i|^{\alpha_i} \sin \frac{1}{\displaystyle\prod_{j=1}^{d} |h_j|^{\beta_j}}. \qquad (7.5)$$

As it appears in (7.2), (7.3), (7.4), and (7.5), the exponents α and β become multi-exponents. In addition, we note that the three forms of oscillations are widely different. In the first one, we speak about a superposition of oscillating behaviors but independent of each other. In the second, it consists of a more complicated type of oscillating signals, where the previous superposition is already contained, provided with other types such as $h_i^{\alpha_i} sin \frac{1}{|h_j|^{\beta_j}}$, for $i \neq j$. The last type is the more complex case as it contains in the singular sine nonseparable powers. More and more complicated forms may be expected, and their theory should be developed.

The chapter is devoted to the study of some oscillating singularities of the types stated above. The aim is principally not to develop a mathematical theory about such singularities, but instead to evoke the existence of such singularities in nature, mainly in the most modern form of physics, the nanomaterials, in many forms such as anisotropic oscillations. We propose to use wavelet theory as a modern mathematical tool to investigate the oscillating singular behavior in nanomaterials.

Indeed, wavelets are applied in many simple cases for the understanding of anisotropic behavior in an image processing framework [49]. In addition, the case of the anisotropic growth has been investigated for ferromagnetic Co-Ni alloy nanowires in [47]. In addition, for a ferroelectric matrix Sr(1-x)Ba(x)TiO(3), the Co-Ni/Sr(1-x)Ba(x)TiO(3) junction is shown to be an interface having a disturbed zone with an observed extent of about 0.5 nm [103]. In addition to the existence of the misfit (parametric disagreement between the two structures at the junction) it does not explain by itself the rupture of ligands at the interfaces. The boundaries of the disturbed zone cannot be observed visually because it persists beyond the junction to propagate up into the Sr(1-x)Ba(x)TiO(3) matrix. This persistence is described in the framework of a low amplitude wave and phase proportional to a singular trigonometric function. Fig. 7.1 shows the eventual link that relates AI methods such as machine learning, nanotechnology, nanomaterials, and thus puts our chapter into the framework of the whole subject/theme of the present book. A fascinating talk due to Anna about ML and nanomaterials may be consulted at [46] https://physicsworld.com/a/machine-learning-puts-nanomaterials-in-the-picture, see also [56,57,86,91].

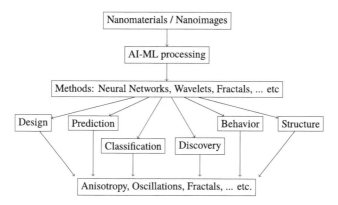

FIGURE 7.1   An approximating flowchart for nanotechnology AI-based methods.

The rest of the chapter will be organized into 7 sections including this introductory one. Section 7.2 is concerned with a review of the regularity of functions in its classical point of view, such as the Hölder and oscillating ones, the two-dimensional case with reviews on both directional and anisotropic regularities. In Section 7.3, some formal calculus is developed for the case of the anisotropic oscillating singularities in the 2-dimensional case. Section 7.4 returns to the wavelet tool for the description of the regularities of functions in the cases reviewed previously. The wavelet characterization of the Hölder regularity, directional regularity, anisotropic, as well as oscillations are provided. In Section 7.5, an application to an example of nanomaterials is provided with the link to the fractal nature of such materials. Section 7.6 is the conclusion. Finally, Section 7.7 is an appendix devoted to the presentation of the wavelet toolkit. We recall that in the whole chapter and, for the sake of simplicity, we restrict ourselves to the case $d = 2$ for the mathematical developments. The higher dimensional case may be deduced by necessary modifications.

7.2   On the regularity of functions

In this section, we review different existing notions of regularity of functions such as the Hölder regularity, oscillating singularity, and directional regularity in higher dimensions. We next introduce our extension of directional oscillating regularity and try to output a wavelet characterization of it as for the classical cases of regularities.

Recall that the Hölder regularity of functions is now widely known and widely studied and has been also widely applied in different domains such as financial time series, medical images, denoising, classification, re-construction of signals, modeling and prediction, etc. This makes the estimation of function regularities an important task.

Definition 7.1. Let F be a real-valued function defined on $\mathbb{R}^d$, $x_0 \in \mathbb{R}^d$ and α a nonnegative real number. F is said to be α-Hölder at x_0 and we write $F \in C^\alpha(x_0)$ if there exists a neighborhood W_0 of x_0, a constant $C > 0$ and a polynomial P of degree less than α such that

$$|F(x) - P(x - x_0)| \leq C|x - x_0|^\alpha, \ \forall \, x \in W_0. \tag{7.6}$$

The point-wise Hölder exponent of F at the point x_0 is

$$\alpha_F(x_0) = \sup\{ \alpha \geq 0 \, ; \ F \in C^\alpha(x_0) \}. \tag{7.7}$$

In the literature, there are many variants of Hölder regularity. For an α non-integer, F is said to be α-Hölder regular at x_0 iff

$$\lim_{h \to 0} \frac{|F(x_0 + h) - P(h)|}{|h|^\gamma} = 0, \ \forall \gamma < \alpha \tag{7.8}$$

and

$$\limsup_{h \to 0} \frac{|F(x_0 + h) - P(h)|}{|h|^\gamma} = +\infty, \ \forall \gamma > \alpha, \tag{7.9}$$

where P is a polynomial of degree less than α, see for example [43,44]. Whenever α is an integer, an equivalent definition may be adopted.

For $0 < \alpha < 1$, the function F is said to be α-Hölder at x_0 iff there exists constants $C, \rho_0 > 0$ such that

$$\sup_{x,y \in B(x_0,\rho)} |F(x) - F(y)| \leq C\rho^\alpha, \ \forall \, \rho < \rho_0. \tag{7.10}$$

As its name indicates, the point-wise Hölder exponent measures the point-wise regularity of the function at a given point. In some cases, it is desirable to take into account the behavior of the function in the neighborhood of the point, and to incorporate this information in the exponent. This induces the so-called local Hölder exponent defined for a function F at a point x_0 by $\tilde{\alpha}_F(x_0)$ as the supremum of real numbers $\tilde{\alpha}$ satisfying

$$\exists \, C, \rho_0 > 0; \ \forall \rho < \rho_0, \quad \sup_{x,y \in B(x_0,\rho)} \frac{|F(x) - F(y)|}{|x - y|^{\tilde{\alpha}}} \leq C. \tag{7.11}$$

Of course, whenever F is higher order differentiable, we have to cut off its Taylor polynomial in Eqs. (7.8), (7.9), and (7.10). Note that the exponents $\alpha_F(x_0)$

and $\tilde{\alpha}_F(x_0)$ may be different. An illustrative example may be given by the function $F(x) = |x|^\alpha \sin(\frac{1}{|x|^\beta})$ with $0 < \alpha < 1$, where we have

$$\alpha_F(0) = \alpha \quad \text{and} \quad \tilde{\alpha}_F(0) = \frac{\alpha}{1+\beta}.$$

Examples, comparisons due to these variants of regularity, may be found, for instance, in [27,66,89,90].

The Hölder exponent of a function F measures its regularity at a given point. F is highly regular as the exponent is high. Consequently, we expect that the Hölder exponent of the kth primitive F_k of a continuous function F near a point x_0 is k times greater than the one of F, i.e., $\alpha_{F_k}(x_0) = \alpha_F(x_0) + k$.

However, it holds in some interesting cases such as $F(x) = |x|^\alpha \sin\left(\frac{1}{|x|^\beta}\right)$, $(\alpha, \beta > 0)$, that the Hölder exponent $\alpha_F(0) = \alpha$, but the Hölder exponent of the primitive F_1 of F at 0 is $\alpha_{F_1}(0) = \alpha + \beta + 1 \neq \alpha + 1$. This reads that the singular trigonometric oscillating term $\sin\left(\frac{1}{|x|^\beta}\right)$, has an essential role in the description of the behavior of the function. In order to take into account the singular behavior corresponding to the oscillating singular sine, one needs to introduce an adopted definition taking both the exterior exponent of regularity $((|x|^\alpha)$ in the last example) and the interior exponent $((\frac{1}{|x|^\beta})$ in the last example) that describes in fact the local power-law divergence of the instantaneous frequency. We deal here with the so-called oscillating singularity. Such a type of singularity has been investigated widely, see [8–10,25,60,62,73,78,79].

To understand the phenomenon and/or the role of the oscillating term, more adequate definitions of regularity are introduced. More precisely, for x_0 fixed, and F a bounded function considers its fractional primitive or order $t > 0$ defined by

$$F_t = (Id - \Delta)^{-t/2}(\phi F), \qquad (7.12)$$

where ϕ is a C^∞ compactly supported function satisfying $\phi(X_0) = 1$. We denote $\alpha_{F_t}(X_0)$ as the Hölder exponent of F_t at X_0. Recall further that the operator $(Id - \Delta)^{-t/2}$ is also obtained as a convolution operator that multiplies the Fourier transform of the function with $(1 + |\xi|^2)^{\frac{-t}{2}}$. We have the following definition.

Definition 7.2. Let $F : \mathbb{R}^d \to \mathbb{R}$ be a bounded function. For $t > 0$, denote F_t the fractional primitive of f of order t. The oscillating singularity exponent of F at a point X_0 is defined by

$$OSE_F(X_0) = (\alpha_F(X_0), \frac{\partial}{\partial t}\alpha_{F_t}(X_0)\Big|_{t=0^+} - 1).$$

In addition, we may also note that due to the possibility of directional-wise behaviors such as anisotropy, the Hölder regularity in its classical form may not be efficient for analyzing anisotropic features. Note that the regularity introduced in Definition 7.1 above is uniform in all directions and did not take into account the directional behavior of the functions. For example, if $F(x_1, x_2) = |x_1|^{\alpha_1} + |x_2|^{\alpha_2}$, where $0 < \alpha_i < 1$, $i = 1, 2$, then $F \in C^{\min \alpha_i}(0, 0)$. Therefore it will be of interest to think about a definition of regularity that considers directional deformations.

Several variants have been proposed, such as the study of the directional regularity $\alpha_{F_{X_0, e}}$ of the trace function $t \in \mathbb{R} \longmapsto F_{X_0, e}(t) = F(X_0 + te)$; X_0 being the point around which the regularity has to be estimated, and e is a fixed direction. For the example just above, $\alpha_{F_{(0,0), e_1}}(0) = \alpha_1$ and $\alpha_{F_{(0,0), e_2}}(0) = \alpha_2$, where (e_1, e_2) is the canonical basis of $\mathbb{R}^2$.

However, this definition has a strong disadvantage for the purposes of a directional multifractal analysis due to the fact that the line $X_0 + te$, $t \in \mathbb{R}$ is a set of vanishing measure. This leads us to think about a directional regularity definition that takes into account a whole neighborhood of the line with a non-vanishing measure. The following definitions have been formulated by many authors [1,19–24,26,28–30,32,33,63,88].

In addition, in [34,35] the authors have formulated a slightly modified definition of the directional regularity. We will see later that such a formulation has two advantages, it permits the wavelet characterization of the directional regularity, and links such a regularity of the anisotropic variant.

Definition 7.3. Let $\alpha = (\alpha_1, \alpha_2) \in \mathbb{R}_2$ be such that $\alpha_1 \geq \alpha_2 > 0$. A polynomial $P(x, y) = \sum_i a_{ij} x^i y^j$ has degree less than α if each monomial component of P of the form $a_{ij} x^i y^j$, with $a_{ij} \neq 0$ satisfies

$$\frac{i}{\alpha_1} + \frac{j}{\alpha_2} < 1.$$

We write $degree\, P < \alpha$.

We now introduce the concept of the anisotropic regularity in the following definition.

Definition 7.4. Let $F : \mathbb{R}^2 \to \mathbb{R}$, $\alpha = (\alpha_1, \alpha_2) \in \mathbb{R}^2$ be such that $\alpha_1 \geq \alpha_2 > 0$. F is said to be regular of order α at $X_0 = (x_0, y_0)$ and we write $F \in C^\alpha(X_0)$ if there exists $C > 0$, a polynomial $P = P(x, y)$ of degree less than α and a neighborhood W_0 of X_0 such that,

$$|F(x, y) - P(x - x_0, y - y_0)| \leq C(|x - x_0|^{\alpha_1} + |y - y_0|^{\alpha_2}), \ \forall (x, y) \in W_0.$$

Definition 7.5. Let $e \in \mathbb{R}^2 \setminus \{0\}$ be a unit vector, $\mathcal{B}$ be an orthonormal basis of $\mathbb{R}^2$ starting with e, $\varepsilon > 0$, and $\alpha = (\alpha_1, \alpha_2) \in \mathbb{R}^2$. The ε-neighborhood of

$X_0 = (x_0, y_0)$ of direction e and exponent α is

$$\mathcal{N}_{e,\alpha}^{\varepsilon}(x_0, y_0) = \left\{(x, y); \; \left(\frac{x - x_0}{\varepsilon^{v_1}}\right)^2 + \left(\frac{y - y_0}{\varepsilon^{v_2}}\right)^2 \leq 1\right\},$$

where $(x - x_0, y - y_0)$ are the coordinates in $\mathcal{B}$ and $v_i = \dfrac{\overline{\alpha}}{\alpha_i}$, with $\overline{\alpha} = \dfrac{2\alpha_1\alpha_2}{\alpha_1 + \alpha_2}$ being the harmonic mean of α.

Corollary 7.6. *Let $X_0 = (x_0, y_0) \in \mathbb{R}^2$, and $F : \mathbb{R}^2 \to \mathbb{R}$. $F \in C^\alpha(X_0)$ iff $\forall \varepsilon > 0$, $\exists c > 0$, $\exists$ a polynomial P, degree$P < \alpha$ and*

$$|F(x, y) - P(x - x_0, y - y_0)| \leq c\varepsilon^{\overline{\alpha}}, \quad \forall(x, y) \in \mathcal{N}_{e,\alpha}^{\varepsilon}(x_0, y_0).$$

The following definition introduces the concept of directional regularity.

Definition 7.7. *Let $e \in \mathbb{R}^2 \setminus \{0\}$ be unitary, $X_0 = (x_0, y_0) \in \mathbb{R}^2$ and $F : \mathbb{R}^2 \to \mathbb{R}$. The Hölder exponent of F in the direction e at X_0 is (denoted by)*

$$\alpha_F(X_0, e) = \sup\left\{\alpha_1; \; \exists \varepsilon > 0, \; F \in C^{\alpha_1, \varepsilon}(X_0, \mathcal{B})\right\},$$

where the coordinates are taken in any orthonormal basis $\mathcal{B}$ of $\mathbb{R}^2$ starting with e.

Definition 7.8. *Let $F : \mathbb{R}^2 \to \mathbb{R}$ be bounded on a neighborhood W_0 of $X_0 = (x_0, y_0)$. Let also $h > 0$ and $\mathcal{B} = (e_1, e_2)$ be an orthonormal basis of $\mathbb{R}^2$. Then, F is h-regular at X_0 according to the direction u, and we write $F \in C_u^h(X_0)$ if there exists $C > 0$, a polynomial $P = P(x, y)$ of u-homogeneous degree less than h and a neighborhood W_0 of X_0 such that,*

$$|F(x, y) - P(x - x_0, y - y_0)| \leq C\left(|x - x_0|^{h/u_1} + |y - y_0|^{h/u_2}\right), \quad \forall(x, y) \in W_0.$$

The u-Hölder exponent of F at X_0 is

$$H_F(X_0, u) = \sup\{h; \; F \in C_u^h(X_0)\},$$

where, as usual, the coordinates are expressed in the basis $\mathcal{B}$.

The following result is proved in [34]. This result may be considered as one of the most important results in regularity description as it guarantees or confirms the existence of a link between the concept of anisotropic regularity to the concept of directional regularity.

Theorem 7.9. *Let $e \in \mathbb{R}^2$ be unitary, and E be the set of vectors $u = (u_1, u_2)$; $0 < u_1 \leq u_2 = 2 - u_1$. The Hölder exponent of F at X_0 in the direction e satisfies*

$$\alpha_F(X_0, e) = \sup_{u \in E}\left(\frac{H_F(X_0, u)}{u_1}\right),$$

where $\mathcal{B}$ is any orthonormal basis of $\mathbb{R}^2$ starting with e.

7.3 Some formal calculus of oscillating singularities

In this section, we review some known cases about the computation of oscillating singularities in order to make readers familiar with such a type of calculus. We apply formal calculus that sometimes needs more assumptions, but they are not pointed out here.

Recall that for $\alpha \in \mathbb{R}$ and $x \in]-1, 1[$, we have the power series development

$$(1 - x)^\alpha = \sum_{n=0}^{+\infty} (-1)^n \frac{\Gamma(\alpha)}{n!\Gamma(\alpha - n)} x^n.$$

For $\alpha = -\frac{t}{2}$, $t > 0$ sufficiently small, this may be transformed to represent formally the operator $(Id - \Delta)^{-t/2}$, where Δ is the Laplace operator (second order derivative in the one-dimensional real case). We obtain a formal representation as

$$(Id - \Delta)^{-t/2} = \sum_{n=0}^{+\infty} (-1)^n \frac{\Gamma(-\frac{t}{2})}{n!\Gamma(-\frac{t}{2} - n)} \Delta^n,$$

with $\Delta^n = \Delta \circ \Delta \circ \cdots \circ \Delta$, is the operator Δ applied n times. Denote next

$$a_n(t) = (-1)^n \frac{\Gamma(-\frac{t}{2})}{n!\Gamma(-\frac{t}{2} - n)}$$

and consider the one-mode oscillating function with oscillating exponent (α, β) at $x = 0$, defined by

$$F(x) = |x|^\alpha \sin(\frac{1}{|x|^\beta}) \quad \text{and} \quad F(0) = 0.$$

We recall that the exponent α reflects the Hölder regularity of F at 0, and β its instantaneous frequency or its oscillating singular behavior at 0, which means that $OSE_F(0) = (\alpha, \beta)$. We obtain formally

$$F_t(x) = (Id - \Delta)^{-t/2} F(x) = \sum_{n=0}^{+\infty} a_n(t) \Delta^n F(x),$$

which in turn permits us to write

$$\widehat{F_t}(\omega) = \sum_{n=0}^{\infty} a_n((i\omega)^2)^n \widehat{F}(\omega) = (1 + |\omega|^2)^{-t/2} \widehat{F}(\omega),$$

where $\widehat{F_t}$ is the Fourier transform of F_t, and $\widehat{F}$ is the Fourier transform of F.

Now, to evaluate the $\Delta^n F(x)$ for all $n \in \mathbb{N}$, we will provide the first four derivatives of f. This is simple, but, to our knowledge, it is yet to be developed. We have for $x > 0$,

$$F'(x) = \alpha x^{\alpha-1} \sin(\frac{1}{|x|^\beta}) - \beta x^{\alpha-\beta-1} \cos(\frac{1}{|x|^\beta}). \tag{7.13}$$

Next,

$$F''(x) = (\alpha)_1 x^{\alpha-2} \sin(\frac{1}{|x|^\beta}) - \beta A_{\alpha,\beta} x^{\alpha-\beta-2} \cos(\frac{1}{|x|^\beta})$$
$$- \beta^2 x^{\alpha-2(\beta+1)} \sin(\frac{1}{|x|^\beta}). \tag{7.14}$$

Similarly,

$$F'''(x) = (\alpha)_2 x^{\alpha-3} \sin(\frac{1}{|x|^\beta}) - \beta B_{\alpha,\beta} x^{\alpha-\beta-3} \cos(\frac{1}{|x|^\beta})$$
$$- 3\beta^2(\alpha - \beta - 1)x^{\alpha-2\beta-3} \sin(\frac{1}{|x|^\beta}) \tag{7.15}$$
$$+ \beta^3 x^{\alpha-3(\beta+1)} \cos(\frac{1}{|x|^\beta})$$

and finally,

$$F^{(4)}(x) = (\alpha)_3 x^{\alpha-4} \sin(\frac{1}{|x|^\beta}) - \beta C_{\alpha,\beta} x^{\alpha-\beta-4} \cos(\frac{1}{|x|^\beta})$$
$$+ 2\beta^3(2\alpha - 3\beta - 3)x^{\alpha-3\beta-4} \cos(\frac{1}{|x|^\beta}) \tag{7.16}$$
$$+ \beta^4 x^{\alpha-4(\beta+1)} \sin(\frac{1}{|x|^\beta}),$$

where

$$(\alpha)_n = \alpha(\alpha - 1)(\alpha - 2)(\alpha - 3)\ldots(\alpha - n), \ n \in \mathbb{N},$$
$$A_{\alpha,\beta} = 2\alpha - \beta - 1, \quad B_{\alpha,\beta} = 3\alpha^2 + \beta^2 - 6\alpha + 3\beta - 3\alpha\beta + 2$$

and

$$C_{\alpha,\beta} = 3\alpha^3 - \beta^3 - 15\alpha^2 - 3\beta^2 + 4\alpha\beta^2 - 6\alpha^2\beta + 18\alpha\beta + 20\alpha - 11\beta - 6.$$

We note from these computations that the terms reflecting the largest contribution of the oscillating behavior of the derivatives are, in fact, the last terms in the right-hand parts of Eqs. (7.13), (7.14), (7.15), and (7.16), respectively. We write this information as

$$F'(x) \simeq -\beta x^{\alpha-\beta-1} \cos(\frac{1}{|x|^\beta}) \text{ as } x \to 0^+,$$

$$F''(x) \simeq -\beta^2 x^{\alpha - 2(\beta + 1)} \sin(\frac{1}{|x|^\beta}) \text{ as } x \to 0^+,$$

$$F'''(x) \simeq \beta^3 x^{\alpha - 3(\beta + 1)} \cos(\frac{1}{|x|^\beta}) \text{ as } x \to 0^+$$

and

$$F^{(4)}(x) \simeq \beta^4 x^{\alpha - 4(\beta + 1)} \sin(\frac{1}{|x|^\beta}) \text{ as } x \to 0^+.$$

Similar computations may be deduced for $x \to 0^-$. We thus obtain by recurrence on the integer parameter n,

$$\Delta^n F(x) \simeq \beta^{2n} |x|^{\alpha - 2n(\beta + 1)} \sin(\frac{1}{|x|^\beta}), \ |x| \to 0,$$

which yields that

$$F_t(x) \simeq |x|^\alpha \sin(\frac{1}{|x|^\beta}) \sum_{n \geq 0} a_n(t) \left(\frac{\beta}{|x|^{\beta + 1}} \right)^{2n}, \ |x| \to 0.$$

Or equivalently,

$$F_t(x) \simeq |x|^{\alpha + t(\beta + 1)} \sin(\frac{1}{|x|^\beta}), \ |x| \to 0.$$

Consequently, the Hölder exponent of F_t at 0 is

$$\alpha_{F_t}(0) = \alpha + t(\beta + 1).$$

We thus join the definition of the singularity exponent β corresponding to the oscillating aspect as

$$\beta = \frac{\partial}{\partial t} \alpha_{F_t}(0) \Big|_{t = 0^+} - 1.$$

In this section, we propose extending the formal calculus developed previously to the higher dimensional case in order to adopt a rigorous definition for the directional oscillating singularity. For the sake of simplicity, we will restrict the calculus to the 2-dimensional case and denote (e_1, e_2) the canonical basis of $\mathbb{R}^2$.

To avoid the use of many notations, we will keep the notation α_F to designate the Hölder regularity (mono-directional, and bi-directional). This means for a 2-variable real-valued function $F = F(x, y)$, the exponent $\alpha_F(X_0)$ will be a couple of exponents, a first one for the Hölder regularity in the x-variable (e_1 direction), and a second exponent reflecting the Hölder regularity in the y-variable (e_2 direction). In terms of the oscillating behavior, we keep the notation

β_F to designate the exponent in the interior of the singular trigonometric function (sin). This means that in the 2-dimensional case, the function $F = F(x, y)$ will be represented by four exponents:

- a first exponent reflecting its Hölder regularity in the x-variable, or the e_1-direction, which is in fact evaluated by α_{FX_0, e_1};
- a second exponent reflecting its Hölder regularity in the y-variable, or the e_2-direction, which is in fact evaluated by α_{FX_0, e_2};
- a third exponent reflecting its singular trigonometric oscillation in the x-variable, or the e_1-direction, which is in fact evaluated by β_{FX_0, e_1};
- a last and fourth exponent reflecting its singular trigonometric oscillation in the y-variable, or the e_2-direction, which is in fact evaluated by β_{FX_0, e_2}.

We immediately note that

$$\beta_{FX_0, e_i}(0) = \frac{\partial}{\partial t} \alpha_{FX_0, e_i}(0) \Big|_{t=0^+} - 1.$$

In the rest of this section we aim to investigate these oscillating exponents for different modes of singular trigonometric oscillations.

7.3.1 Case 1

Consider the first model of oscillating singularities

$$F(x, y) = |x|^\alpha \sin(\frac{1}{|x|^\beta}) + |y|^\gamma \sin(\frac{1}{|y|^\gamma}) = G(x) + H(y).$$

The Hölder regularities of F at 0 according to x and y are given by the couple

$$\alpha_F(0) = (\alpha, \gamma).$$

Now, for t as previously consider the operator $(Id - \Delta)^{-t/2}$, where Δ stands for the Laplace operator in the two-dimensional case,

$$\Delta = \frac{\partial^2}{\partial x^2} + \frac{\partial^2}{\partial y^2} = \Delta_x + \Delta_y.$$

Then, analogous calculus as for the one-dimensional case yields two separate and independent parts such as

$$F_t(x, y) = (Id - \Delta)^{-t/2} F(x, y) = G_t(x) + H_t(y),$$

where

$$G_t(x) = (Id - \Delta)^{-t/2} G(x)$$

and

$$H_t(y) = (Id - \Delta)^{-t/2} H(y)$$

are the one-dimensional case respective transforms of the one-dimensional functions $G(x)$ and $H(y)$ as above. Consequently, the Hölder exponents of F at 0 according to the direction x and y are summarized in the couple

$$\alpha_{F_t}(0) = (\alpha + t(\beta + 1), \gamma + t(\delta + 1)), \tag{7.17}$$

or equivalently

$$(\alpha_{F_t}(0, e_1), \alpha_{F_t}(0, e_2)) = (\alpha + t(\beta + 1), \gamma + t(\delta + 1)). \tag{7.18}$$

Consequently,

$$\frac{\partial}{\partial t}\alpha_{F_t}(0)\Big|_{t=0^+} = (\beta + 1, \delta + 1), \tag{7.19}$$

or equivalently,

$$\begin{cases} \beta = \dfrac{\partial}{\partial t}\alpha_{F_t}(0, e_1)\Big|_{t=0} - 1, \\[2mm] \delta = \dfrac{\partial}{\partial t}\alpha_{F_t}(0, e_2)\Big|_{t=0} - 1. \end{cases} \tag{7.20}$$

We now examine a different type of operator to understand more the directional behavior of functions. Instead of the operator $(Id - \Delta)^{-t/2}$ applied above, we apply the operator

$$(Id - \Delta_x)^{-t/2}(Id - \Delta_y)^{-t/2}.$$

In this way, we obtain already for the same function F above:

$$F_t(x, y) = F_t(x) + G_t(y).$$

As a consequence, we obtain the same results as in (7.17), (7.19), and (7.20), which is coherent with Definition 7.10.

Note here that we may use a quite different form of the differential operator by considering

$$(Id - \Delta_x)^{-t/2}(Id - \Delta_y)^{-s/2},$$

for two independent parameters t and s. Denote $\mathbf{t} = (t, s)$ and

$$F_{\mathbf{t}}(x, y) = F_{t,s}(x, y) = (Id - \Delta_x)^{-t/2}(Id - \Delta_y)^{-s/2}F(x, y) = F_t(x) + G_s(y).$$

As a consequence, the directional Hölder exponent at 0 will be

$$\alpha_{F_{\mathbf{t}}}(0) = \alpha_{F_{t,s}}(0) = (\alpha + t(\beta + 1), \gamma + s(\delta + 1)). \tag{7.21}$$

This yields a slightly modified version of (7.20), but already with the gradient of $H_t(F, 0)$ such as

$$
\begin{cases}
\beta = \dfrac{\partial}{\partial t}\alpha_{F_t}(0, e_1)\Big|_{t=0^+} - 1, \\[3mm]
\gamma = \dfrac{\partial}{\partial s}\alpha_{F_t}(0, e_2)(0)\Big|_{s=0^+} - 1.
\end{cases}
\tag{7.22}
$$

The last formulations are somewhat coherent even formally with (7.20), and with Definition 7.10 even with the modification.

We continue to examine different forms of the regularizing differential operator and consider now

$$
(Id - \Delta_x)^{-t/2} + (Id - \Delta_y)^{-t/2}.
$$

Applying such an operator, we obtain

$$
F_t(x, y) = F_t(x) + G_t(y), \quad \text{as } |x|, |y| \to 0.
$$

As a consequence, we re-obtain the same formulation as in (7.17), (7.19), and (7.20), which is also coherent with Definition 7.10 above.

Similarly, if we consider now two independent parameters t, s and the operator

$$
(Id - \Delta_x)^{-t/2} + (Id - \Delta_y)^{-s/2},
$$

we obtain by applying such an operator

$$
F_{t,s}(x, y) = F_t(x) + G_s(y), \quad \text{as } |x|, |y| \to 0.
$$

As a consequence, we obtain here also the same result as in (7.22).

From this example, we deduce the computation of the exponent of directional oscillating singularities due to the superposition of several oscillating and independent signals, each one depends on different variables from the others, as described in (7.2).

7.3.2 Case 2

In this section, we investigate a different type of oscillating singularities by considering the two-variable function

$$
F(x, y) = |x|^\alpha \sin(\frac{1}{|x|^\beta})|y|^\gamma \sin(\frac{1}{|y|^\gamma}) = G(x)H(y).
$$

Similarly to previous cases, we may write that

$$
F_t(x, y) = (Id - \Delta)^{-t/2} F(x, y) = G_t(x)H(y) + G(x)H_t(y).
$$

Therefore for t small enough we obtain a directional Hölder exponent of F_t as in (7.17), (7.19), and (7.20).

We now consider as in the previous case the product operator

$$(Id - \Delta_x)^{-t/2}(Id - \Delta_y)^{-t/2}.$$

We obtain

$$F_t(x, y) = (Id - \Delta_x)^{-t/2}(Id - \Delta_y)^{-t/2}F(x, y) = G_t(x)H_t(y),$$

which yields the same developments as above.

Consider next the different form due to a couple of independent parameters $\mathbf{t} = (t, s)$ as

$$(Id - \Delta_x)^{-t/2}(Id - \Delta_y)^{-s/2}.$$

This time, we obtain

$$F_{t,s}(x, y) = (Id - \Delta_x)^{-t/2}(Id - \Delta_y)^{-s/2}F(x, y) = G_t(x)H_s(y).$$

As a consequence, the directional Hölder exponent at 0 will be as in (7.21) or equivalently (7.22).

Continuing to examine the summing operator different form

$$(Id - \Delta_x)^{-t/2} + (Id - \Delta_y)^{-t/2},$$

we obtain

$$F_t(x, y) = G_t(x)H(y) + G(x)H_t(y),$$

which yields the same results as in the first step. Similarly, if we consider the operator

$$(Id - \Delta_x)^{-t/2} + (Id - \Delta_y)^{-s/2},$$

we obtain in the same way

$$F_{\mathbf{t}}(x, y) = F_{t,s}(x, y) = F_t(x)H(y) + G(x)H_s(y).$$

For $t, s > 0$ small enough, we obtain, as previously, the directional regularity as in (7.21) and (7.22).

Compared to the previous case, the present one describes a mixed case (product) resulting from correlated oscillating signals, which are independent, and, as previously, each one depends on different variables. This case may be subscribed under the whole class described in (7.3). We also deduce that the computation of the exponent of directional oscillating singularities is compatible with Definition 7.10.

7.3.3 Case 3

Consider the model function

$$F(x, y) = |x|^\alpha |y|^\gamma \sin(\frac{1}{|x|^\beta |y|^\delta}).$$

As previously, we will examine the different forms for the regularizing operator in order to obtain a coherent definition for the directional oscillating singularities.

The application of the first form operator gives at 0, the estimation

$$F_t(x, y) = (Id - \Delta)^{-t/2} F(x, y)$$

$$= \left[C_1 |x|^{\alpha + t(\beta+1)} |y|^{\gamma+t\delta} + C_2 |x|^{\alpha+t\beta} |y|^{\gamma+t(\delta+1)} \right] \sin(\frac{1}{|x|^\beta |y|^\gamma}),$$

where $C_1 = C(\beta, t)$ is a constant depending only on β and t, and similarly, $C_2 = C(\delta, t)$ is a constant depending only on δ and t. Consequently, the directional Hölder exponent of F_t at 0 is

$$\alpha_{F_t}(0) = (\alpha + t(\beta + 1), \gamma + t(\delta + 1)),$$

which joins the previous cases in (7.17).

We now investigate different variants for the regularizing operator. Consider

$$F_t(x, y) = (I - \Delta_x)^{-t/2}(I - \Delta_y)^{-t/2} F(x, y).$$

Simple computations yield that

$$F_t(x, y) = C |x|^{\alpha + t(\beta+1)} |y|^{\gamma + t(\delta+1)} \sin \left(\frac{1}{|x|^\beta |y|^\gamma} \right),$$

where $C = C_{\alpha, \beta, \gamma, \delta}$ is a constant depending on α, β, γ, δ. As a result, the directional Hölder exponent of F_t at 0 is here also the same as in (7.17), which leads to (7.19) and (7.20).

Consider next the two-parameter formulation

$$F_{t,s}(x, y) = (I - \Delta_x)^{-t/2}(I - \Delta_y)^{-s/2} F(x, y).$$

Near 0, this yields that

$$F_{t,s}(x, y) = C |x|^{\alpha + t(\beta+1)+s\beta} |y|^{\gamma+t\delta+s(\delta+1)} \sin \left(\frac{1}{|x|^\beta |y|^\gamma} \right),$$

where $C = C_{\alpha, \beta, \gamma, \delta}$ is a constant depending on α, β, γ, δ. Consequently, the directional Hölder exponent will be

$$\alpha_{F_t}(0) = (\alpha + t(\beta + 1) + s\beta, \gamma + t\delta + s(\delta + 1)). \tag{7.23}$$

This new form yields here also the same result as in (7.22), and joins again Definition 7.10.

As we note, this example is more complicated when compared to the previous cases. The singular sine contains a mixed case (product) resulting from correlated Hölderian signals, which are independent, and as previously, each one depends on different variables and/or directions. This case may be subscribed under the whole class described in (7.5). We also deduce that the computation of the exponent of directional oscillating singularities is compatible with Definition 7.10.

7.3.4 Case 4

Consider now the last model

$$F(x, y) = |x|^{\alpha} |y|^{\gamma} sin(\frac{1}{|x|^{\beta} + |y|^{\delta}}).$$

By applying the first variant of the regularizing operator $(Id - \Delta)^{-t/2}$, we obtain the directional Hölder exponent of F_t at 0 as in (7.17), (7.19), and (7.20).

Consider now the expression

$$F_t(x, y) = (I - \Delta_x)^{-t/2}(I - \Delta_y)^{-t/2}F(x, y).$$

Whenever $x, y \to 0$, similar techniques as in the previous cases yield that

$$F_t(x, y) = |x|^{\alpha+t(\beta+1)}|y|^{\gamma+t(\delta+1)} \sin\left(\frac{1}{|x|^{\beta}|y|^{\gamma}}\right).$$

Consequently, we obtain here also (7.17), and thus (7.19) and (7.20).

Finally, consider the transform

$$F_t(x, y) = F_{t,s}(x, y) = (I - \Delta_x)^{-t/2}(I - \Delta_y)^{-s/2}F(x, y).$$

Analogous calculus as above permits us to obtain

$$F_{t,s}(x, y) = |x|^{\alpha+t(\beta+1)}|y|^{\gamma+s(\delta+1)} \sin\left(\frac{1}{|x|^{\beta}|y|^{\gamma}}\right), \quad |x|, |y| \to 0.$$

As a result, the directional Hölder exponent of $F_{t,s}$ will be the same as estimated in (7.21) and (7.22).

This example describes a case of oscillating singular signals where the singular sine contains a superposition of Hölderian signals, which are independent, and as previously, each one depends on different variables and/or directions. This case may be subscribed under the whole class described in (7.4). We here also deduce the compatibility with Definition 7.10.

It results from all computations that a rigorous definition for directional oscillating singularities may be formulated as follows. Therefore a definition of the directional oscillating singularity may be formulated as follows.

Definition 7.10. Let $F : \mathbb{R}^2 \to \mathbb{R}$ be a bounded function. The directional oscillating singularity exponent of F at a point X_0 is defined by

$$
OSE_F(X_0) = \begin{pmatrix} \alpha_{F_{X_0,e_1}}(0) & \frac{\partial}{\partial t}\alpha_{F_t X_0,e_1}(0)\Big|_{t=0^+} - 1 \\[2mm] \alpha_{F_{X_0,e_2}}(0) & \frac{\partial}{\partial t}\alpha_{F_t X_0,e_2}(0)\Big|_{t=0^+} - 1 \end{pmatrix}
$$

$$
= \begin{pmatrix} \alpha_{F_{X_0,e_1}}(0) & \beta_{F_{X_0,e_1}}(0) \\[2mm] \alpha_{F_{X_0,e_2}}(0) & \beta_{F_{X_0,e_2}}(0) \end{pmatrix}.
$$

7.4 Wavelet characterization of the regularity

The present section is concerned with the development of the wavelet characterization of the regularity of function, in the different form exposed above. The idea of using wavelets in this context rests on the idea that instead of estimating the Hölder regularity of a function directly via the definition, we estimate its wavelet coefficients near the point in question.

This is motivating for many reasons, especially in relation to the application in images analysis, such as nano-cases. Indeed, the presence of irregularities, rupture, deformations, and singularity behavior or aspects in images, surfaces, and so forth is well known nowadays. To investigate these aspects, many tools have been used, such as fractal images and box dimensions. However, the majority of cases are based on observations, and an experimental approach. Rigorous mathematical proofs are always difficult and need more effort and also tools. Wavelets are one of the powerful tools in analyzing irregular objects and in discovering their hidden structures. Nanomaterials are one of the objects in nature that present many hidden aspects such as scaling laws, and fractal and multifractal behaviors.

The characterization of Hölder regularity, both uniform and point-wise by wavelets has been investigated by many authors, see [20–25,51,52,58,59,61]. The fundamental result established in [59] is stated in the following theorem.

Proposition 7.11. *[52], [59]. Let ψ be a $C^r(\mathbb{R}^d)$ analyzing wavelet on $\mathbb{R}^d$. For a real-valued function F on $\mathbb{R}^d$ let*

$$
C_{a,b}(F) = \frac{1}{a^d} \int_{\mathbb{R}^d} F(t)\bar{\psi}\left(\frac{t-b}{a}\right) dt \tag{7.24}
$$

be its wavelet transform at the position $b \in \mathbb{R}$ and the scale $a > 0$. The following assertions hold:

- $F \in C^\alpha(\mathbb{R}^d)$ *if and only if* $|C_{a,b}(F)| \leq Ca^\alpha$ *for all b and $0 < a < 1$;*

- If $F \in C^\alpha(x_0)$, then for $0 < a < 1$ and $|b - x_0| \leq 1/2$,

$$|C_{a,b}(F)| \leq Ca^\alpha \left(1 + \frac{|b - x_0|}{a}\right)^\alpha ; \qquad (7.25)$$

- If (7.25) holds and if $F \in C^\varepsilon(\mathbb{R}^d)$ for an $\varepsilon > 0$, then there exists a polynomial P of degree less than α such that, if $|x - x_0| \leq 1/2$,

$$|F(x) - P(x - x_0)| \leq C|x - x_0|^\alpha \log\left(\frac{2}{|x - x_0|}\right). \qquad (7.26)$$

Analogous results as in Proposition 7.11 have been established by several authors, including [1,17–24,26,28–30,32–35,63–65].

The oscillating singularity exponent has been characterized by means of the wavelet transform in [9,10,25,60,62,78,79] as follows.

Proposition 7.12. Let $F \in C^\varepsilon(\mathbb{R}^d)$ for some $\varepsilon > 0$, α, β and x be fixed real numbers. Then, the oscillating singularities exponent $H_F(x) = (\alpha, \beta)$ if and only if the following assertions hold:

1.

$$\liminf_{j \to +\infty, k2^{-j} \to x} \frac{\log|C_{j,k}(F)|}{\log(2^{-\alpha j} + |x - k2^{-j}|^\alpha)} \geq 1;$$

2. There exists a sequence $(j_n, k_n 2^{-j_n})$ satisfying $k_n 2^{-j_n} \to x$ as $n \to \infty$,

$$\liminf_{n \to +\infty} \frac{\log(2^{-j_n} + |x - k_n 2^{-j_n}|)}{-j_n \log 2} = \frac{1}{1 + \beta} \qquad (7.27)$$

and

$$\liminf_{n \to +\infty} \frac{\log|C_{j_n,k_n}(F)|}{\log(2^{-\alpha j_n} + |x - k_n 2^{-j_n}|^\alpha)} = 1; \qquad (7.28)$$

3. The exponent β is the lower bound of all real numbers satisfying (7.27) and (7.28).

The sequence $(j_n, k_n 2^{-j_n})_n$ is said to be a minimizing sequence for F at the point x.

For the case of the 2-dimensional case, the various types of regularity exposed previously have been also investigated by means of wavelets such as in [1,17–24,26,28–35,54,55,61,63–65,88,92–94]. For the sake of simplicity, we will recall briefly the basic results in the 2-dimensional case. The following results concern the anisotropic regularity.

Definition 7.13. Let ϕ be a function in the Schwartz class such that $\widehat{\phi}(0) = 1$ and $Support(\widehat{\phi}) \subset B(0, 1)$. For $a > 0$, $u = (u_1, u_2) \in \mathbb{R}^2$ such that $u_1, u_2 > 0$,

$u_1 + u_2 = 2$, denote

$$\phi_{a,u}(x, y) = \frac{1}{a^2} \phi \left(\frac{x}{a^{u_1}}, \frac{y}{a^{u_2}} \right).$$

Let also $\mathcal{B}$ be an orthonormal basis of $\mathbb{R}^2$. The anisotropic wavelet transform of $F \in L^2(\mathbb{R}^2)$ relative to u and $\mathcal{B}$ is defined by

$$d_{a,b}(F, \mathcal{B}, u, \lambda) = \int_{\mathbb{R}^2} F(x, y) e^{-i\lambda . X} \phi_{a,u}(\Omega_{\mathcal{B}}(X - b)) dX,$$

where $\Omega_{\mathcal{B}}$ is the $(2, 2)$-matrix mapping the canonical basis of $\mathbb{R}^2$ to the basis $\mathcal{B}$.

Proposition 7.14. *[63] Let F be a locally bounded function with slow growth. If $F \in C^\alpha(x_0)$ and $(a^{u_1}|\lambda_1|)^2 + (a^{u_2}|\lambda_2|)^2 \geq 1$, then there exists $C > 0$ such that*

$$|d_{a,b}(F, \mathcal{B}, u, \lambda)| \leq C \left[a^{\overline{\alpha}} + |x_0 - b_1|^{\alpha_1} + |y_0 - b_2|^{\alpha_2} \right],$$

whenever $a \leq 1$ and $|X_0 - b| \leq 1$.

To estimate the directional-wise oscillating regularity of a function $F = F(x, y)$, we should pass as for the mono-directional case by the directional Hölder regularity of its fractional primitive $F_t = (Id - \Delta)^{-t/2} F$. To deduce or evaluate the oscillating behavior of the function F by means of wavelets, we need to apply the equivalent of Proposition 7.14 and Proposition 7.12 to the function $F_t = F_t(x, y)$ to estimate first its directional chirp exponents.

Denote by φ and ψ some suitable scaling (father) and mother wavelets (in the Schwartz class, smooth enough, compactly supported, enough or infinitely vanishing moments). It is well known that the set $(\varphi_k, \psi_{j,k})$, $j, k \in \mathbb{Z}$, where

$$\varphi_k = \varphi(. - k), \text{ and } \psi_{j,k} = 2^{-j/2} \psi(2^j . - k)),$$

is an orthonormal basis in $L^2(\mathbb{R})$.

Let $u = (u_1, u_2) \in \mathbb{R}^2$ be such that $0 < u_1 \leq u_2 = 2 - u_1$. For $j \in \mathbb{N}$, we write $I_{j,u}$ the set of pairs (G, l), with $G = (G_1, G_2) \in \{F, M\}^2$ has at least one component equal to M and $l = (l_1, l_2) \in \mathbb{N}^2$, where

$$l_i = [ju_i] \text{ if } G_i = F, \tag{7.29}$$

$$[ju_i] \leq l_i < [(j + 1)u_i] \text{ if } G_i = M \text{ and } (j + 1)u_i > [ju_i] \tag{7.30}$$

and

$$l_i = [ju_i] \text{ if } G_i = M \text{ and } (j + 1)u_i = [ju_i]. \tag{7.31}$$

Note that $I_{j,u}$ is bounded independently of j. Consider next the functions

$$\phi_k(x) = \prod_{i=1}^{2} \psi_F(x_i - k_i)$$

and

$$\psi_{j,k,u}^{(G,l)}(x) = \prod_{i=1}^{2} \psi_{G_i}(2^{l_i} x_i - k_i).$$

We obtain here an orthonormal basis of $L^2(\mathbb{R}^2)$ composed of the set of functions

$$(\phi_k, 2^{|l|/2} \psi_{j,k,u}^{(G,l)}), \; j \in \mathbb{N}, (G,l) \in l_{j,u}, \; k \in \mathbb{Z}^2, \; |l| = l_1 + l_2,$$

in such a way that any function $F \in L^2(\mathbb{R}^2)$ may be expressed as

$$F(x) = \sum_{k \in \mathbb{Z}^2} C_k \phi_k(x) + \sum_{j=0}^{\infty} \sum_{k \in \mathbb{Z}^2} \sum_{(G,l) \in l_{j,u}} d_{j,k,u}^{(G,l)} \psi_{j,k,u}^{(G,l)}(x),$$

where, as usual, the coefficients C_k and $d_{j,k,u}^{(G,l)}$ are evaluated as

$$C_k = \int_{\mathbb{R}^2} F(x)\phi_k(x)\,dx$$

and

$$d_{j,k,u}^{(G,l)} = 2^{|l|} \int_{\mathbb{R}^2} F(x)\psi_{j,k,u}^{(G,l)}(x)\,dx.$$

Examples of well-known bases in this form are due to Triebel ([92,93]). The following definition is formulated in [34].

Theorem 7.15. *Let* $e \in \mathbb{R}^2$ *with* $|e| = 1$. *Let* E *be the set of all* $u = (u_1, u_2) \in \mathbb{R}^2$ *satisfying* $0 < u_1 \leq u_2 = 2 - u_1$, *and* $F \in C^\epsilon(\mathbb{R}^2)$ *for* $\epsilon > 0$. *The Hölder exponent of* F *in the direction* e *at* X_0 *is given by*

$$\alpha_F(X_0, e) =$$

$$\sup_{u \in E} \left(\lim_{j \to \infty} \inf_{k \in \mathbb{Z}, (G,l) \in I_{j,u}} \frac{\log |d_{j,k,u}^{G,I}|}{\log(2^{-j u_1}) + |x_{0,1} - \frac{k_1}{2^{l_1}}| + |x_{0,2} - \frac{k_2}{2^{l_2}}|^{u_{12}}} \right),$$

where $u_{12} = \dfrac{u_1}{u_2}$, *and the coordinates are on any orthonormal basis* $\mathcal{B}$ *starting with* e, *and where the set* $I_{j,u}$ *is given in* (7.29), (7.30), *and* (7.31).

For the next one, denote $R_u = (2^{-j_1} k_1, 2^{-j_2} k_2) + [0, 2^{-j_1}] \times [0, 2^{-j_2})$ known as a u-dyadic cube, and denote W_{R_u} the associated wavelet coefficient of the function F relative to the pair (j, k) due to R_u. Write also

- $d_{j,u}(X_0) = \displaystyle\max_{R'_u \in Adj(R_{j,u}(X_0))} d_{R'_u}$;
- $R_{j,u}(X_0)$ the u-dyadic cube containing x_0 at the scale j;
- $Adj(R_{j,u}(X_0))$ the set of all u-dyadic cubes adjacent to R_u;

- $d_{R_u} = \sup\limits_{R'_u \subset R_u} |C_{R'_u}|$ the dominant u-wavelet coefficient.

Theorem 7.16. *With the same notations as in Definition 7.7, we have*

$$\alpha_F(X_0, e) = \sup_{u \in E}\left(\liminf_{j \to \infty} \frac{\log |d_{j,u}(X_0, \mathcal{B})|}{\log 2^{-ju_1}}\right),$$

where $\mathcal{B}$ is any orthonormal basis of $\mathbb{R}^2$ starting with the vector e.

To achieve wavelet characterizations, we recall that the result of Theorem 7.9 may serve in the wavelet characterization of the anisotropic or directional regularity in some cases. Indeed, whenever this upper bound is reached for some $\mathcal{U} = (u_1, u_2)$, the exponent $\alpha_F(X_0, e)$ will be the result of the anisotropic wavelet estimation of $H_F(X_0, \mathcal{U})$ as estimated in Definition 7.8, and relative to the anisotropy $\mathcal{U}$. In this case, a wavelet characterization may be based on the anisotropic wavelet introduced in Definition 7.13 with respect to the anisotropy $\mathcal{U}$. The following result gives a wavelet characterization of the u-Hölder regularity.

Theorem 7.17. *The u-Hölder exponent of F at X_0 in the direction e is given by*

$$H_F(X_0, u) = \liminf_{j \to \infty} \frac{\log d_{j,u}(X_0)}{-j \log 2}.$$

7.5 An application on nanomaterials

7.5.1 AI ML and nanotechnology brief history

Among the many artificial intelligence methods and models, machine learning has become one of the popular methods in the nanotechnology area in the last decades. AI machine learning methods and nanotechnology meet, for instance, in the discovery of the design of nanomaterials, development of new nanomaterials, and their involvement in hardware, and also in the area of data processing such as images, big data sets, etc.

In [87], for example, an AI platform has been applied to visualize nanoscale patterns due to images of nanomaterials based on Deep Learning neural networks. An illustrative diagram is proposed in [87], and may be consulted to explain visually the platform, and provides a general idea on the link between AI, machine learning, and nanofrontiers. Readers may also refer to https://sites.google.com/view/ aifornanotechnology.

One of the popular mathematical tools nowadays in AI fractals occupies a great place. Nowadays, fractal analysis and geometry, especially fractal dimensions are widely applied in image processing, such as in materials-issued images in order to characterize, for example, the breaking up zones in materials.

Indeed, different behaviors observed in nanomaterials involve, and/or necessitate, fractal models to be explored, and exploited, such as the porous structure, and oscillating behavior, which are known phenomena in materials.

Fractal analysis/geometry constitutes an excellent mathematical tool also in detecting irregularities in images. These irregularities may be due to different causes, such as, experimental conditions in laboratories, natural effects, etc.

Oscillating behavior is observed in many phenomena in nature such as sea waves as the most famous example. Mathematical investigations of the behavior of materials has been widely studied. Even though the use of modern and sophisticated mathematical tools such as wavelets, fractional calculus in nano-materials, is somewhat recent, see [12–15,37,39,40].

Experimental studies, however, are somewhat more developed. In the field of nanomedicine, which is nowadays a central topic for researchers from both theory and applications, we may cite the implementation of plasmonic material nanoparticles for cancer diagnosis. Such implementation may induce noisy signals, which is why scientists search to apply noble materials that may be more noiseless, such as gold (see, for example, [53]). Oscillating behavior may also be observed for the giant magnetoresistance behavior of multi-layer nanofilms (see [75,102]).

Recent investigations have also been carried out for fluidization of nanoparticles in the presence of an oscillating magnetic field. By applying an oscillating AC magnetic field for exciting large permanent magnetic particles mixed in with the nanoparticle agglomerates, and the fluidization behavior of the nanoagglomerates, including the fluidization regime, the minimum fluidization velocity, the bed pressure drop, and the bed expansion have been investigated in [100]. It is shown that the bed of nanoparticle agglomerates may be smoothly fluidized, and that the minimum fluidization velocity may be significantly reduced relative to the frequencies of the oscillating magnetic field. In addition, channeling or slugging of the bed disappears and the bed expands uniformly without bubbles, and with negligible elutriation for low instantaneous oscillating frequencies of the magnetic field.

Nano is originally extracted from the Greek word dwarf to designate one thousand millionth of a unit, any unit of measure. Next, the prefix nano is added to the words technology to yield what we call now nanotechnology. It is also applied with the word mechanics to designate the branch of science called now nanomechanics. However, nanotechnology itself goes back to Richard Feynman since the end of the 1940s when introducing his famous principle of nanotechnology by stating that the principles of physics may allow us to maneuver things atom by atom.

The technology has been developed to provide us effectively with the maneuvering tool in nanotechnology by discovering the electron microscope, giving researchers the ability to analyze nano-structures. The first scanning electron microscope became therefore available not very much later from the postulate of Feynman about atom-by-atom handling of particles, effectively in the 1960s.

In the 1980s, Binnig and his collaborators (Rohrer, 1981, and Quate, and Gerber, 1986) created the scanning atomic-force microscope and the scanning tunneling microscope, allowing for the investigation of the surface of a material at nanometer scales.

Nanotechnology has currently allowed scientists to understand what is happening in both phenomena and materials at the nanoscale, and also to create and apply structures, devices, and systems, such as nano-structure composites as well as nanoformation polymers.

Currently, with the technological development, and implementation of modern instruments, such as the nano-microscopes, the physical properties of materials have become more and more comprehensive. Moreover, new materials and new properties have been discovered. The fractal structure is one of the physical properties expressed at the supermolecular level, which is explored further by nano-microscopes.

In the present section, we mainly propose conducting some concrete examples from nanomaterial imaging in order to illustrate the presence of anisotropic oscillating singularities and computing the eventual multifractal parameters for such images.

The use of 2D images may be somewhat justified by the fact that in nanomaterials, we have not to look only to the sizes of particles, but also to the way that these particles are localized. Indeed, it is observed that for many nanomaterials, the majority of their atoms localize on the surface of a particle, in contrast to ordinary materials where atoms occur over the volume of a particle [69,83].

7.5.2 Fractal processing of nanomaterials

In the present part, we propose conducting some discussions about the use of fractal dimensions for nanomaterial images. This will permit us to reach at the same time the use of fractals in both images/signals and nanomaterial applications, thus, we propose the description of how fractals are used for understanding nanoimages in addition to those images being issued from nanomaterials.

Estimating the fractal dimension of images will permit us to describe well the structural properties. The most (and even simple) used method to estimate the fractal dimension of an image is the so-called Box-dimension, known also the Bouligand–Minkowski dimension. It is based on a simple mathematical formula that computes the log-log slope of the maximum number of boxes (squares, balls, cubes, circles, etc.) used to cover the image (black pixels, for example, when the image is converted to black and white), by the size of such boxes.

In the present section, we consider the image presented in Fig. 7.2, which designates an image of Titanium dioxide TiO_2 nanoparticles. Such nanomaterials are well known and have many important properties. In terms of industrial aims, they are capable of adding new functionalities to infrastructures, such as self-cleaning properties. They also serve to remove air pollutants through photocatalysis. In building constructions, TiO_2 are used to degrade organic pollutants, without affecting the aesthetic characteristics of concrete structures. Titanium dioxide nanoparticles are particles of TiO_2, with diameters less than 100 nm. In terms of aims related to health, these nanoparticles are applied, for example, in ultrafine form. Indeed, TiO_2 are used in sunscreens due to its ability to block UV radiation while remaining transparent on the skin. The health risks of ultrafine TiO_2 from dermal exposure on intact skin are considered extremely low, and it is considered safer than other substances used for UV protection. In fact, the discovery and the use of fractal structures for the TiO_2 material were not pointed out first in the present work. It was instead investigated by many authors such as [41], where the authors studied the formation of the fractal nano-structures for the TiO_2.

FIGURE 7.2 Titanium dioxide TiO_2 nanoparticles.

In Table 7.1, estimations of the fractal dimension due to the box-counting method are provided. A log-log model relating the number of boxes to cover the image, denoted here by N_s, to the size of boxes is applied in the form

$$\log(N_s) = -a \log(Size) + b,$$

where the size is computed by means of dyadic cubes 2^{-s}, and triadic 3^{-s}, and b is a fitting coefficient, assuring the regression. The coefficient (slope) a will be eventually the estimated dimension. We note from Table 7.1 a mean value of the dimension estimated as

$$dim \approx 1.9646.$$

7.5.3 The (anisotropic) oscillations in nanomaterials

To emphasize more and more the fractal/multifractal nature of the TiO_2 nanomaterial, a zooming or a higher level multiresolution representation yielded Fig. 7.3 where the oscillating nature or phenomenon starts to appear. In addition, the directional oscillations appear in different directions, which explains the presence of anisotropy in the oscillating behavior.

TABLE 7.1 Estimation of the fractal dimension of TiO_2 illustrated in Fig. 7.2.

Scale	Iteration	Fitting coefficients	Dimension
2	5	0.0534	1.9615
2	6	0.0615	1.9594
2	9	0.0583	1.9689
3	5	0.0734	1.9623
3	6	0.0639	1.9674
3	9	0.0323	1.9778

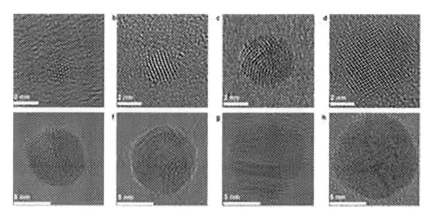

FIGURE 7.3   Some 2D illustrations of Titanium dioxide, TiO_2, nanoparticles.

In Table 7.2, we provide some oscillating regularity's exponents' estimations relative to some directional cases for Fig. 7.3 at the center point of each figure. We used the following illustrative directions

$$e_1 = (1, 0), \quad e_2 = (0, 1), \quad e_3 = (\frac{1}{\sqrt{2}}, \frac{1}{\sqrt{2}}).$$

Finally, in Table 7.3, we estimate the oscillating exponent in many directions already relative to Fig. 7.3.

Finally, it is worth noting that nanoparticles occupy space in several forms such as spheres and cylinders. This last form is at the heart of the connection between anisotropic and directional regularities. Indeed, at the nanoscale, these cylinders have very small thicknesses, almost null, or negligible in the mathematical sense. This is expressed by the fact that the most influential variable is the height, which means here a mono-one-dimensional aspect where the anisotropic character appears as un-influencing. These remarks make the study of nanoimages by a strong tool such as wavelets and fractals or mixed models of both of them of great importance. In fact, at the nanoscale, the thickness, which

TABLE 7.2 Estimation of the directional Hölder regularity of TiO_2 illustrated in Fig. 7.3.

Direction	Sub-figure	Directional Hölder regularity
e_1	Fig. 7.3-a	0.2331
e_2	Fig. 7.3-a	0.2345
e_3	Fig. 7.3-a	0.8561
e_1	Fig. 7.3-b	0.2344
e_2	Fig. 7.3-b	0.2412
e_3	Fig. 7.3-b	0.8354
e_1	Fig. 7.3-d	0.2532
e_2	Fig. 7.3-d	0.2431
e_3	Fig. 7.3-d	0.8532

TABLE 7.3 Estimation of the directional oscillating exponent of TiO_2 illustrated in Fig. 7.3.

Direction	Sub-figure	Oscillating exponent
e_1	Fig. 7.3-a	(0.2331, 0.3345)
e_2	Fig. 7.3-a	(0.2345, 0.3412)
e_3	Fig. 7.3-a	(0.8561, 0.5123)
e_1	Fig. 7.3-b	(0.2344, 0.3281)
e_2	Fig. 7.3-b	(0.2412, 0.3521)
e_3	Fig. 7.3-b	(0.8354, 0.5242)
e_1	Fig. 7.3-d	(0.2532, 0.3302)
e_2	Fig. 7.3-d	(0.2431, 0.3282)
e_3	Fig. 7.3-d	(0.8532, 0.5125)

may be assumed to be negligible, may have a great influence on the material. Mathematically speaking, this motivates and proves further that regularity or directional behavior is, in fact, different from the anisotropic one. As we have noted in the application above, the fractal (box) dimension is evaluated independently of wavelets, and shows at least the possible fractal structure of the material, and the regularities are evaluated by the wavelets independently from the fractal dimensions. We may thus ask what to expect when mixing both tools?

7.6   Conclusions

In this chapter, we have essentially revised the notion of regularity or oscillating singularity, precisely in higher dimensions. The goal has been to prove or point out to researchers that this notion can be extended to non-isotropic cases as was done for the classical Hölder regularity. Formal definitions and calculus have been developed. Specific applications to nanoimages extracted from nanomaterials have proven that directional oscillations can indeed occur in nature.

We believe that this chapter is a step forward and a first basis for developing a fundamental mathematical and physical study for rigorous definitions and characterizations by wavelets of these directional oscillations, as well as applications on other image cases from several fields.

7.7 Appendix – Wavelet toolkit review

This appendix is devoted to a review of the basics of the wavelet toolkit. Wavelet analysis has its origins in the mid-1980s, introduced in the context of signal analysis and exploration of petroleum. Next, wavelet analysis has been applied also to analyze different signals, such as seismic ones, more sensitively than Fourier techniques, leading to the first appearance of the continuous wavelet transform formula. Thus Wavelet theory has become an active area of research in many fields, including electrical engineering, mathematical analysis (harmonic analysis, operator theory), physics, and so forth [36]. Like Fourier analysis, wavelet analysis deals with decomposition of functions in terms of a set of functional bases. Unlike Fourier analysis, the wavelet one expands functions in terms of wavelets that are generated in the form of translations and dilations of a fixed function called the mother wavelet.

The wavelets obtained in this way have special scaling properties. They are localized in both time and frequency, offering an advantage in many cases and overcoming many difficulties raised by Fourier analysis theory. Wavelet theory permits the representation of L_2-functions in a functional Riesz basis well localized in time and in frequency spaces.

For more details, we may refer the reader to [12–14,45,50,92,93].

Definition 7.18. In $L^2(\mathbb{R})$, a wavelet is a function $\psi \in L^2(\mathbb{R})$ that satisfies the following conditions:

- Admissibility,

$$C_\psi = \int_0^\infty \frac{|\widehat{\psi}(\xi)|^2}{\xi} d\xi < \infty.$$

- Localization in time,

$$\int_{-\infty}^{+\infty} |\psi(t)|^2 dt = 1.$$

- Vanishing moments,

$$i = 0, \dots, N-1, \quad \int_{\mathbb{R}} \psi(t) t^i dt = 0.$$

To analyze a function by wavelets, one passes, as for Fourier analysis, by its wavelet transform or its continuous wavelet transform (CWT). The Continuous wavelet transform is based first on the introduction of a scale (dilation) parameter $a > 0$ and another parameter $b \in \mathbb{R}$ known as the position or the translation, respectively, to the analyzing wavelet ψ that plays the role of Fourier sine and cosine and will be subsequently called the mother wavelet. The translation parameter determines the position or the time around which we want to assess the behavior of the analyzed function, while the scale factor is used to assess the behavior around the position.

Definition 7.19. The continuous wavelet transform of a function $f \in L^2(\mathbb{R})$ at the scale a and the position b is given by

$$C_f(a, b) = \int_{-\infty}^{+\infty} f(t) \, \overline{\psi}_{a,b}(t) \, dt,$$

where $\psi_{a,b}$ is an L^2-normalized copy of ψ,

$$\psi_{a,b}(t) = \frac{1}{\sqrt{a}} \, \psi(\frac{x-b}{a}).$$

The wavelet transform $C_f(a, b)$ has several properties:

1. It is linear, in the sense that $\forall \alpha, \beta \in \mathbb{R}$ and $f_1, f_2 \in L^2(\mathbb{R})$,

$$C_{(\alpha f_1 + \beta f_2)}(a, b) = \alpha C_{f_1}(a, b) + \beta C_{f_2}(a, b).$$

2. It is translation-invariant:

$$C_{(\tau_{b'} f)}(a, b) = C_f(a, b - b'),$$

where $\tau_{b'}$ refers to the translation of f by b' given by

$$(\tau_{b'} f)(x) = f(x - b').$$

3. It is dilation-invariant, in the sense that, for some $\lambda, r > 0$ fixed, then

$$C_f(a, b) = \lambda C_f(ra, rb),$$

where $f(x) = \lambda f(rx)$.

As in Fourier or Hilbert analysis, wavelet analysis provides a Plancherel type relation that permits itself the reconstruction of the analyzed function from its wavelet transform. The function f can be reconstructed from its wavelet transform as follows (see [5–7,45,50].

Theorem 7.20. For $f \in L^2(\mathbb{R})$, we have the L^2-equality

$$f(x) = \frac{1}{C_\psi} \int_{a>0} \int_{b \in \mathbb{R}} C_f(a, b) \, \psi(\frac{x-b}{a}) \, \frac{da \, db}{a^2}.$$

Proof. We first show that for all $f, g \in L^2(\mathbb{R})$, we have

$$\int \int C_f(a, b) \, \overline{C_g(a, b)} \, \frac{da \, db}{a} = C_\psi \int f(x) g(x) \, dx. \qquad (7.32)$$

Indeed, we observe that

$$C_f(a,b) = \frac{1}{\sqrt{a}} \int f(x)\psi(\frac{x-b}{a})dx$$

$$= \frac{1}{2\pi}\mathcal{F}(\widehat{f}(y)\overline{\widehat{\psi}(ay)}e^{-iby}).$$

Consequently,

$$\int_b C_f(a,b)\overline{C_g(a,b)}db = \frac{1}{2\pi}\int_y \widehat{f}(y)\overline{\widehat{g}(y)}|\widehat{\psi}(ay)|^2 dy.$$

By application of Fubini's theorem, it follows that

$$\int_{a>0}\int_b C_f(a,b)\overline{C_g(a,b)}\frac{dadb}{a} = \frac{1}{2\pi}\int_{a>0}\int_y \widehat{f}(y)\overline{\widehat{g}(y)}|\widehat{\psi}(ay)|^2\frac{dady}{a}$$

$$= \frac{1}{2\pi}C_\psi \int_y \widehat{f}(y)\overline{\widehat{g}(y)}dy$$

$$= C_\psi \int_y f(y)\overline{g(y)}dy.$$

Now, by applying Eq. (7.32) we may write that

$$\|f(x) - \frac{1}{C_\psi}\int_{\frac{1}{A}\leq a\leq A}\int_{|b|\leq B} C_f(a,b)\psi(\frac{x-b}{a})\frac{dadb}{a^2}\|_{L^2}$$

$$= \sup_{\|g\|=1}\left(f(x) - \frac{1}{C_\psi}\int_{\frac{1}{A}\leq a\leq A}\int_{|b|\leq B} C_f(a,b)\psi(\frac{x-b}{a})\frac{dadb}{a^2}\right)\overline{g(x)}dx$$

$$= \sup_{\|g\|=1}\left(\int f(x)\overline{g(x)}dx - \frac{1}{C_\psi}\int_{\frac{1}{A}\leq a\leq A}\int_{|b|\leq B} C_f(a,b)\overline{C_g(a,b)}\frac{dadb}{a}\right)$$

$$= \sup_{\|g\|=1}\frac{1}{C_\psi}\int_{(a,b)\notin[1/A,A]\times[-B,B]} C_f(a,b)\overline{C_g(a,b)}\frac{dadb}{a}$$

$$\leq \frac{1}{C_\psi}\left[\int_{(a,b)\notin[1/A,A]\times[-B,B]}|C_f(a,b)|^2\frac{dadb}{a}\right]^{\frac{1}{2}}$$

$$\cdot\left[\sup_{\|g\|=1}\int_{(a,b)\notin[1/A,A]\times[-B,B]}|C_f(a,b)|^2\frac{dadb}{a}\right]^{\frac{1}{2}}.$$

The first equality is an application of the Riesz lemma, the second comes from Fubini's theorem and the Cauchy–Schwartz inequality. At this point, the previous lemma shows that the last quantity tends to 0. □

Now, we will recall the basic properties and the history of a multidimensional wavelet basis.

The first constructed bases were separable, and their construction focused on an analogy with the Haar one. In the one dimensional case, the Haar basis is defined by

$$\begin{cases} \psi_{j,k}(x) = 2^{j/2}\psi(2^j x - k); \ j,k \in \mathbb{Z} \\ \psi = \xi_{[0,1/2[} - \xi_{[1/2,1[}. \end{cases}$$

In the multi-dimensional case, it is defined as follows:

$$\begin{cases} \psi^{\varepsilon}_{j,k}(x) = 2^{j/2}\psi^{\varepsilon}(2^j x - k); \ ; j,k \in \mathbb{Z} \\ \varepsilon \in \{0,1\}^d \setminus \{(0,...,0)\} \\ \psi^{\varepsilon} = \psi^{\varepsilon_1} \otimes \psi^{\varepsilon_2} \otimes ... \otimes \psi^{\varepsilon_d} \\ \psi^1 = \psi \ , \ \psi^0 = \varphi = \xi_{[0,1]}. \end{cases}$$

We generally speak about Triebel wavelets, see, for example, [34,35,92,93].

The next concept in wavelet theory that is essentially applied in signal/image processing and in practical and applied fields in general is the concept of multi-resolution analysis ([67,72,76,82]).

Definition 7.21. A multi-resolution analysis (multi-scale) on $\mathbb{R}^d$ is a sequence $\{V_j\}_{j\in\mathbb{Z}}$ of closed subspaces in $L^2(\mathbb{R}^d)$ satisfying the following properties

$$V_j \subset V_{j+1}, \quad \forall j \in \mathbb{Z},$$

$$\overline{\bigcup_{j\in\mathbb{Z}} V_j} = L^2(\mathbb{R}^d) \quad , \quad \bigcap_{j\in\mathbb{Z}} V_j = \{0\},$$

$$f \in V_j \Longleftrightarrow f(2^j.) \in V_0,$$

$$f \in V_0 \Longleftrightarrow f(.-k) \in V_0, \ \forall k \in \mathbb{Z}^d,$$

$$\exists \ \varphi \in V_0; \ \{\varphi(.-k)\}_{k\in\mathbb{Z}^d} \text{ is a Riesz basis in } V_0.$$

In the one dimensional case, we denote by P_j the orthogonal projection on V_j. The main idea in the multi-resolution analysis is the ability to construct an orthonormal wavelet basis $\{\psi_{j,k}; \ j,k \in \mathbb{Z}\}, \ \psi_{j,k}(x) = 2^{j/2}\psi(2^j x - k)$, such that

$$P_{j+1}f = P_j f + \sum_k < f, \psi_{j,k} > \psi_{j,k}, \ \forall \ f \in L^2(\mathbb{R}).$$

We will describe in brief the construction of such a basis. Let W_j be the orthogonal supplementary of V_j in V_{j+1}. The first properties of the multi-resolution analysis allow that

$$L^2(\mathbb{R}) = \oplus_{j\in\mathbb{Z}} W_j.$$

It was proved ([45], [50]) that for a fixed $j \in \mathbb{Z}$, $\{\psi_{j,k}; k \in \mathbb{Z}\}$ is an orthonormal basis of W_j and that $\{\psi_{j,k} ; \ j,k \in \mathbb{Z}\}$ is an orthonormal basis in the hole space

$L^2(\mathbb{R})$. To construct ψ, we need to set

$$\varphi = \sum_{k \in \mathbb{Z}} h_k \varphi_{1,k},$$

which is a trivial consequence of the fact that $\varphi \in V_0 \subset V_1$. The last one is generated by $\{\varphi_{1,k}; k \in \mathbb{Z}\}$. We write then

$$\psi(x) = \sum_{k \in \mathbb{Z}} (-1)^k h_{1-k} \varphi_{1,k}(x).$$

Let (V_j^1) be a multi-resolution analysis of $L^2(\mathbb{R})$ with a scaling function φ, a wavelet ψ and an orthogonal projections P_j^1. Consider then the orthogonal projection P_j^d in $L^2(\mathbb{R}^d)$ defined as the tensor product of d copies of P_j^1,

$$P_j^d = P_j^1 \otimes P_j^1 \otimes \ldots \otimes P_j^1.$$

Denote $V_j^d = P_j^d(L^2(\mathbb{R}^d))$. We have $V_j^d = V_j^1 \otimes V_j^1 \otimes \ldots \otimes V_j^1$. The closure in $L^2(\mathbb{R}^d)$ of V_j^d has an orthonormal basis

$$\begin{cases} \varphi_{j,k}^d = 2^{jd/2} \varphi^d(2^j x - k) \, ; \, j \in \mathbb{R} \, , \, k \in \mathbb{R}^d \\ \varphi^d = \varphi \otimes \varphi \otimes \ldots \otimes \varphi. \end{cases}$$

Then, the description of the orthogonal supplementary W_j^d of V_j^d in V_{j+1}^d follows immediately. We write

$$W_j^d = \bigoplus_{\varepsilon \neq (0, \ldots, 0)} V_j^{\varepsilon_1} \otimes V_j^{\varepsilon_2} \otimes \ldots \otimes V_j^{\varepsilon_d}.$$

This yields an orthonormal basis of $L^2(\mathbb{R}^d)$ associated to P_j^d

$$\begin{cases} \psi_{j,k}^\varepsilon(x) = 2^{jd/2} \psi^{\varepsilon_1}(2^j x_1 - k_1) \ldots \psi^{\varepsilon_d}(2^j x_d - k_d) \, ; \, j, k_i \text{ integers} \\ \psi^1 = \psi \, , \, \psi^0 = \varphi = \xi_{[0,1]}. \end{cases}$$

This last formula appears to be better than the one defined by a tensor product. The separable wavelets present are simple. In contrast, nonseparable wavelets remain difficult to use and to reconstruct. However, in analysis either in nature one can speak about propagation in privileged directions, i.e., signals whose regularity varies according to the direction of propagation. One plans to study their regularity by means of the well adapted wavelets. An important example of directional phenomena is supplied by spirals, such as the domain between the two curves of equations (in polar coordinates)

$$r = \theta^{-\alpha} \quad \text{and} \quad r = (\theta + \pi)^{-\alpha}.$$

Another example that bears similarities with the spirals is the set

$$C_\alpha = \bigcup_n \left[\frac{1}{(2n+1)^\alpha}, \frac{1}{(2n)^\alpha} \right].$$

If the aim is a point-wise analysis, without particular emphasis on directions, then the isotropic wavelet, such as the isotropic Mexican hat, will be more economical. However, if the signal to be detected has a preferred direction, then one needs a wavelet with good angular selectivity, see [2–4,76].

Acknowledgment

The authors would like to thank the editors of the book for providing us with the opportunity of contributing. Accordingly, a series of studies on mathematical tools for the exploration and exploitation of nanosystems as well as their improvement have been presented in our chapter.

References

[1] P. Abry, M. Clausel, S. Jaffard, S.G. Roux, B. Vedel, Hyperbolic wavelet transform: an efficient tool for multifractal analysis of anisotropic textures, Rev. Mat. Iberoam. 31 (1) (2015) 313–348.

[2] J.-P. Antoine, R. Murenzi, P. Vandergheynst, Directional wavelets revisited: Cauchy wavelets and symmetry detection in patterns, Appl. Comput. Harmon. Anal. 6 (1999) 314–345.

[3] J.-P. Antoine, R. Murenzi, Two-dimensional directional wavelets and the scale-angle representation, Signal Process. 52 (1996) 259–281.

[4] J.-P. Antoine, R. Murenzi, P. Vandergheynst, Two-dimensional directional wavelets in image processing, Int. J. Imaging Syst. Technol. 7 (1996) 152–165.

[5] S. Arfaoui, I. Rezgui, A. Ben Mabrouk, Wavelet Analysis on the Sphere, Spheroidal Wavelets, Degryuter, ISBN 978-3-11-048188-4, 2017.

[6] S. Arfaoui, A. Ben Mabrouk, C. Cattani, Fractal Analysis Basic Concepts and Applications, Series on Advances in Mathematics for Applied Sciences, vol. 91, World Scientific, 2022.

[7] S. Arfaoui, A. Ben Mabrouk, C. Cattani, Wavelet Analysis, Basic Concepts and Applications, 1st ed., CRC Taylor-Francis, Chapman & Hall, Taylor & Francis, Boca Raton, April 21, 2021.

[8] A. Arneodo, F. Argoul, E. Bacry, J. Elezgaray, J.F. Muzy, Ondelettes, multifractales et turbulences: de l'ADN aux croissances cristallines, Diderot Editeur, Arts et Sciences, Paris, 1995.

[9] A. Arneodo, E. Bacry, S. Jaffard, J.F. Muzy, Oscillating singularities on Cantor sets: a grand-canonical multifractal formalism, J. Stat. Phys. 87 (1/2) (1997) 971–998.

[10] A. Arneodo, E. Bacry, S. Jaffard, J.F. Muzy, Singularity spectrum of multifractal functions involving oscillating singularities, J. Fourier Anal. Appl. 4 (2) (1998) 159–174.

[11] M. Baldoni, C. Baroglio, D. Cavagnino, L. Saitta, Towards automatic fractal feature extraction for image recognition. Chapter 22, in: Huan Liu, Hiroshi Motada (Eds.), Feature Extraction Construction and Selection: A Data Mining Perspective, in: The Kluwer International Series in Engineering and Computer Science, vol. SECS 453, 1998, pp. 357–373.

[12] D. Baleanu, Wavelet Transforms and Their Recent Applications in Biology and Geoscience, InTechOpen, 2012.

[13] D. Baleanu, Advances in Wavelet Theory and Their Applications in Engineering, Physics and Technology, InTechOpen, 2012.

[14] D. Baleanu, Wavelet Transform and Some of Its Real-World Applications, InTechOpen, 2015.

[15] D. Baleanu, Z.B. Guvenc, J.A. Tenreiro Machado, New Trends in Nanotechnology and Fractional Calculus Applications, Springer, 2010.

[16] A. Banerjee, S.S. Banerjee, Growing gold fractal nano-structures and studying changes in their morphology as a function of film growth rate, Mater. Res. Express 3 (2016) 105016, https://doi.org/10.1088/2053-1591/3/10/105016.

[17] M. Ben Abid, M. Ben Slimane, I. Ben Omrane, B. Halouani, Mixed wavelet leaders multifractal formalism in a product of critical Besov spaces, Mediterr. J. Math. 14 (2017) 176.

[18] M. Ben Abid, M. Ben Slimane, I. Ben Omrane, M. Turkawi, Multivariate wavelet leaders Rényi dimension and multifractal formalism in mixed Besov spaces, Int. J. Wavelets Multiresolut. Inf. Process. (2022), https://doi.org/10.1142/S0219691321500478.

[19] H. Ben Braiek, M. Ben Slimane, Critère de la régularité directionnelle, C. R. Acad. Sci. Paris, Ser. I 349 (2011) 385–389.

[20] A. Ben Mabrouk, Multifractal analysis of some non isotropic quasi-self-similar functions, Far East J. Dyn. Syst. 7 (1) (2005) 23–63.

[21] A. Ben Mabrouk, Wavelet analysis of some non isotropic quasi-self-similar functions, PhD in mathematics, Faculty of Sciences, University of Monastir, Tunisia, 2007.

[22] A. Ben Mabrouk, Wavelet analysis of anisotropic quasi-self-similar functions in a nonlinear case, in: Colloque WavE 2006, Polytechnic Federal School of Lausanne, Switzerland, 10-14 July 2006.

[23] A. Ben Mabrouk, On some nonlinear non isotropic quasi-self-similar functions, Nonlinear Dyn. 51 (2008) 379–398.

[24] A. Ben Mabrouk, An adapted group dilation anisotropic multifractal formalism for functions, J. Nonlinear Math. Phys. 15 (1) (2008) 1–23.

[25] A. Ben Mabrouk, Wavelet analysis of nonlinear self-similar distributions with oscillating singularity, Int. J. Wavelets Multiresolut. Inf. Process. 6 (3) (2008) 1–11.

[26] A. Ben Mabrouk, Directionlets and some generalized nonlinear self-similarities, Int. J. Math. Anal. 5 (26) (2011) 1273–1285.

[27] A. Ben Mabrouk, J. Aouidi, Lecture Note on Wavelet Multifractal Analysis of Self Similarities, Lampert Academic Publishing, Verlag, ISBN 978-3-8465-9646-3, 2012.

[28] M. Ben Slimane, Etude du formalisme multi-fractal pour les fonctions, Thèse de Doctorat en Mathématiques Appliquées, Ecole Nationale des Ponts et Chaussées, 1996.

[29] M. Ben Slimane, Anisotropic two-microlocal spaces and regularity, J. Funct. Spaces (2014) 505796.

[30] M. Ben Slimane, Multifractal formalism and anisotropic selfsimilar functions, Math. Proc. Camb. Philos. Soc. 124 (1998) 329–363.

[31] M. Ben Slimane, M. Ben Abid, I. Ben Omrane, B. Halouani, On wavelet and leader wavelet based large deviation multifractal formalisms for non-uniform Hölder functions, J. Fourier Anal. Appl. 25 (2019) 506–522.

[32] M. Ben Slimane, M. Ben Abid, I. Ben Omrane, B. Halouani, Criteria of pointwise and uniform directional Lipschitz regularities on tensor products of Schauder functions, J. Fourier Anal. Appl. 460 (2018) 496–515.

[33] M. Ben Slimane, M. Ben Abid, I. Ben Omrane, M. Turkawi, Multifractal analysis of rectangular pointwise regularity with hyperbolic wavelet bases, J. Fourier Anal. Appl. 27 (2021) 90.

[34] M. Ben Slimane, H. Ben Braiek, Directional and anisotropic regularity and irregularity criteria in Triebel wavelet bases, J. Fourier Anal. Appl. 18 (2012) 893–914.

[35] M. Ben Slimane, H. Ben Braiek, On the gentle properties of anisotropic Besov spaces, J. Math. Anal. Appl. 396 (2012) 21–48.

[36] E.B. Bouchereau, Analyse d'images par transformées en ondelettes Application aux images sismiques, Thèse de Doctorat en Mathématiques Appliquées, Université Joseph Fourier, Grenoble I, 1997.

[37] L. Byoungsang, Y. Seokyoung, L.J. Woong, K. Yunchul, C. Junhyuck, Y. Jaesub, R.J. Chul, L. Jong-Seok, L.J. Heon, Statistical characterization of the morphologies of nanoparticles through machine learning based electron microscopy image analysis, ACS Nano 14 (2020) 17125–17133.

[38] D. Calitoiu, J.B. Oommen, D. Nussbaum, Neural network-based chaotic pattern recognition - Part 2: stability and algorithmic issues, in: Marek Kurzynski, Edward Puchata, Michat Wozniak, Andrzej Zolnierek (Eds.), Computer Recognition Systems, Proceedings of the 4th International Conference on Computer Recognition Systems CORES '05, Springer-Verlag Berlin Heidelberg, 2005, pp. 3–16.

[39] C. Cattani, J. Rushchitsky, Wavelet and Wave Analysis as Applied to Materials with Micro or Nanostructure, Series on Advances in Mathematics for Applied Sciences, vol. 74, 2007.

[40] C. Cattani, Y. Karaca, Computational Methods for Data Analysis, De Gruyter, 2018.

[41] G.L. Celardo, D. Archetti, G. Ferrini, L. Gavioli, P. Pingue, E. Cavaliere, Evidence of diffusive fractal aggregation of TiO2 nanoparticles by femtosecond laser ablation at ambient conditions, Mater. Res. Express 4 (2017) 015013, https://doi.org/10.1088/2053-1591/aa50e9.

[42] D. Chetverikov, R. Peteri, A brief survey of dynamic texture description and recognition, in: Marek Kurzynski, Edward Puchata, Michat Wozniak, Andrzej Zolnierek (Eds.), Computer Recognition Systems, Proceedings of the 4th International Conference on Computer Recognition Systems CORES '05, Springer-Verlag Berlin Heidelberg, 2005, pp. 17–26.

[43] K. Daoudi, Généralisations des Systèmes de Fonctions Itérées, Thèse de Mathématiques Appliquées, Paris 9 Dauphine, 1996.

[44] K. Daoudi, J.L. Véhel, Y. Meyer, Construction of continuous functions with prescribed local regularity, Constr. Approx. 14 (3) (1998) 349–385.

[45] I. Daubechies, Ten Lectures on Wavelets, Society for Industrial and Applied Mathematics, Philadelphia, PA, USA, 1992.

[46] A. Demming, Machine learning puts nanomaterials in the picture. Characterization and modelling research update, https://physicsworld.com, May 13, 2019.

[47] U. Diane, Nanoparticules métalliques anisotropes synthétisées par voie chimique: fils, plaquettes et particules hybrides de cobalt-nickel, propriétés structurales et magnétiques; fils d'argent auto-organisés, Matériaux, Université Paris-Diderot-Paris VII, 2005 (Français).

[48] U. Frisch, G. Parisi, Fully developed turbulence and intermittency, in: Proc. Int. Summer School Phys. Enrico Fermi, North Holland, 1985, pp. 84–88.

[49] C. Germain, Contribution à la caractérisation multi-échelle de l'anisotropie des images texturées, Thèse de Doctorat de l'Université de Bordeaux I. Numéro d'ordre 1808, 1997.

[50] M. Holschneider, Wavelets an Analysis Tool, Mathematical Monographs, Clarendon Press, Oxford, 1995.

[51] M. Holschneider, P.H. Tchamitchian, Régularité Locale de la Fonction Non-Differentiable de Riemann, Lecture Notes in Mathematics, vol. 1438, 1990, pp. 102–124.

[52] M. Holschneider, P. Tchamitchian, Pointwise analysis of Riemann's nondifferentiable function, Invent. Math. 105 (1) (1991) 157–175.

[53] X. Huang, M.A. El-Sayed, Gold nanoparticles: optical properties and implementations in cancer diagnosis and photothermal therapy, J. Adv. Res. 1 (1) (2010) 13–28.

[54] L. Huang, D. Yang, On function spaces with mixed norms - a survey, J. Math. Study 54 (2021) 262–336.

[55] I. Iglewska-Nowak, Directional wavelets on n-dimensional spheres, Appl. Comput. Harmon. Anal. 44 (2) (2018) 201–229.

[56] G. Iovane, A.V. Nasedkin, F. Passarella, Fundamental solutions in antiplane elastodynamic problem for anisotropic medium under moving oscillating source, Eur. J. Mech. A, Solids 23 (2004) 935–943.

[57] G. Iovane, A.V. Nasedkin, F. Passarella, Moving oscillating loads in 2D anisotropic elastic medium: plane waves and fundamental solutions, Wave Motion 43 (2005) 51–66.

[58] S. Jaffard, Exposants de Hölder en des points donnés et coefficients d'ondelettes, C. R. Acad. Sci. Paris, Sér. I Math. 308 (1989) 79–81.

[59] S. Jaffard, Pointwise smoothness, two-microlocalization and wavelet coefficients, Publ. Mat. Barc. 35 (1) (1991) 155–168.

[60] S. Jaffard, Oscillation spaces: properties and applications to fractal and multifractal functions, J. Math. Phys. 39 (8) (1998) 4129–4141.

[61] S. Jaffard, Pointwise regularity criteria, C. R. Acad. Sci. Paris, Ser. I 339 (2004) 757–762.

[62] S. Jaffard, Beyond Besov spaces Part 2: oscillation spaces, Constr. Approx. 21 (2004) 29–61.

[63] S. Jaffard, Pointwise and directional regularity of nonharmonic Fourier series, Appl. Comput. Harmon. Anal. 28 (2010) 251–266.

[64] S. Jaffard, C. Melot, R. Leonarduzzi, H. Wendt, P. Abry, S.G. Roux, M.E. Torres, p-exponent and p-leaders, Part I: negative pointwise regularity, Physica A 448 (2016) 300–318.

[65] S. Jaffard, C. Melot, R. Leonarduzzi, H. Wendt, P. Abry, S.G. Roux, M.E. Torres, p-exponent and p-leaders, Part II: multifractal analysis. Relations to detrended fluctuation analysis, Physica A 448 (2016) 319–339.

[66] S. Jaffard, Y. Meyer, Wavelet methods for pointwise regularity and local oscillations of functions, Mem. Am. Math. Soc. 123 (587) (1996).

[67] J.-P. Kahane, P.G. Lemarié-Rieusset, Séries de Fourier et ondelettes, Cassini, 1998.

[68] H. Kataoka, K. Okayasu, A. Matsumoto, et al., Pre-training without natural images, Int. J. Comput. Vis. 130 (2022) 990–1007.

[69] I. Khan, K. Saeed, I. Khan, Nanoparticles: properties, applications and toxicities, Arab. J. Chem. 12 (7) (2019) 908–931.

[70] W. Klonowski, E. Olejarczyk, R. Stepien, SEM image analysis for roughness assessment of implant materials, in: Marek Kurzynski, Edward Puchata, Michat Wozniak, Andrzej Zolnierek (Eds.), Computer Recognition Systems, Proceedings of the 4th International Conference on Computer Recognition Systems CORES '05, Springer-Verlag Berlin Heidelberg, 2005, pp. 553–560.

[71] S. Kockentiedt, K. Toennies, E. Gierke, N. Dziurowitz, C. Thim, S. Plitzko, Automatic detection and recognition of engineered nanoparticles in SEM images, in: M. Goesele, T. Grosch, B. Preim, H. Theisel, K. Toennies (Eds.), Vision, Modeling, and Visualization, 2012, 8 pp.

[72] P.G. Lemarié-Rieusset, Sur l'existence des analyses multi-résolutions en théorie des ondelettes (On the existence of multiresolution analyses in wavelet theory), Rev. Mat. Iberoam. 8 (3) (1992) 457–474 (in French).

[73] H.F. Liu, X. Gong, Z.H. Dai, Z.H. Yu, A new method to estimate the oscillating singularity exponents in locally self-similar functions, Phys. Lett. A 310 (1) (2003) 30–39.

[74] S. Liu, C. Cattani, Y. Zhang, Introduction of fractal based information processing and recognition, Appl. Sci. 9 (2019) 1297, https://doi.org/10.3390/app9071297.

[75] J. Liu, Y. Zhao, C. Lian, Z. Dai, J.-T. Sun, S. Meng, Ab initio study on anisotropic thermoelectric transport in ternary pnictide KZnP, J. Phys. Mater. 2 (2019) 024001.

[76] S.G. Mallat, Une exploration des signaux en ondelettes, Editions Ecole Polytechnique, 2000.

[77] M.M. Mansor, F.L. Mohd. Isa, D.A. Green, A.V. Metcalfe, Modelling directionality of paleoclimatic time series, ANZIAM J. 57 (EMAC2015) (2016) C66–C81.

[78] C. Mélot, Sur les singularité oscillantes et le formalisme multifractal, Thesis in Mathematics, University Paris 12, 2002.

[79] C. Melot, Oscillating singularities in Besov spaces, J. Math. Pures Appl. 83 (2004) 367–416.

[80] S. Jaffard, C. Mélot, Wavelet analysis of fractal boundaries. Part 1: local exponents, Commun. Math. Phys. 258 (2005) 513–539.

[81] S. Jaffard, C. Mélot, Wavelet analysis of fractal boundaries. Part 2: multifractal analysis, Commun. Math. Phys. 258 (541–565) (2005) 541–565.

[82] Y. Meyer, Oudelettcs er Operateus, Hermann, 1988.

[83] M. Mostafa, A. Alrowaili, M.M. AlShehri, M. Mobarak, A.M. Abbas, Structural and optical properties of calcium titanate prepared from gypsum, J. Nanotechnol. (2022) 6020378.

[84] S. Nojiri, S.D. Odintsov, The oscillating dark energy: future singularity and coincidence problem, Phys. Lett. B 637 (3) (2006) 139–148.

[85] H.S. Oh, H. Kim, S.J. Lee, The numerical methods for oscillating singularities in elliptic boundary value problems, J. Comput. Phys. 170 (2) (2001) 742–763.

[86] A.G. Okunev, M.Y. Mashukov, A.V. Nartova, A.V. Matveev, Nanoparticle recognition on scanning probe microscopy images using computer vision and deep learning, Nanomaterials 10 (2020) 1285, https://doi.org/10.3390/nano10071285.

[87] A. Rajagopal, V. Nirmala, J. Andrew, A.M. Vedamanickam, AI visualization in nanoscale microscopy, arXiv:2201.00966.

[88] J. Sampo, S. Sumetkijakan, Estimations of Hölder regularities and direction of singularity by Hart Smith and curvelet transforms, J. Fourier Anal. Appl. 15 (2009) 58–79.

[89] S. Seuret, J.L. Véhel, A time domain characterization of 2-microlocal spaces, J. Fourier Anal. Appl. 9 (5) (2003) 473–495.

[90] S. Seuret, J.L. Véhel, The local Hölder function of a continuous function, Appl. Comput. Harmon. Anal. 13 (3) (2002) 263–276.

[91] M.S. Swapna, S. Sankararaman, Fractal applications in bio-nanosystems, Advancements Bioequiv Availab. 2 (4) OABB.000541.2019, https://doi.org/10.31031/OABB.2019.02. 000541.

[92] H. Triebel, Wavelet bases in anisotropic function spaces, in: Proc. Conf. Function Spaces, Differential Operators and Nonlinear Analysis, Milovy, 2004, Math. Inst. Acad. Sci., Czech Republic, Prague, 2005, pp. 370–387.

[93] H. Triebel, Theory of Function Spaces III, Monographs in Mathematics, vol. 78, Birkhäuser, Basel, 2006.

[94] H. Triebel, Interpolation Theory, Function Spaces, Differential Operators, North Holland Mathematical Library, vol. 18, 1978, Amsterdam-New York.

[95] V. Tvergaard, B.N. Legarth, Interface crack growth for anisotropic plasticity with non-normality effects, Int. J. Solids Struct. 44 (22–23) (2007) 7357–7369.

[96] V. Tvergaard, B.N. Legarth, Effect of anisotropic plasticity on mixed mode interface crack growth, Eng. Fract. Mech. 74 (16) (2007) 2603–2614.

[97] V.M. Volchuk, O.V. Uzlov, O.V. Puchikov, D.S. Zotov, V.I. Sokoliuk, Fractal model of mechanical properties evaluation of C-Mn-Al-Ti-N steel with acicular ferrite structure for railway freight cars, AIP Conf. Proc. 2389 (2021) 080002, https://doi.org/10.1063/5.0063496, Published Online: 23 September 2021.

[98] R. Yamada, K. Okayasu, A. Nakamura, H. Kataoka, Fractal geometry-based automatic generation of large-scale image database for pre-training in 3D object recognition, J. Jpn. Soc. Precis. Eng. 87 (4) (2021) 374–379.

[99] X. Yang, W. Yang, J. Li, X. Zhang, Oscillatory singularity behaviors near interface crack tip for mode II of orthotropic bimaterial, J. Appl. Math. 2013 (2013) 716768.

[100] Q. Yu, R.N. Dave, C. Zhu, J.A. Quevedo, R. Pfeffer, Enhanced fluidization of nanoparticles in an oscillating magnetic field, AIChE J. 51 (7) (2005) 1971–1979.

[101] J. Yuying, H. Xuan, W. Zhongwei, H. Xiangang, Machine learning boosts the design and discovery of nanomaterials, ACS Sustain. Chem. Eng. 9 (2021) 6130–6147.

[102] Zhang, Physical Fundamentals of Nanomaterials, Micro & Nano Technologies Series, Chemical Industry Press, William Andrew Applied Sciences Publisher, Elsevier, 2018.

[103] http://www.insp.upmc.fr/Microscopie-Electronique-a.html.

Chapter 8

Comparative analysis of approaches to optimize fractal image compression

Rakesh Garg and Richa Gupta

Amity University, Noida, Uttar Pradesh, India

8.1 Introduction

The term fractal encoding was introduced by Barnsley and Sloan in 1988 [1]. Fractal-based compression is a lossy image-compression approach that exploits redundancy present in natural images. Researchers of a US-based company also compressed a two-dimensional random fractal curve using partitioning and segmentation of the curve [2]. This was the first scheme that was made public. Further, Barnsley and Hurd [3] proposed using the group of affine transformations for representation of images. This knowledge led to the birth of FIC for the image of characteristics.

Fractal image compression extended its reputation after the introduction of more suitable encoding for natural images based on a partitioned iterated function system (PIFS) [4]. An impediment in encoding images using FIC is the requirement of the enormous time spent in searching the domain pool for a suitable match. The computation complexity of the algorithm diverges with the size of the search space, partitioning scheme, and the metrics used for mapping. These complications associated with FIC have encouraged numerous researchers to overcome the pressing concerns of the algorithm and obtain a good-quality decoded image. Currently, not only real-time images but hyperspectral images are also using fractal transforms. The capability of fractals to handle massive data associated with a hyperspectrum and characteristics such as resolution independence makes it apposite for encoding hyperspectral images [5–7]. The chapter aims to address the following questions through the literature survey. RQ1: Classification of the approaches that are used to improve the fractal compression techniques. RQ2: Find the potential strategies for future optimization. RQ3: Find various performance measures that are used in the various algorithms. The organization of the chapter is as follows. Section 8.2 is a detailed description of the theoretical framework required to understand the underlying concepts of fractal image encoding. Section 8.3 offers critical is-

Copyright © 2024 Elsevier Inc. All rights reserved, including those for text and data mining, AI training, and similar technologies.

sues or parameters affecting FIC. Section 8.4 contains different strategies for accelerating the encoding and decoding speed of compression based on a fractal approach. The studied literature is analyzed and summarized in Section 8.5 as a quick review and conclusions are drawn in Section 8.6.

8.2 Theoretical framework

Well-known fractals like a fern and an apollonian gasket possess global self-similarity. A property of global self-similarity makes images easy to reproduce using IFS. An image having self-similarity is composed of a scaled version of itself or a subpart of itself. An iterated function system is used to create fractal images. The next subsection discusses the compression of a grayscale image using a baseline method.

Fractal image encoding of grayscale image

Grayscale images do not have self-similarity that is present in fractal images. Jacquin [4] brought forward the idea of IFS and introduced a (PIFS) partitioned iterated function system for encoding of grayscale images.

Partitioned iterated function systems (PIFS)

Jacquin [4] suggested using a set of transformation to store the image. The subsequent expression gives a fractal affine transformation of the image block in the pool. Parameters a and b in Eq. (8.1) are called scaling and offset coefficients, respectively:

$$W \begin{bmatrix} i \\ j \\ u(i, j) \end{bmatrix} = \begin{bmatrix} a_{11} & a_{12} & 0 \\ a_{21} & a_{22} & 0 \\ 0 & 0 & a \end{bmatrix} \begin{bmatrix} i \\ j \\ u(i, j) \end{bmatrix} + \begin{bmatrix} x \\ y \\ b \end{bmatrix}, \qquad (8.1)$$

where W_i is a set of the affine transformation. W_i can stretch, rotate, zoom, and translate the original image. Parameters a_{11} to a_{22} are called the transform coefficients of W_i. Coefficients x and y are the locations of the domain block u. Also, it is remarkable that the contraction W_i is composed of two linear affines transform $w_{i,1}$ and $w_{i,2}$, $w_{i,1}$. This changed the spatial coordinates of the image and named geometric transform as shown in Eq. (8.2):

$$w_{i,1} \begin{pmatrix} i \\ j \end{pmatrix} = \begin{pmatrix} a_{11} & a_{12} \\ a_{21} & a_{22} \end{pmatrix} \begin{pmatrix} i \\ j \end{pmatrix} + \begin{pmatrix} e \\ f \end{pmatrix}. \qquad (8.2)$$

Transform $w_{i,2}$ is called the massic, and it works on the intensity levels of the image. Eq. (8.3) defines $w_{i,2}$, which adjusts the contrast and brightness of the domain block:

$$w_{i,2} = a * u + b. \qquad (8.3)$$

Further, scaling a and offset b in Eq. (8.3) can be calculated by minimizing the distance between the range and the domain block. The distance between domain block u_k and range block v is determined by the following distance metric, as shown in Eq. (8.4):

$$d(a_i.u_k + b_i, v) = \|a_i.u_k + b_i - v\|, \tag{8.4}$$

where k, $0 < k \leq 7$ represents eight dihedral transformations of domain block u_k. The next expression calculates the value of a and b on minimizing $d(,)$.

$$a_i = \frac{\left[N \langle u_k, v \rangle - \left\langle u_k, \overrightarrow{I} \right\rangle \left\langle v, \overrightarrow{I} \right\rangle \right]}{\left[N \langle u_k, u_k \rangle - \left\langle u_k, \overrightarrow{I} \right\rangle^2 \right]}, \tag{8.5}$$

$$b_i = \frac{1}{N} \left[\left\langle v * \overrightarrow{I} \right\rangle - a \left\langle u_k * \overrightarrow{I} \right\rangle \right].$$

In the above expression, $N = 64$ and I is the identity matrix. The fractal code of the range block is constituted by the location of the mapped domain block, scaling, offset, and type of dihedral transformation. The length of the code depends on the bit used to represent these parameters.

Finally, for the reconstruction of the image, all (number of range blocks) the affine transformation or PIFS $(w_1, w_2 \ldots, w_N)$ is determined. The affine transform is applied to a random initial image recursively until it converges. The encoder of FIC is computationally extensive. Encoding of an image involves various parameters. These parameters affect the performance of FIC. The following section discusses the performance issues of FIC.

8.3 Factor affecting FIC

The calculation difficulty associated with the encoding of FIC can better be understood using the following practical example. Let f signify the image of size 512×512. Let us assume image domain block of 16×16 in a pool. The domain pool comprised of $(512 - 16 + 1) \times (512 - 16 + 1) = 247\,009$ numbers of blocks. The range pool is defined by nonoverlapping blocks of size half of the domain block, which aggregates to 4096 blocks. Mapping of the range to the domain block is an integral process of FIC. In the mapping process the size of blocks to be mapped plays an important role. The major drawback of the encoding algorithm proposed above is its difficulty in mapping large-size range blocks. The smaller-size block is easy to match but suffers from a large overhead.

A fractal transform is found for each range block v in the range pool by mapping to the most similar domain block in the domain pool. The domain block is subsampled and resized according to the size of range pool for each mapping. The correspondence between the subsampled domain block u and the range block v is determined by the mean square error (MSE).

Each domain block is subjected to eight dihedral transformations. Hence, each block needed $247\,009 \times 8 = 1\,976\,072$ MSE computation. Hence, a whole image encoded using an exhaustive search required $4096 \times 1\,976\,072 = 8\,093\,990\,912$ calculations. Another factor that impacts the encoding time is a comprehensive search to determine the best geometrical and luminance transform needed for mapping. For each range block the contrast, a, and brightness b of the subsampled domain is adjusted to minimize the difference between the blocks. Finally, the length of the code is determined by the bits assigned to transform the parameter and the location of domain block. High-level quantization of these transform parameters leads to a weak class of the reconstructed image. The decoder uses more approximations in the case of a high threshold that sequentially results in artifacts in the image. These blocking effects or image degradation due to quantization of coefficients is preventable by using decorrelating parameters.

Lastly, reconstruction majorly is determined by the iterative decoding process. The complexity of decoding depends on the initial image used. The subsequent section deliberates various approaches measure used to speed up the encoding process.

8.4 Approaches for accelerate FIC

8.4.1 Block classification

Jacquin [4] suggested creating classes based on the quantitative properties of the block. His work is founded on the classification scheme proposed by Ramamurthi and Gerso [8]. In [4], classification of the domain block is based on perceptual geometric properties of the block. Jacquin's approach presented three types of blocks: edge block, shade, and midrange block and presented the elaborated scheme. Fisher, Jacobs, and Boss [2,9] introduced the idea of subdivision of square blocks into four quadrants. The authors used pixel intensity and the order of variance to classify image subblocks. Fisher et al. [9] presented the use of the quadtree in FIC. In [2], three main classes are further divided into subclasses based on variance and average pixel intensity, as shown in Eq. (8.6):

$$class\ 1 : a_1 \geq a_2 \geq a_3 \geq a_4,$$
$$class\ 2 : a_1 \geq a_2 \geq a_4 \geq a_3, \qquad (8.6)$$
$$class\ 3 : a_1 \geq a_4 \geq a_2 \geq a_3.$$

In the above expressions, a_i is the pixel intensity of the four quadrants. Here, parameter i varies as $i = 1, 2, 3, 4$. In total, 72 classes are formed for the block, but only two are searched for the mapping purpose. Boss and Jacobs [10] achieved a higher quality of the image with much less searching using archetype classification. Archetype classification is a more elaborate arrangement that creates classes in advance using training images based on experimental studies. For a given set of domain blocks D_i an archetype classification gives a domain block D_k that has a minimum least-square difference. Eq. (8.7) finds a particular domain block that covers the set:

$$D_k = \arg\min_{D_k} \sum_{i \neq k} \min_{a,b} \| D_i - (a D_k + b B) \|. \tag{8.7}$$

Here, a, b, and B are parameters of the best transform w to be found for the set of domain block. Wang et al. [11] proposed hybrid partial fractal mapping for image coding. The properties of edges were used to classify the domain blocks in the presented framework for a partial fractal image coder. Analogous to the Wang approach, Duh et al. [12] used edge properties for classification of blocks into three classes. Computationally competent classes were formed using only vertical and horizontal DCT coefficients. Analogous to Fisher's classification method [2], a faster classification scheme is proposed by [13]. The idea presented in [13] used two parameters, named as the approximate first derivative (AFD), and the normalized root mean square error RMSE for sorting image blocks. Tong et al. [14] used the standard deviation for the classification of the image block to hasten the mapping process. Later, [15] also used standard deviation to enhance the results of Tong's method by applying an intelligent classification algorithm to all domain blocks. The standard deviation algorithm improved results of the search as it searches only for domain blocks with similar-valued range blocks.

8.4.2 Partitioning of image blocks

The performance of the FIC algorithm depends on the sort and size of partitioning implemented. The proposal of automated FIC using PIFS [4] is based on quadtree partitioning followed by many image encoding schemes [16–18]. These algorithms have been presented to improve the performance and reduce the complexity of the algorithm. Fisher et al. [16] emphasized the requirement and procedures to partition the original image into a range and a domain pool. The authors recommended the practice of quadtree partitioning as a generalization of a fixed-size range block. The outsized range block is challenging to map, while smaller-sized partitioning suffers from overhead due to there being more blocks to map. Also, the authors discussed the pros and cons of using HV and triangular partitioning to create an image pool. Reusens [19] gave an understanding on adaptive partitioning about IFS and presented the idea of polygon-based segmentation of image blocks. A vital point of the paper was the balance of the information between transformations of blocks and partition representation used in the encoding of the image. Reusens proposed polygon segmentation based on piecewise constant approximation [20]. The author exploited the optimal degree of adaptation about size as well as the shape of image blocks for better results. Fisher and Menlove [17] introduced the novel idea of HV partitions. At this juncture, a rectangular range block is segmented either vertically or horizontally to produce smaller rectangles. The location of the partition is estimated using indices i and j for which horizontal and vertical differences had a maximum value. Biased horizontal h_j and vertical v_i differences are calculated using Eq. (8.8):

$$h_j = \min(j, M - j - 1) \Big/ (M - 1) \left(\sum_i r_{i,j} - \sum_i r_{i,j+1} \right),$$

$$v_i = \min(i, N - i - 1) \Big/ (N - 1) \left(\sum_j r_{i,j} - \sum_j r_{i+1,j} \right).$$

$$(8.8)$$

In Eq. (8.8), the parameters i and j are indices of pixel values with r_{ij} varying. The terms $\min(j, M - j - 1)/(M - 1)$ and $\min(i, N - i - 1)/(N - 1)$ are biases being multiplied by horizontal and vertical sums, respectively. The performance of the algorithm is further optimized using quadrant classification, encoding based on block size, and the restricting ratio of the domain range. Square and rectangle partitions suffered from block artifacts around the edges. Devoine et al. [21] proposed a Delaunay triangulation-based FIC to overcome the drawback of previous algorithms [4,17]. Mixed triangular and quadrilateral partitions showed better results regarding storage, as a triangle can have any orientation. Consequently, it reduces the blocking effect. Quadrilaterals shaped by the grouping of adjacent triangles further reduces the number of mappings between the range and the domain blocks. Liaobtc, Chernc, and Tsaod [22], proposed an adaptive side coupling quadtree (ASCQ) structure that kept the domain and the range pool collectively. Duh, Jeng, and Chen [12] introduced an evolutionary computing centered FIC to find a better partitioning of the image as compared to quadtree partitioning. Further, in Saupe [28], four algorithms based on a derivative chain code were discussed to store the partitioning. Edge-orientation-based adaptive shape partitioning was presented by Kuo et al. [23]. Here, in shape-adaptive fractal coding the initial image block is partitioned into rectangular or triangular subblocks. The authors took advantage of the fact that edge orientation is a byproduct of the isometry calculation done in FIC, hence, did not increase the computational cost of the algorithm. Although the authors proposed many partitioning schemes, a quadtree is most commonly used.

Many types of research used quadtree partitioning due to its convenient use [24–26]. Progressive FIC using Lagrange optimization along with quadtree partitioning were presented by Kopilovic et al. [24]. He et al. [25] aimed to reduce the convergence of the reconstruction image using a progressive decoding scheme combined with adaptive quadtree partitioning. The depth of the quadtree has an impact on the MSE calculation for given scaling and offset values. Lower quadtree depth affects more image regions and consequently the error calculations. In [26] a progressive quadtree structure is proposed to deal with the dependency of the depth of the quadtree structure in fractal image coding. Moreno and Otazu [27] presented a computationally efficient algorithm for fractal image encoding based on Hilbert scanning of an embedded quadtree structure that allowed to represent images as an embedded bitstream for the fractal function.

8.4.3 Nearest-neighbor search

Saupe [28], in 1995 proposed use of a multidimensional nearest-neighbor search in fractal image compression. The author suggested combining the classification scheme proposed in [9] and a fast searching method. Notation $dx(,)$ is used to represent Euclidean distance in Eq. (8.9):

$$e(R, D) = \langle D, \phi(D) \rangle^2 g(\Delta(R, D))$$
$$\text{where } g(\Delta) = \Delta^2 \left(1 - \frac{\Delta^2}{4}\right). \tag{8.9}$$

Here, operator $\langle,\rangle$ represents the inner product. Also, the least-square error is proportional to the Euclidean distance between the normalized projection of the range and the domain vector. Saupe's approach [28] was an extension of the kd-tree algorithm proposed by Arya et al. [29]. The authors proposed to find the approximate nearest neighbor rather than the nearest neighbor. This idea introduced a search parameter into the kd-tree structure. Tong [30] presented an adaptive method for [29] using error modeling to find the optimal search parameter. The method used the fractal coefficient to facilitate the faster search.

8.4.4 Clustering

Lepsøy and Øien [31] used a cluster centroid of the range block, opted-in cooperation with a block classification scheme. Wein and Blake [32] presented the methodology of clustering of the domain block using a kd-tree and a pairwise nearest-neighbor algorithm. The authors proposed creating a cluster of any two domain blocks that have $RMS(D1, D2) = \varepsilon$. The parameter ε is an acceptable small difference value. Eq. (8.10) is the rationale behind the performance of the clustering algorithm used to speed up the encoding process for any domain block D_1 and D_2 from a cluster:

$$\delta(R, D_2) = RMS(R, a_2 D_2 + b_2) \leq RMS(R, a_1 D_1 + b_1) \text{ then}$$
$$RMS(R, a_2 D_2 + b_2) = RMS(R, a_1 D_1 + b_1) + a_1 \varepsilon. \tag{8.10}$$

The algorithm presented in [33] applied Kohonen's self-organizing map (SOM) to create clusters of image blocks. Belloulata et al. [34] presented the idea of reducing computation complexity in hybrid wavelet-based FIC by applying a progressive, constructive clustering algorithm on subband domain blocks. Similarly, Chunmei Wang and Qiansheng Cheng [35] suggested usage of an attribute cluster network. A fuzzy classification is a commonly used method for clustering applications. Loe et al. [36] introduced a fuzzy classifier to accelerate encoding of FIC. The authors applied a shape-oriented similarity fuzzy set measure to define a group of three classes of the image block. A more flexible and accurate fuzzy-classifier-based FIC is presented in [37]. The algorithm in [37] used six feature parameters established on edge blocks of the image to represent the textured image. In the case of a mismatch, within-cluster occurs, the algorithm allowed extension of the search with a higher degree of membership function or to partition the block into four subblocks. Jaferza deh et al. [38] proposed to accelerate the encoding process by using fuzzy c means clustering for categorizing image blocks using a new metric based on a 1D-DCT.

8.4.5 Vector quantization

Sitaram et al. [39] presented an algorithm to design an efficient codebook for vector quantization (VQ) regarding complexity and size. VQ is suitable for lossy data-compression techniques like FIC, as the frequency of occurrence of data is inversely related to the rate of error. Many authors proposed hybrid schemes to improve the performance of FIC, but not much effort has been put into the VQ perspective of a FIC coder. In an early work, Jacquin [4] suggested using VQ for all blocks except sharp-edge blocks to code using fractals. Gharavi-Alkhansari and Haung [40] proposed VQ only as a particular case of block transform coding techniques. Further, a comparative study of performance of a fractal image coder and product code VQ is discussed by Lepsøy [41]. None of the above methods worked on the VQ of the encoder. Hamzaoui et al. [42] proposed a hybrid method that took advantage of the classification method [43] and the preestimated centroid of groups using the algorithm [44]. This hybrid approach initially classifies and designs a set of a fixed cluster centroid from training images. In the clustering process of the feature vector $\phi(I_{Di}(D_i))$ of each domain, the block is mapped to the nearest centroid of the cluster. Improvement in fractal image compression using VQ is introduced in the algorithm by forcing two conditions on the least-square error $e(R, D)$ discussed as shown in Eq. (8.11):

$$\text{if } \frac{1}{\sqrt{n}}e(R, I_R^{-1}(m_c)) \leq \delta \quad \text{OR}$$

$$e(R, I_R^{-1}(m_c)) \leq (1 + \varepsilon)e(R, I_R^{-1}(I_{Di}(Di))). \tag{8.11}$$

According to the proposed hybrid algorithm [33], if either of the conditions is valid for all domain to block Di in a group with centroid m_c, then the centroid is used for encoding of the range block using VQ. Both the conditions ensure that in the case of least-square or collage error approximation of the range block to the nearest centroid is acceptable then cluster centroid m_c is used as a VQ domain block. In the above Eq. (8.10), parameters δ and ε have scalar values used along with quadtree partitioning. Similar to [33], Davoine et al. [21] presented a novel method based on VQ and adaptive Delaunay triangulation. The authors extended the work of [31] and [10] that exercised VQ classification and archetype classification on square-shaped image blocks, whereas [21] proposed to use standard VQ based classification scheme to enhance the speed of matching by reducing the number of triangular image blocks. Hamzaoui et al. [45] used square isometries along with a classification scheme to increase the size of mean shape-gain vector quantization (MSGVQ). Further, the authors extended the use of MSGVQ to accelerate the searching of the domain block and decoding speed of a conventional fractal coder using a VQ codebook [46]. Also, the authors suggested using fixed vector design based on a clustering algorithm to improve the rate-distortion performance as compared to other hybrid FIC methods [47,48]. A study by Yang and Lin [49], applied self-transformation by IFS. The authors proposed a mutable VQ embedding method established on the locally adaptive compression algorithm.

8.4.6 Discrete cosine transforms

Barthel and Voy'e [50] initially proposed speedup FIC in the frequency domain in 1993. The authors proposed a modification in the luminance transform so that it works in the frequency domain. This proposition allowed the majorities of domain blocks to be mapped using a lower-order DCT transform [51–53]. This alternative method allows every fractal transform to be carried out in the DCT domain. Range and domain blocks of size 8×8 were transformed into the DCT domain using Eq. (8.12):

$$F(u, v) = \frac{1}{16}c(u)c(v)\sum_{x=0}^{7}\sum_{y=0}^{7} f(x, y)$$
$$\times \cos\frac{(2x + 1)u\pi}{16}\cos\frac{(2y + 1)v\pi}{16} \qquad (8.12)$$
$$\text{where } c(u), c(v) = \left\{ \begin{array}{ll} \frac{1}{\sqrt{2}}u, & v = 0, \\ 1 & \text{otherwise.} \end{array} \right\}$$

For better results, the transformed range block is classified into two classes based on their complexity. Eqs. (8.13a) and (8.13b) are used to estimate the error image and quantize it with the help of the quantization table:

$$E(u, v) = F_R(u, v) - \tau o\phi(F_D(u, v)), \qquad (8.13a)$$
$$E_q(u, v) = E(u, v)/Q(u, v), \qquad (8.13b)$$

where in the above equations functions $E(u, v)$ and $E_q(u, v)$ are the error and quantized image, respectively. Parameters τ, ϕ are the compound transformation and counter activity factor, respectively. The quantization table $Q(u, v)$ mentioned in the equation is designed according to human perception properties of

the image. Cutis et al. [54] proposed to use a pruned DCT to accelerate fractal image coding calculations. The pruned DCT coefficients are high-frequency
components, hence setting them to zero is difficult to recognize due to the HVS.
Further, Duh et al. [12] presented the idea to reduce search time. The classification scheme limits the search space since range blocks are searched corresponding to the domain block in the same class only. Similar to this scheme, a unified
feature and DCT-coder-based FIC was proposed by Zhou [55]. A unified feature
reduces the search space by eliminating the inappropriate image subblock from
the pool before mapping the range and the domain block. Also, as compared to
the coding process in JPEG, FIC has an asymmetric coding process.

8.4.7 Variance-based FIC

Saupe and Jacob [56] presented the idea of using a block variance-based condition for partitioning nodes of a quadtree structure in FIC. A quadtree approach
suffers a large number of vain attempts to map the domain block to the range
block. The variance σ^2 of the block I of size $n \times n$ is calculated using Eq. (8.14),
where X is a pixel value at location i:

$$\sigma_I^2 = 1/n^2 \left(\sum_{i=1}^{n \times n} X_I^2 \right) - \left(1/n^2 \sum_{i=1}^{n \times n} X_I \right)^2. \qquad (8.14)$$

A variance-based test condition to split the quadtree structure removed all un-
successful searches and consequently speeded up the encoding process. Lee and
Lee [57] used the local variance of the domain block to speed the encoding
process by up to tenfold for an image, depending upon the complexity. The
following condition limits the search by excluding all the inadmissible domain
blocks from the pool.

$$\left| \sigma_{d_m}^2 - \sigma_{R_n}^2 \right| \leq T_{th}. \qquad (8.15)$$

Eq. (8.15) states that if the difference in the local variance σ^2 of the mth domain block and the nth range block is more than the threshold T_{th}, blocks do
not match, but vice versa is not correct. Similar to the idea of using ordered
domain blocks according to variance in [57], Lee [58] analyzed the partial domain block search exploiting the relationship between square variance distance
(SVD) and small Euclidean distance (SED). The author claimed that the probability of finding SED is higher if the search starts with a domain block having
minimum SVD to the next-neighboring domain block arranged in descending
order according to the squared variance. Ponomarenko et al. [59] presented a
histogram-based analysis of local variance of the range and domain block to
accelerate the search process. Wu et al. [60] implemented variance- and mean-
based classes of range and domain block along with the simplification of eight
transformations to overcome the long encoding time. Further, a partial distance

search is used to reduce the computational cost of calculating the root mean square error between the blocks. He et al. [61] performed an empirical study on the choice of the control parameter threshold for a shade range block, codebook, and size of the search window. The algorithm proposed in [61] reduced the encoding time as compared to [57] due to the prior exclusion of the range block based on a threshold for the shade of the block. Zhou, Zhang, and Zhang [62] further modified [57] using a neural-network-based FIC [63]. The approach increases the performance of the algorithm regarding reconstruction quality and encoding time. A neural-network-based FIC has the advantage of parallel working and better matching between subblocks that helped to address issues of the enormous encoding of FIC. Shen et al. [64] presented the idea of reducing the search space based on localized searching of the domain block in neighboring boxes concerning the range block. Also, the authors used local self-similarity and its relationship with an affine transformation. Han [65] proposed the usage of variance to divide the range and domain blocks into six classes and further applied a genetic search scheme to speed up the FIC.

8.4.8 Spatial correlation

Vector-quantization-based encoding [66] removed interblock correlation to reduce the computational complexity of the process. Wang and Hsieh [67] proposed a faster FIC inspired by range block correlation and a searched ordering technique to encode position and isometry parameters. Further, to accelerate encoding, the authors used the search scheme presented in [57]. Truong et al. [68] created a limited search space on the rationale that correlation between neighboring domains as well as range blocks in an image is present. A search space for a range block r_i is created using vertical, horizontal, and diagonal neighboring domain blocks in relative directions. Every domain block d was expanded into four direct neighbors in the corresponding direction as $\{d_0^H, d_1^H, d_2^H, d_3^H\}$. Wu et al. [69] proposed two-stage spatial correlation based on the genetic algorithm in which initially the local optima is explored to find the appropriate map in a limited search space. If the first stage fails, the second stage of exploiting sufficient similarity in the whole image is proposed. Xing-yuan et al. [70] presented a similar feature of spatial correlation to accelerate fractal image coding based on PIFS. A speeded-up encoding of FIC using a correlation information feature is introduced in [71] that depicts the structural characteristics of a block. Wang et al. [72] reduced the encoding time using a spatial texture feature based on cooccurrence matrices to represent an image. Also, the authors applied an intelligent classification algorithm as a search strategy to enhance the performance of FIC as compared to the baseline algorithm. A novel scheme to improve matching probability based on the classification of blocks using an absolute Pearson's correlation coefficient was proposed by Wang and Zheng [73]. The proposed method increases encoding speed by sorting blocks within each class formed using Pearson's correlation coefficient. Many other highly cited region functionality based FIC worked to improved performance of encoding algorithms [74–76].

8.4.9 Iteration-free convergence

Another scheme [77] showed improvement in conventional FIC by using a noniterative paradigm. The algorithm designed a better domain pool with less redundancy than the same domain block utilized at the encoder and decoder ends. The encoding process proposed in [77] is shown below. Wang [78] used an iteration-free fractal codec to identify tempering in the image and proposed an algorithm to restore the image.

Belloulata [79] introduced directional subband coding based on noniterative FIC and claimed to reduce the computational time. Kamal et al. [80] attempted to increase the coding speed using a genetic algorithm. GA, being a mathematically motivated search method, allowed better mapping between a domain and range blocks. Further, Kamal et al. [80] proposed an iteration-free FIC using VQ, GA, and simulated annealing. The algorithm showed improvement in image quality and coding speed.

8.4.10 Local search or no-search method

The initial studies of Jacquin [4] and Fisher et al. [2] are the most practical FIC schemes utilizing exhaustive search and block-based transformation. Contrary to a profound search, Monro and Dudbridge [81] showed the use of a first-order polynomial Bath Fractal Transform (BFT) to encode the range block without searching. BFT has various implementation options for the degree of searching and order of approximation. Commonly employed BFT are zero-order, bilinear, and biquadratic mapping of grayscales [82]. Further, Monro and Woolley [83] examined use of a zero searching variant of BFT to accelerate encoding of FIC. In the BFT a domain portion is taken from iterated function system. An attractor in BFT defined as a nonoverlapping tiling of the image with contraction mapping w_k chosen. A fractal function was defined on a unique attractor for each tile k to an approximate grayscale image. The authors proposed a recursive set of mapping v_k that is a contractive of image fragment $f(x, y)$. Eq. (8.16) shows a grayscale mapping of an image using v_k as a polynomial function.

$$\begin{aligned} v_k(x, y, f_{fractal}) &= a + b_x x + b_y y + c_x x^2 \\ &+ c_y y^2 + d_x x^3 + d_y y^3 + ef(x, y). \end{aligned} \tag{8.16}$$

In FIC based on BFT, evaluates lower-order movements of image blocks and solution of the linear equation shown. In the idea presented by Monro and Woolley [83], the examined two-dimensional fractal transformed by keeping all coefficients equal to zero, except a and e. Also, the work investigated the first order and second order of BFT. Dudbridge [84] claimed linear cost both for encoding and decoding processes using a no search scheme. The author suggested using one common domain block for the union of four range blocks. In 2004, Furao and Hasegawa [85] proposed a faster alternative algorithm for FIC without searching the domain pool. The authors presented a redesigned compression algorithm to deal with every time-consuming part of the fractal image encoding process. Computational complexity due to finding an appropriate scaling coefficient and illumination offset and domain block to map to range block is handled using the redesigned error in Eq. (8.17):

$$
\begin{aligned}
e(R, \tilde{D}) &= \min_{D \in pool} e(R, D) \\
&= \min_{D \in pool} \min_{D \in i} \left\| a_i(D - \bar{d}I) - (R - \bar{r}I) \right\|^2.
\end{aligned}
\tag{8.17}
$$

Here, only scaling parameter a_i and the mean of range block $\bar{r}$ are considered as IFS coefficients. In conventional FIC there is a quadruple loop, Furao, and Hasegawa [85] suggested only searching for $a_i = 1, 2, \ldots, m$. A further similarity index is used to reduce the number of candidate domain blocks for mapping to the range block. Also, the similarity index utilized to fix the domain block location concerning the range block. The domain block at the location $(row - D/2, col - D/2)$ is fixed concerning the range block located at (row, col) used to avoid searching of the pool. Ongwattanakul et al. [86] claimed to improve the fidelity criterion to use the fixed location of the domain block to map to the appropriate range block. To relax the computation complexity of FIC Wu et al. [87] proposed a searchless encoding method based on a quadtree structure. Further, in [88] the authors proposed a parallel hardware architecture to accelerate the encoding process. Wang and Wang [89] presented a modified gray-level transform to deal with the low-fidelity outcome of the Furao and Hasegawa [85] approach. Wang and Wang [89] used an adaptive plane, so that the number of transform coefficients present is used to encode an image block. In [89] Wang and Wang modified the error function proposed by Tong and Pi [14] using adaptive planes P_1 and P_2 constructed from the domain and range blocks, respectively. Here, in Eq. (8.18) a_i is a scaling parameter that is varied from $1, 2, \ldots, m$:

$$
e(R, D)^2 = \min_i \left\| a_i(D - P_1) - (R - P_2) \right\|^2.
\tag{8.18}
$$

Wang et al. [90] extended their work to overcome the drawback of the modified gray-level transform by using fitting planes to improve the transform. Jiang and Jiang [91] showed the use of wavelet decomposition to improve the encoding process of no-search FIC. In an extension of Furao's method [85] Salarian and Hassanpour [92] proposed a no-search FIC in the DCT domain. The algorithm proposed accelerates the encoding process as compared to FIC in the spatial and wavelet domains. de Lima, Schwartz, and Pedrini [93] showed an improvement in bit rate by utilizing a searchless FIC for 3D video encoding.

8.4.11 Parallel algorithm

Ancarani et al. [94] provided a parallel ASIC architecture that is an add-on board for a PC platform. Acken et al. [95] proposed an efficient FIC based on a parallel approach that performed a full search and used a quadtree partitioning scheme in comparison to the previous algorithm. Hufnagl and Uhl [96] presented the idea of using massively parallel SIMD arrays for FIC. The authors compared different algorithms to find an appropriate algorithm for the architecture proposed. In [97] an implementation and comparison of the performance of different strategies of serialization were carried out. The authors focused on a low-budget workstation cluster for FIC implementation. Gomes et al. [98] proposed using a multicore processor due to its high processing power to encode the image using FIC in parallel. Similarly, in [99], a discrete wavelet transform is used alongside the parallel implementation of FIC. Jackson et al. [88] exercised a searchless approach to encode the image using FIC and claimed to achieve a higher compression ratio. Vidya et al. [100] provided an architecture for FIC and utilized control parallelism by implementing ASIC. A full search with fixed block size partitioning was employed in [101]. Panigrahy et al. [101] claimed a reduction in computation time by eliminating multiplication and division required to compute parameters for the domain block. Saad and Abdullah [102] implemented a faster FIC in a low-cost Altera Cyclone II FPGA. To speed up the encoding process of FIC the authors of [102] used a parallel-pipeline implementation.

8.4.12 Wavelet fractal hybrid encoder

Pentland and Horowitz [103] proposed to link wavelets with the fractal coder. The approach discussed in [103] lightly allied to Jacquin's method and was used within a subband fixed vector quantizer. Wavelet transforms can partition the signal into a different resolution that gives a high compression ratio. A coder proposed in [104] utilized blocks as the vector codebook from low-frequency image subbands to quantize into higher subbands. Similarly, many researchers also stated the links between wavelets and fractals along with extension and generalization of the image-compression algorithm [105,106]. A combination of FIC with a wavelet transform was proposed to take advantage of the wavelet coefficient's ability to classify the domain block efficiently to reduce encoding time [106]. Davis [47] proposed a wavelet-based framework to address issues of convergence of the fractal scheme and finding the contraction mapping to attractor for a given image. Biorthogonal and symmetric basis function used in place of Eqs. (8.19a) and (8.19b) to achieve generalization of Haar:

$$\left\langle \tilde{\phi}_\Gamma^{N-R}, F \right\rangle \approx \frac{g_\Gamma}{2^{D-R}} \left\langle \tilde{\phi}_{\Pi(\Gamma)}^{N-D}, F \right\rangle + h_\Gamma = \frac{g_\Gamma}{2^{D-R}} \left\langle W \tilde{\phi}_\Gamma^{N-R}, F \right\rangle + h_\Gamma, \quad (8.19a)$$

$$\left\langle \tilde{\psi}_\gamma^j, F \right\rangle \approx \frac{g_\Gamma}{2^{D-R}} \left\langle \tilde{\psi}_{\gamma'}^{j-(D-R)}, F \right\rangle = \frac{g_\Gamma}{2^{D-R}} \left\langle W \tilde{\psi}_\gamma^j, F \right\rangle. \quad (8.19b)$$

Andreopoulos et al. [107] claimed better compression performance using a hybrid scheme. The proposed method encoded a wavelet-transformed image using fractal coding and a lossy-prediction method. All high-frequency coefficients obtained from a wavelet transformation of the image coded using conventional FIC, whereas the low-frequency component was subjected to a binary tree predictive coder. In [108], an encoding scheme different from that of Li and Kuo [47,104] was presented whereby the only partial image is coded using fractal prediction. A fractal contraction map is used to obtain the interscale wavelet coefficient, and a bit plane wavelet encoder is applied to the prediction residual. Wu et al. [109] extended the previous hybrid schemes and proposed a searchless and noniterative decoding algorithm. Wu and associates exploited the correlation between parent and child coefficients at a finer scale with the same orientation at the same spatial location. Similarity characteristics were used by many researchers [110–112] in wavelet-fractal coding integration. Iano et al. [113] proposed an algorithm that used a modified set partitioning in hierarchical tree coding and a classification method proposed by Fisher [9] and improved the performance of the encoder. Song et al. [114] proposed an adaptive wavelet-fractal encoder using a four-fork tree structure to exploit the energy stored in the low-frequency sublevel band and imaged features that previous algorithms [47,104] failed to do. Moreno and Otazu [27] presented a computationally efficient algorithm for fractal image encoding based on Hilbert scanning of an embedded quadtree structure in the wavelet domain. Zang and Wang [115] offered a faster fractal image encoder using a wavelet transform with a diamond search. A novel idea of a multidescription fractal-wavelet coder was proposed by Yang [116] that aimed to exploit the fractal's ability to represent various resolution-scale similarity present in wavelet transform coefficient. Wu [117] used a genetic algorithm along with a wavelet transform in FIC to reduce the number of MSE computations.

8.4.13 Genetic algorithm

The intrinsic parallelism of the genetic algorithm and characteristics like adaptability, self-organizing, and learning makes it suitable for applications such as FIC. A genetic algorithm is a widely used stochastic optimization tool due to its various characteristics. In [36] Lee, Gu, and Phua used GA to optimize a fuzzy-classifier-based fractal image encoder. The authors distinguished different classes by a membership grade function defined for image blocks. The algorithm used an optimized set of parameters of the membership function to achieve better classifications. The authors of [118,119] took advantage of a GA search and proposed chromosomes to contain PIFS of all the blocks. In the proposed algorithm, the fitness of a chromosome is measured as the distance between the attractor and the original image. Wu et al. [69] proposed spatial correlation and a GA-based two-stage FIC algorithm. The proposed method maintained the quality of the image with speeded up encoding as compared to [118]. Zheng et al.

[120] presented the extensible search space and modified GA coefficient, and a manifold mutation operator to improve the performance of the algorithm. In [121] the authors proposed a faster fractal encoding method using a genetic schema algorithm. The operator in the evolutionary process performed on blocks used the adaptive operator by the schema theory. Tseng et al. [122] presented the idea of the accelerated encoding process using particle-swarm optimization (PSO) based on visual information of the edge property. The authors suggested using a directional map rather than a full search to find a candidate of higher similarity in the swarm. Xing-yuan et al. [70] proposed a two-stage hybrid genetic scheme to improve the performance of FIC. A simulated annealing genetic algorithm (SAGA) is adopted to find global optima, particularly in the case of failure of local optima. The authors also compared the results of SAGA with the approach presented in [119]. The authors of the paper compared the results with [68] and [119]. Wu and Lin introduced a genetic algorithm based on FIC using a hybrid-select mechanism. The authors claimed 130 times faster encoding speed as compared to conventional GA-based FIC. [123] used PSO to improve the encoding speed, while maintaining the quality of the image. Wu [117] embedded DWT in GA for faster encoding speed while keeping the quality of the reconstructed image. In [124], the authors aimed to find a high-speed and better-quality decoder using VQ, GA, and simulated annealing.

8.5 Summary and analysis of review

In this section, a quick review of the surveyed literature is presented in a tabular fashion. Table 8.1 illustrates the study of various methods to accelerate the encoding of FIC. Fundamental papers on approaches have been presented along with some citations in the table. Papers, in summary, have been selected by the year of publication. Table 8.2 shows the comparative analysis of different approaches by various parameters.

8.6 Conclusions

The chapter presented various vital issues affecting the performance of the fractal encoding algorithm. The theoretical background of FIC was reported in the initial section. We then presented a discussion on a theory of IFS, partitioned iterative function systems, and the contractive theorem. The core contribution of this chapter is it presents the various techniques applied to overcome the enormous time-consumption problem of FIC. In the chapter, illustrative techniques from the literature were selected for analysis and study. The working of each approach was discussed along with its advantages and disadvantages. Factors affecting the performance measure of every approach was covered in the chapter. Tables 8.1 and 8.2 cover various important approaches and their comparative analysis with each other by critical factors. The frequency domains have been reported and used in combination of FIC to improve the performance

TABLE 8.1 Summarized literature of approaches used to improve encoding performance of fractal image compression.

Citation	Approach	Authors, Year of publication	Metric	Fidelity criteria	Codebook	Decoding process	Partitioning scheme	Key concerns
358	Block Classification	E.W. Jacobs, Y. Fisher, R.D. Boss, 1992	RMS	PSNR BPP	3 bits to save orientation, 12 bits for range block identity and location of domain block. Fixed number of bits used for scaling and offset. Number of classes	Any initial image is subjected to all the transforms iteratively until the image converges	Quadtree	A large number of parameters influence compression ratio, speed, and accuracy of the algorithm
58	DCT	K.U. Barthel and T. Voyé, 1993	RMSE	PSNR BPP	Search class, geometric index, and isometry of range block, codebook index of luminance transformation, the partition of range blocks of higher hierarchy	Value of scaling chosen such that lower error at the decoder is obtained and less iteration	Transform coding, Quadree	The high number of transform parameters restrict coding efficiency
21	Nearest Neighbor	D. Saupe, 1995	MSE	PSNR, BPP, drop speed	24 classes and other transform parameters	Any initial image is subjected to all the transforms iteratively until the image converges	Quadtree	Slight degradation of image quality and compression ratio at a gain of speed
39	Partitioning of range and domain blocks	E. Reusens, 1994	MSE	PSNR, rate (bit/pixel)	Type of partition decided the length of the code	Any initial image is subjected to all the transforms iteratively until the image converges	Polygon	
68	Partitioning of range and domain blocks	D. Saupe, M. Ruhl, 1996	MSE	PSNR, CR, Time (CPUs), optimization level	Range position, partition type, and partition location, scaling and offset, one bit to specify whether the range is potential or not, domain location using domain index, rotation information (2 bits per transform)	Two optimizations applied to the decoding process. Named as pixel referencing and low-dimensional fixed-point approximation. The approximation method reduces decoding time significantly. Domain subsampling is done rather than averaging	HV partition	Decoding is performed by ordering blocks according to size

continued on next page

TABLE 8.1 (continued)

Citation	Approach	Authors, Year of publication	Metric	Fidelity criteria	Codebook	Decoding process	Partitioning scheme	Key concerns
156	Vector Quantization	F. Davoine, M. Aritonini, Jean-Marc Chassery, and M. Barlaud, 1996	RMS	PSNR, CR	log[M] bits required to code the address of the triangle. N Bits used for the number of the triangle	Contractive mapping W applied iteratively to any arbitrary image until converges. In every iteration, W is applied to each triangle	Delaunay triangulation	Some triangles created during partitioning. Overhead of bits to code triangle and its location
89	Wavelet	Y. Iano et al., 2006	MSE	Compression ratio, PSNR, BPP, time	Scaling and offset 5 and 7 bits, respectively,	* Decoding end divides the bitstream into approximation subband and detailed subband. * Approximation subband and detailed subband subjected PIFS and modified SPIHT, respectively.	Modified set partitioning	Value of PSNR becomes constant with an increasing number of quantization bits. It further reduces the bit budget that affects the performance of the coder
28	Spatial Correlation	Chou-Chen Wang, 2001	MSE	Time, BPP, PSNR, Navg	Two bits to indicate search order, one bit to differentiate between parameters and search order. Uniform quantizer for all the IFS parameter	* At the decoder, raster scan order is applied to reconstruct the image	Quadtree	Quantization of parameter affects the quality of the decoded image
114	Spatial correlation	T.K. Truong, C.M Kung, J. Hjeng, M. I.hsieh, 2004	MSE	PSNR, no. of blocks hit, time, BPP, speed up	Codebook contained the parameters of all the eligible domain blocks associated with range block	Any initial image is subjected to all the transforms iteratively until the image converges	Quadtree	The significant difference between local and global minima
66	Clustering	C. J. Wein and Ian F. Blake, 1996	MSE	PSNR, BPP, time		Any initial image is subjected to all the transforms iteratively until convergence	k-dimensional tree structure used to partition domain blocks	The significant difference between local and global minima

continued on next page

TABLE 8.1 (continued)

Citation	Approach	Authors, Year of publication	Metric	Fidelity criteria	Codebook	Decoding process	Partitioning scheme	Key concerns
21	Noniterative, Block Averaging	Hsuan T., Chang, 2000	MSE	Bitrate, PSNR	One bit to identify parent and child block. Two extra bits to distinguish splitting of the block	The decoder used information in the header to reconstruct the image. The image is reconstructed either using mean information or contractive affine transformation	Quadtree	Range blocks can be decoded in parallel. Useful for hardwire applications
88	Variance	C. He, S.X. Yang and X. Huang, 2004	Minimum (RMSE)	PSNR, BPP	Collection of the resized domain block	The codebook is restricted searched as compared to block classification where each class is used to find the best match	Quadtree	The algorithm does not guarantee best mapping but gives good computational complexity
99	Searchless	S. Furao, O. Hasegawa, 2004	MSE	PSNR, BPP, Time	Domain block is not kept in the codebook. The average intensity is saved	Any initial image is subjected to all the transforms iteratively until the image converges	Quadtree	The high value of threshold results in the poor quality of image, while a low threshold leads to a long encoding time
24	Genetic Algorithm	Ming-Sheng Wu, Yih-Lon Lin, 2010	MSE	Time, number of MSE computations, PSNR	Search space is created using three classes depending on their DCT coefficient value.	Any initial image is subjected to all the transforms iteratively until the image converges		The significant difference between local and global minima
3	Parallel Approach	M. Panigrahy, I. Chakrabarti, and A. S. Dhar, 2015	SAD (sum of absolute difference)	PSNR, clock frequency, CR, the threshold		Any initial image is subjected to all the transforms iteratively until the image converges	Fixed partitioning	The method used full search and fixed partitioning that amounts to the encoding time

TABLE 8.2 Analysis of different approaches to various parameters.

Method	Base Factors	Criteria	Mapping Criteria	Searching Technique
Block Classification	Quadrant variance, quadrant brightness	PSNR, BPP	RMS	Exhaustive Search
Nearest Neighbor	Size of the neighborhood	PSNR, BPP	MSE	Nearest-Neighbor Search
Partitioning scheme	Shape of image block and quantity of blocks, overlapping present	PSNR, BPP	RMS	Exhaustive Search, no search
DCT	Size of image block	PSNR, BPP, ET	MSE	Exhaustive Search, no search
DWT	Decomposition level in wavelet, subband level	PSNR, BPP, ET	MSE	Exhaustive Search
Vector quantization	Codebook design, vector quantization, and its complexity	Number of clusters, PSNR, BPP, ET	RMSE	Codebook Search
Variance	Size of window, the no. of hit blocks	PSNR, BPP, ET	RMSE, SSIM	Exhaustive search, no search, restricted search
No Search	Image type, range block size, used threshold	PSNR, BPP, ET	RMSE	No Search
Spatial correlation	Range block size, size of the search window size and used the threshold	PSNR, BPP, ET	MSE	Fast Search [57]
Parallel	Range block size, used threshold, partitioning applied	PSNR, clock frequency, CR, the threshold	SAD	Exhaustive Search, Searchless
Clustering	Number of clusters and quantity of blocks present in the cluster	PSNR, encoding time	MSE	Exhaustive, search, no search
Noniterative convergence	Size of domain pool	Bitrate PSNR	MSE	Exhaustive Search
Genetic algorithm-based FIC	Crossover, the initial population chosen, mutation function, fitness function	PSNR, BPP	Acceptance criteria	Exhaustive Search, no search

of the algorithm. It is observed that a high value of the threshold results in the poor quality of image, while a low threshold leads to long encoding times in searchless FIC. The literature is finally summarized for quick study and analysis. Much work [125–127] has been carried out by researchers for improvement of FIC in various fields of application However, the future still holds many advanced approaches that can accelerate the encoding process with an excellent quality reconstructed image.

References

[1] M.F. Barnsley, A.D. Sloan, A better way to compress images, Byte (1988) 215–223.

[2] Y. Fisher, E.W. Jacobs, R.D. Boss, Fractal Image Compression Using Iterated Transforms, 1992, pp. 35–61.

[3] M.F. Barnsley, L.P. Hurd, Fractal Image Compression, vol. 5, 1993.

[4] A.E. Jacquin, Image coding based on a fractal theory of iterated contractive image transformations, IEEE Trans. Image Process. 1 (1) (1992) 18–30, https://doi.org/10.1109/83.128028.

[5] E.R. Vrscay, D. Otero, D. La Torre, A simple class of fractal transforms for hyperspectral images, Appl. Math. Comput. 231 (2014) 435–444, https://doi.org/10.1016/j.amc.2014.01.007.

[6] S. Zhu, D. Zhao, F. Wang, Hybrid prediction and fractal hyperspectral image compression, Math. Probl. Eng. 2015 (2015), https://doi.org/10.1155/2015/950357.

[7] D. Zhao, S. Zhu, F. Wang, Lossy hyperspectral image compression based on intra-band prediction and inter-band fractal encoding, Comput. Electr. Eng. 54 (2016) 494–505, https://doi.org/10.1016/j.compeleceng.2016.03.012.

[8] B. Ramamurthi, A. Gersho, Classified vector quantization of images, IEEE Trans. Commun. 34 (11) (1986) 1105–1115, https://doi.org/10.1109/TCOM.1986.1096468.

[9] Y. Fisher, Fractal Encoding - Theory and Applications to Digital Images, Springer-Verlag, New York, 1994.

[10] R.D. Boss, E.W. Jacobs, Archetype classification in an iterated transformation image compression algorithm, in: Fractal Image Compression Theory Appl. to Digit. Images, 1995, pp. 79–90.

[11] Z. Wang, D. Zhang, Y. Yu, Hybrid image coding based on partial fractal mapping, Signal Process. Image Commun. 15 (9) (2000) 767–779, https://doi.org/10.1016/S0923-5965(99)00018-1.

[12] D.J. Duh, J.H. Jeng, S.Y. Chen, DCT based simple classification scheme for fractal image compression, Image Vis. Comput. 23 (13) (2005) 1115–1121, https://doi.org/10.1016/j.imavis.2005.05.013.

[13] T. Kovács, A fast classification based method for fractal image encoding, Image Vis. Comput. 26 (8) (2008) 1129–1136, https://doi.org/10.1016/j.imavis.2007.12.008.

[14] C.S. Tong, M. Pi, Fast fractal image encoding based on adaptive search, IEEE Trans. Image Process. 10 (9) (2001) 1269–1277, https://doi.org/10.1109/83.941851.

[15] X. Wu, D.J. Jackson, H.C. Chen, A fast fractal image encoding method based on intelligent search of standard deviation, Comput. Electr. Eng. 31 (2005) 402–421.

[16] Y. Fisher, A discussion of fractal image compression, in: D. Saupe, P. H.O., J. H. (Eds.), Chaos and Fractals, Springer-Verlag, New York, 1992, pp. 903–919.

[17] Y. Fisher, S. Menlove, Fractal encoding with HV partitions, in: Y. Fisher (Ed.), Fractal Image Compression: Theory and Application to Digital Images, Springer New York, New York, NY, 1995, pp. 119–136.

[18] A. Boukhelif, Accelerating fractal image compression by domain pool reduction adaptive partitioning and structural block classification, Adv. Model. Anal. B 48 (3–4) (2005) 27–42, https://doi.org/10.1109/ISCC.2004.1358601.

[19] E. Reusens, Partitioning complexity issue for iterated functions systems based image coding, in: VII European Signal Processing Conference, vol. 1, 1994, pp. 171–174, [Online]. Available: https://pdfs.semanticscholar.org/40b6/16d8b96939d77d44d83a1567e97a2bcd6829. pdf. (Accessed 9 July 2017).

[20] X. Wu, C. Yao, Image coding by adaptive tree-structured segmentation, in: Proceedings. Data Compression Conference, 1991, pp. 73–82.

[21] F. Davoine, M. Antonini, J.M. Chassery, M. Barlaud, Fractal image compression based on Delaunay triangulation and vector quantization, IEEE Trans. Image Process. 5 (2) (1996) 338–346, https://doi.org/10.1109/83.480769.

[22] M. Liaobtc, M. Chernc, C. Tsaod, Fractal image coding system based on an adaptive side-coupling quadtree structure, vol. 14, 1996, pp. 401–415.

[23] C.J. Kuo, W.J. Huang, T.G. Lin, Isometry-based shape-adaptive fractal coding for images, J. Vis. Commun. Image Represent. 10 (4) (Dec. 1999) 307–319, https://doi.org/10.1006/jvci. 1999.0422.

[24] I. Kopilovic, D. Saupe, R. Hamzaoui, Progressive fractal coding, in: Proceedings 2001 International Conference on Image Processing (Cat. No. 01CH37205), vol. 1, 2001, pp. 86–89.

[25] C. He, S.X. Yang, X. Huang, Progressive decoding method for fractal image compression, IEE Proc., Vis. Image Signal Process. 151 (3) (2004) 207, https://doi.org/10.1049/ip-vis: 20040316.

[26] C.H. Yuen, O.Y. Lui, K.W. Wong, Hybrid fractal image coding with quadtree-based progressive structure, J. Vis. Commun. Image Represent. 24 (8) (Nov. 2013) 1328–1341, https:// doi.org/10.1016/j.jvcir.2013.09.002.

[27] J. Moreno, X. Otazu, Image compression algorithm based on Hilbert scanning of embedded quadTrees: an introduction of the Hi-SET coder, in: 2011 IEEE International Conference on Multimedia and Expo, Jul. 2011, pp. 1–6.

[28] D. Saupe, Accelerating fractal image compression by multi-dimensional nearest neighbor search, in: Proceedings DCC '95 Data Compression Conference, 1995, pp. 222–231.

[29] S. Arya, D.M. Mount, N.S. Netanyahu, R. Silverman, A.Y. Wu, An optimal algorithm for approximate nearest neighbor searching fixed dimensions, J. ACM 45 (6) (1998) 891–923, https://doi.org/10.1145/293347.293348.

[30] C.S. Tong, M. Wong, Adaptive approximate nearest neighbor search for fractal image compression, IEEE Trans. Image Process. 11 (6) (2002) 605–615, https://doi.org/10.1109/TIP. 2002.1014992.

[31] S. Lepsøy, G.N. Øien, Fast attractor image encoding by adaptive codebook clustering, in: Y. Fisher (Ed.), Fractal Image Compression-Theory and Application, Springer-Verlag, New York, 1995.

[32] C.J. Wein, I.F. Blake, On the performance of fractal compression with clustering, IEEE Trans. Image Process. 5 (3) (1996) 522–526, https://doi.org/10.1109/83.491325.

[33] R. Hamzaoui, M. Muller, D. Saupe, VQ-enhanced fractal image compression, in: Proceedings of 3rd IEEE International Conference on Image Processing, vol. 1, 1996, pp. 153–156.

[34] K. Belloulata, A. Baskurt, R. Prost, Fast directional fractal coding of subbands using decision-directed clustering for block classification, in: 1997 IEEE International Conference on Acoustics, Speech, and Signal Processing, vol. 4, 1997, pp. 3121–3124.

[35] Chunmei Wang, Qiansheng Cheng, Attribute cluster network and fractal image compression, in: WCC 2000 - ICSP 2000. 2000 5th International Conference on Signal Processing Proceedings. 16th World Computer Congress 2000, vol. 3, 2000, pp. 1613–1616.

[36] K.F. Lee, W.G. Gu, K.H. Phua, Speed-up fractal image compression with a fuzzy classifier 5965 (96) (1997).

[37] J. Han, Fast fractal image compression using fuzzy classification, in: 2008 Fifth International Conference on Fuzzy Systems and Knowledge Discovery, Oct. 2008, pp. 272–276.

[38] K. Jaferzadeh, K. Kiani, S. Mozaffari, Acceleration of fractal image compression using fuzzy clustering and discrete-cosine-transform-based metric, IET Image Process. 6 (7) (2012) 1024, https://doi.org/10.1049/iet-ipr.2011.0181.

[39] V.S. Sitaram, Chien-Min Huang, P.D. Israelsen, Efficient codebooks for vector quantization image compression with an adaptive tree search algorithm, IEEE Trans. Commun. 42 (11) (1994) 3027–3033, https://doi.org/10.1109/26.328984.

[40] M. Gharavi-Alkhansari, T.S. Huang, Fractal-based techniques for a generalized image coding method, [Online]. Available: https://pdfs.semanticscholar.org/2094/503e2de4dcfed86ad43f22fcb637d3535bf4.pdf. (Accessed 26 June 2017).

[41] S. Lepsøy, Attractor Image Compression—Fast Algorithms and Comparisons to Related Techniques, Norwegian Institute of Technology, Norway, 1993.

[42] R. Hamzaoui, M. Muller, D. Saupe, Enhancing fractal image compression with vector quantization, in: Digit. Signal Process. Work. Proceedings, 1996, IEEE, 1996, pp. 231–234, [Online]. Available: papers//b6c7d293-c492-48a4-91d5-8fae456be1fa/Paper/p3524%5Cnfile:///C:/Users/Serguei/OneDrive/Documents/Papers/Enhancing fractal image compression with-1996-02-26.pdf.

[43] E.W. Jacobs, Y. Fisher, R.D. Boss, Image compression: a study of the iterated transform method, Signal Process. 29 (3) (1992) 251–263, https://doi.org/10.1016/0165-1684(92)90085-B.

[44] R. Hamzaoui, Codebook clustering by self-organizing maps for fractal image compression, Fractals 5 (Supplementary Issue) (1995) 27–38, [Online]. Available: https://pdfs.semanticscholar.org/5ff2/61f2f0d86efeb1a9017a968d60200d78a749.pdf. (Accessed 26 June 2017).

[45] R. Hamzaoui, B. Ganz, D. Saupe, Quadtree based variable rate oriented mean shape-gain vector quantization, in: Proceedings DCC '97. Data Compression Conference, 1997, pp. 327–336.

[46] R. Hamzaoui, D. Saupe, Combining fractal image compression and vector quantization, IEEE Trans. Image Process. 9 (2) (2000) 197–208, https://doi.org/10.1109/83.821730.

[47] G.M.G.M. Davis, A wavelet-based analysis of fractal image compression, IEEE Trans. Image Process. 7 (2) (1998) 141–154, https://doi.org/10.1109/83.660992.

[48] J. Li, J. Kuo, Hybrid wavelet-fractal image compression based on a rate-distortion criterion, [Online]. Available: https://www.microsoft.com/en-us/research/wp-content/uploads/2016/12/vcip97_fwt.pdf. (Accessed 28 June 2017).

[49] C.-H. Yang, Y.-C. Lin, Fractal curves to improve the reversible data embedding for VQ-indexes based on locally adaptive coding, J. Vis. Commun. Image Represent. 21 (4) (May 2010) 334–342, https://doi.org/10.1016/j.jvcir.2010.02.008.

[50] K.U. Berthel, T. Voye, Adaptive fractal image coding in the frequency domain, in: Internal Workshop on Image Processing, 1994, pp. 1–10.

[51] Y. Zhao, B. Yuan, A hybrid image compression scheme combining block-based fractal coding and DCT, Signal Process. Image Commun. 8 (2) (1996) 73–78, https://doi.org/10.1016/0923-5965(95)00036-4.

[52] Y. Zhao, B. Yuan, Image compression using fractals and discrete cosine transform, Electron. Lett. 30 (6) (1994) 474–475.

[53] B.E. Wohlberg, G. de Jager, Fast image domain fractal compression by DCT domain block matching, Electron. Lett. 31 (May) (1995) 869, https://doi.org/10.1049/el:19950582.

[54] K.M. Curtis, C. Neil, V. Foropoulod, A hybrid fractalldct image compression method, in: 14th International Conference on Digital Signal Processing, 2002, pp. 1337–1340.

[55] Y.-M. Zhou, C. Zhang, Z.-K. Zhang, An efficient fractal image coding algorithm using unified feature and DCT, Chaos Solitons Fractals 39 (4) (2009) 1823–1830, https://doi.org/10.1016/j.chaos.2007.06.089.

[56] D. Saupe, S. Jacob, Variance-based quadtrees in fractal image compression, Electron. Lett. 33 (1) (1997) 46, https://doi.org/10.1049/el:19970052.

[57] C.K. Lee, W.K. Lee, Fast fractal image block coding based on local variances, IEEE Trans. Image Process. 7 (6) (1998) 888–891, https://doi.org/10.1109/83.679437.

[58] Sang-Moon Le, A fast variance-ordered domain block search algorithm for fractal encoding, IEEE Trans. Consum. Electron. 45 (2) (May 1999) 275–277, https://doi.org/10.1109/30.793409.

[59] N.N. Ponomarenko, K. Egiazarian, V.V. Lukin, J.T. Astola, Lossless acceleration of fractal compression using domain and range block local variance analysis, Time (2001) 419–422.

[60] Y.G. Wu, M.Z. Huang, Y.L. Wen, Fractal image compression with variance and mean, in: Proc. - IEEE Int. Conf. Multimed. Expo, vol. 1, 2003, pp. I353–I356.

[61] C. He, S.X. Yang, X. Huang, Variance-based accelerating scheme for fractal image encoding, Electron. Lett. 40 (2) (2004) 115, https://doi.org/10.1049/el:20040084.

[62] Y. Zhou, C. Zhang, Z. Zhang, Improved Variance-Based Fractal Image Compression Using Neural Networks, 2006, pp. 575–580.

[63] K.T.T. Sun, S.J.J. Lee, P.Y.Y. Wu, Neural network approaches to fractal image compression and decompression, Neurocomputing 41 (1–4) (Oct. 2001) 91–107, https://doi.org/10.1016/S0925-2312(00)00349-0.

[64] J.-L. Shen, M.-Q. Zhou, Y.-L. Luo, X. Zheng, Y. Wang, Fractal image coding based on local similarity, in: 2007 International Conference on Machine Learning and Cybernetics, Aug. 2007, pp. 350–355.

[65] J. Han, Fast fractal image encoding based on local variances and genetic algorithm, in: Proceedings - 5th International Conference on Wireless Communications, Networking and Mobile Computing, WiCOM 2009, 2009, pp. 6–9.

[66] Chaur-Heh Hsieh, Jyi-Chang Tsai, Lossless compression of VQ index with search-order coding, IEEE Trans. Image Process. 5 (11) (1996) 1579–1582, https://doi.org/10.1109/83.541428.

[67] C.C. Wang, C.H. Hsieh, An efficient fractal image-coding method using interblock correlation search, IEEE Trans. Circuits Syst. Video Technol. 11 (2) (2001) 257–261, https://doi.org/10.1109/76.905992.

[68] T. Truong, C. Kung, J. Jeng, M. Hsieh, Fast fractal image compression using spatial correlation, Chaos Solitons Fractals 22 (5) (Dec. 2004) 1071–1076, https://doi.org/10.1016/j.chaos.2004.03.015.

[69] M.-S. Wu, W.-C. Teng, J.-H. Jeng, J.-G. Hsieh, Spatial correlation genetic algorithm for fractal image compression, Chaos Solitons Fractals 28 (2) (Apr. 2006) 497–510, https://doi.org/10.1016/j.chaos.2005.07.004.

[70] W. Xing-yuan, L. Fan-ping, W. Shu-guo, Fractal image compression based on spatial correlation and hybrid genetic algorithm, J. Vis. Commun. Image Represent. 20 (8) (2009) 505–510, https://doi.org/10.1016/j.jvcir.2009.07.002.

[71] Q. Wang, D. Liang, S. Bi, Fast fractal image encoding based on correlation information feature, in: 2010 3rd International Congress on Image and Signal Processing, Oct. 2010, pp. 540–543.

[72] X.-Y. Wang, Y.-X. Wang, J.-J. Yun, An improved fast fractal image compression using spatial texture correlation, Chin. Phys. B 20 (10) (Oct. 2011) 104202, https://doi.org/10.1088/1674-1056/20/10/104202.

[73] J. Wang, S. Member, N. Zheng, A novel fractal image compression scheme with block classification and sorting based on Pearson's correlation coefficient, IEEE Trans. Image Process. 22 (9) (2013) 3690–3702.

[74] H. Hartenstein, M. Ruhl, D. Saupe, Region-based fractal image compression, IEEE Trans. Image Process. 9 (7) (2000) 1171–1184, https://doi.org/10.1109/83.847831.

[75] K. Belloulata, et al., Fractal image compression with region-based functionality, IEEE Trans. Image Process. 11 (4) (2002) 351–362, https://doi.org/10.1109/TIP.2002.999669.

[76] S. Zhu, Y. Hou, Z. Wang, K. Belloulata, Fractal video sequences coding with region-based functionality, Appl. Math. Model. 36 (11) (Nov. 2012) 5633–5641, https://doi.org/10.1016/j.apm.2012.01.025.

[77] H.T. Chang, C.J. Kuo, Iteration-free fractal image coding based on efficient domain pool design, IEEE Trans. Image Process. 9 (3) (Mar. 2000) 329–339, https://doi.org/10.1109/83.826772.

[78] C. Te Wang, T.S. Chen, S.H. He, Detecting and restoring the tampered images based on iteration-free fractal compression, J. Syst. Softw. 67 (2) (2003) 131–140, https://doi.org/10.1016/S0164-1212(02)00094-8.

[79] K. Belloulata, Fast fractal coding of subbands using a non-iterative block clustering, J. Vis. Commun. Image Represent. 16 (1) (Feb. 2005) 55–67, https://doi.org/10.1016/j.jvcir.2004. 02.001.

[80] A.R.N.B. Kamal, S.T. Selvi, H. Selvaraj, Iteration-free fractal coding for image compression using genetic algorithm, Int. J. Comput. Intell. Appl. 07 (04) (Dec. 2008) 429–446, https:// doi.org/10.1142/S1469026808002399.

[81] D.M. Monro, F. Dudbridge, Fractal approximation of image blocks, in: [Proceedings] ICASSP-92 1992 IEEE Int. Conf. Acoust. Speech, Signal Process., vol. 3, 1992, pp. 485–488.

[82] D.M. Monro, A hybrid fractal transform, in: IEEE Int. Conf. Acoust. Speech, Signal Process., vol. 5, 1993, pp. 1–4.

[83] D.M. Monro, S.J. Woolley, Fractal image compression without searching, in: Proc. ICASSP '94. IEEE Int. Conf. Acoust. Speech Signal Process., vol. v, 1994, pp. V/557–V/560.

[84] F. Dudbridge, Least squares block coding by fractal functions, in: Fractal Image Compression: Theory and Application to Digital Images, Springer US, New York, 1994, pp. 231–244.

[85] S. Furao, O. Hasegawa, A fast no search fractal image coding method, Signal Process. Image Commun. 19 (2004) 393–404, https://doi.org/10.1016/j.image.2004.02.002.

[86] S. Ongwattanakul, X.W.X. Wu, D.J. Jackson, A new searchless fractal image encoding method for a real-time image compression device, in: 2004 IEEE Int. Symp. Circuits Syst. (IEEE Cat. No. 04CH37512), vol. 3, 2004.

[87] X. Wu, D.J. Jackson, H. Chen, A new searchless two-level IFS fractal image encoding method, in: The 19th International Conference on Computers and Their Applications, 2004, pp. 6–10.

[88] D.J. Jackson, H. Ren, X. Wu, K.G. Ricks, A hardware architecture for real-time image compression using a searchless fractal image coding method, J. Real-Time Image Process. 1 (3) (Mar. 2007) 225–237, https://doi.org/10.1007/s11554-007-0024-2.

[89] X.-Y. Wang, S.-G. Wang, An improved no-search fractal image coding method based on a modified gray-level transform, Comput. Graph. 32 (4) (Aug. 2008) 445–450, https://doi.org/ 10.1016/j.cag.2008.02.004.

[90] X.Y. Wang, Y.X. Wang, J.J. Yun, An improved no-search fractal image coding method based on a fitting plane, Image Vis. Comput. 28 (8) (2010) 1303–1308, https://doi.org/10.1016/j. imavis.2010.01.008.

[91] J. Mingyan, J. Zheng, A new searchless fractal image encoding method based on wavelet decomposition, in: Proceedings of the World Congress on Intelligent Control and Automation (WCICA), vol. 2, 2006, pp. 9583–9586.

[92] M. Salarian, H. Hassanpour, A new fast no search fractal image compression in DCT domain, in: 2007 Int. Conf. Mach. Vis., 2007.

[93] V. de Lima, W.R. Schwartz, H. Pedrini, 3D searchless fractal video encoding at low bit rates, J. Math. Imaging Vis. 45 (3) (Mar. 2013) 239–250, https://doi.org/10.1007/s10851-012-0357-8.

[94] F. Ancarani, D. De Gloria, M. Olivieri, C. Stazzone, Design of an ASIC architecture for high speed fractal image compression, in: Proceedings Ninth Annual IEEE International ASIC Conference and Exhibit, 1996, pp. 223–226.

[95] K.P. Acken, M.J. Irwin, R.M. Owens, A parallel ASIC architecture for efficient fractal image coding, J. VLSI Signal Process. 19 (2) (1998) 97–113, https://doi.org/10.1023/A: 1008005616596.

[96] C. Hufnagl, A. Uhl, Algorithms for fractal image compression on massively parallel SIMD arrays, Real-Time Imaging 6 (4) (Aug. 2000) 267–281, https://doi.org/10.1006/rtim.1998. 0164.

[97] K. Qureshi, S.S. Hussain, A comparative study of parallelization strategies for fractal image compression on a cluster of workstations, Int. J. Comput. Methods 05 (03) (Sep. 2008) 463–482, https://doi.org/10.1142/S0219876208001534.

[98] R. de Q. Gomes, V. Guerreiro, R. da R. Righi, L.G. da Silveira, J. Yang, Analyzing performance of the parallel-based fractal image compression problem on multicore systems, AASRI Proc. 5 (2013) 140–146, https://doi.org/10.1016/j.aasri.2013.10.070.

[99] U.B. Kodgule, B.A. Sonkamble, Discrete wavelet transform based fractal image compression using parallel approach, Int. J. Comput. Appl. 122 (16) (Jul. 2015) 18–22, https://doi.org/10.5120/21785-5068.

[100] D. Vidya, R. Parthasarathy, T.C. Bina, N.G. Swaroopa, Architecture for fractal image compression, J. Syst. Archit. 46 (14) (Dec. 2000) 1275–1291, https://doi.org/10.1016/S1383-7621(00)00018-7.

[101] M. Panigrahy, I. Chakrabarti, A.S. Dhar, VLSI design of fast fractal image encoder, in: 18th International Symposium on VLSI Design and Test, Jul. 2014, pp. 1–2.

[102] A.-M.H.Y. Saad, M.Z. Abdullah, High-Speed Implementation of Fractal Image Compression in Low Cost FPGA, vol. 47, Elsevier, 2016, pp. 429–440.

[103] A. Pentland, B. Horowitz, A practical approach to fractal-based image compression, in: [1991] Proceedings. Data Compression Conference, 1991, pp. 176–185.

[104] R. Rinaldo, G. Calvagno, Image coding by block prediction of multiresolution subimages, IEEE Trans. Image Process. 4 (7) (Jul. 1995) 909–920, https://doi.org/10.1109/83.392333.

[105] G. Davis, Adaptive self-quantization of wavelet subtrees: a wavelet-based theory of fractal image compression, July 1995.

[106] A. van de Walle, Merging fractal image compression and wavelet transform methods, Fractals 05 (supp01) (Apr. 1997) 3–15, https://doi.org/10.1142/S0218348X97000590.

[107] I. Andreopoulos, Y.A. Karayiannis, T. Stouraitis, A hybrid image compression algorithm based on fractal coding and wavelet transform, in: 2000 IEEE International Symposium on Circuits and Systems. Emerging Technologies for the 21st Century. Proceedings (IEEE Cat No. 00CH36353), vol. 3, 2000, pp. 37–40.

[108] J. Li, C.C.J. Kuo, Image compression with a hybrid wavelet-fractal coder, IEEE Trans. Image Process. 8 (6) (1999) 868–874, https://doi.org/10.1109/83.766863.

[109] Meng Wu, H.O. Ahmad, M.N.S. Swamy, A new fractal zerotree coding for wavelet image, in: 2000 IEEE International Symposium on Circuits and Systems. Emerging Technologies for the 21st Century. Proceedings (IEEE Cat No. 00CH36353), vol. 3, 2000, pp. 21–24.

[110] X. Xie, Z.M. Ma, Fractal predictive image coding based on zerotrees of wavelet coefficients, J. Image Graph. 5A (11) (2000) 920–924.

[111] Y.H. Xie, D.S. Fu, A fractal image coding algorithm research based on wavelet transformation, J. Image Graph. 8A (7) (2003) 839–842.

[112] D.F. Wang, W. Jiang, Fractal image coding combined with wavelet subtree, Syst. Eng. Electron. 27 (6) (2005) 1120–1122.

[113] Y. Iano, F.S. da Silva, A.L.M. Cruz, A fast and efficient hybrid fractal-wavelet image coder, IEEE Trans. Image Process. 15 (1) (2006) 98–105, https://doi.org/10.1109/TIP.2005.860317.

[114] S. Chun-lin, F. Rui, L.I.U. Fu-qiang, C. Xi, A novel fractal wavelet image compression approach, J. China Univ. Min. Technol. 17 (1) (Mar. 2007) 121–125, https://doi.org/10.1016/S1006-1266(07)60026-1.

[115] Y. Zhang, X. Wang, Fractal compression coding based on wavelet transform with diamond search, Nonlinear Anal., Real World Appl. 13 (1) (2012) 106–112, https://doi.org/10.1016/j.nonrwa.2011.07.017.

[116] J. Yang, Multiple description wavelet-based image coding using iterated function system, Math. Probl. Eng. 2013 (Mar. 2013) 1–12, https://doi.org/10.1155/2013/924274.

[117] M.-S. Wu, Genetic algorithm based on discrete wavelet transformation for fractal image compression, J. Vis. Commun. Image Represent. 25 (8) (Nov. 2014) 1835–1841, https://doi.org/10.1016/j.jvcir.2014.09.001.

[118] S.K. Mitra, C.A. Murthy, M.K. Kundu, Technique for fractal image compression using genetic algorithm, IEEE Trans. Image Process. 7 (4) (Apr. 1998) 586–593, https://doi.org/10.1109/83.663505.

[119] L. Vences, I. Rudomin, Genetic algorithms for fractal image and image sequence compression, in: Computacion Vis., 1997, [Online]. Available: http://citeseerx.ist.psu.edu/viewdoc/download?doi=10.1.1.30.9233&rep=rep1&type=pdf. (Accessed 24 October 2017).

[120] Y. Zheng, G. Liu, X. Niu, An improved fractal image compression approach by using iterated function system and genetic algorithm, Comput. Math. Appl. 51 (11) (Jun. 2006) 1727–1740, https://doi.org/10.1016/j.camwa.2006.05.010.

[121] M.-S. Wu, J.-H. Jeng, J.-G. Hsieh, Schema genetic algorithm for fractal image compression, Eng. Appl. Artif. Intell. 20 (4) (Jun. 2007) 531–538, https://doi.org/10.1016/j.engappai.2006.08.005.

[122] C-C. Tseng, J-G. Hsieh, J-H. Jeng, Fractal image compression using visual-based particle swarm optimization, Image Vis. Comput. 26 (8) (Aug. 2008) 1154–1162, https://doi.org/10.1016/J.IMAVIS.2008.01.003.

[123] A. Muruganandham, R.S.D. Wahida Banu, Adaptive fractal image compression using PSO, Proc. Comput. Sci. 2 (2010) 338–344, https://doi.org/10.1016/j.procs.2010.11.044.

[124] A.R. Nadira, B. Kamal, Iteration free fractal image compression for color images using vector quantization, genetic algorithm and simulated annealing, Online J. Sci. Technol. 5 (1) (2015), [Online]. Available: https://www.tojsat.net/journals/tojsat/articles/v05i01/v05i01-05.pdf. (Accessed 22 October 2017).

[125] B. Wohlberg, G. De Jager, A review of the fractal image coding literature, IEEE Trans. Image Process. 8 (12) (1999) 1716–1729, https://doi.org/10.1109/83.806618.

[126] B. Cui, P. Huang, W. Xie, Fractal dimension characteristics of wind speed time series under typhoon climate, J. Wind Eng. Ind. Aerodyn. 229 (Oct. 2022) 105144, https://doi.org/10.1016/j.jweia.2022.105144.

[127] B.V. Prithvi, S.K. Katiyar, Interpolative operators: fractal to multivalued fractal, Chaos Solitons Fractals 164 (Nov. 2022) 112449, https://doi.org/10.1016/j.chaos.2022.112449.

Part III

Fractals in disease identification and control

Chapter 9

Alzheimer disease (AD) medical image analysis with convolutional neural networks

AD-CNN

Ayesha Sohail[a], Muddassar Fiaz[b], Alessandro Nutini[c], and M. Sohail Iqbal[d]

[a]*School of Mathematics and Statistics, The University of Sydney, Camperdown, NSW, Australia,* [b]*Comsats University Islamabad, Islamabad, Pakistan,* [c]*Centro Studi Attività Motorie – Biology and Biomechanics Dept., Lucca, Italy,* [d]*Department of General Medicine, Shahdara Hospital, Lahore, Pakistan*

9.1 Introduction

Brain amyloidosis, as reported by Raghavan et al. [1], is one of the deadly brain diseases. Amyloid is a protein that results from a protein misfolding process that takes place in parallel to physiological folding and generates toxic and insoluble proteins that are deposited in the tissues in the form of tangles of fibrillary proteins in the form of a β-sheet. Aβ comprises extracellular sequences and parent protein transmembrane region. The spontaneous conversion into fibrillary aggregates of Aβ monomers are found to be correlated with AD growth. The neurodegenerative effects of AD are hypothesized to derive from Aβ. Two different pathways [2] drive the fibrillization of Aβ, by adjacent sAβ to amyloid fibers, the initial nucleation and the secondary nucleation. Secondary nucleation has been recognized as the primary pathway in AD for the formation of amyloid plaque. These studies attempt to model a sort of "amyloid precursor" that emerges from these interactions. It is important to develop a vision to associate the amyloid structure with the pathology that it creates. The equation of the changes in Aβ concentrations in the ODE model is shown under the above premises [3].

There is currently no appropriate cure for Alzheimer's disease, just symptomatic treatment with a minimal result. There is a huge demand for more reliable and earlier diagnosis. More reliable medical and prognostic knowledge is constantly being sought by clinicians, patients, and their relatives. Researchers aim to test new treatments before patients develop dementia, when they are most

Intelligent Fractal-Based Image Analysis. https://doi.org/10.1016/B978-0-44-318468-0.00017-9

Copyright © 2024 Elsevier Inc. All rights reserved, including those for text and data mining, AI training, and similar technologies.

likely to be successful. Clinical diagnosis is only somewhat reliable and involves the occurrence of dementia, but specific biomarkers for AD-related pathologic modifications, such as amyloid imaging, will allow for more accurate detection and early diagnosis, even though patients are only marginally symptomatic.

The convolutional neural networks (CNN) are used in the recent literature for the analysis of images. In the field of medical imaging, the cost-effective deep learning algorithms are evolving rapidly, however, there are still open problems in this domain that still need to be addressed. In a human brain, there are approximately 86 billion neurons. The neurons work as the basic building bricks for the central nervous system (CNS). The neurons are interconnected at points called synapses. The synapses work as junctions at the site to transmit "electric nerve impulses". Synapses connect neurons or a neuron and a gland/ muscle cell. Currently, different researchers are working in this domain to use and improve the algorithms of CNN [4].

A method for studying and analyzing such physical events is mathematical modeling. The notion of these reciprocal effects linked systems and unusual behavior involves mathematical models with their nonlinear equations.

9.2 Brain amyloidosis and medical imaging

Amyloid is a protein that results from a protein misfolding process that takes place in parallel to physiological folding and generates toxic and insoluble proteins that are deposited in the tissues in the form of tangles of fibrillary proteins in the form of $A\beta$-sheets. White matter, gray matter, and cerebral spinal fluid are the three main elements that are investigated in the AD. Depending on these materials, physicians and radiologists will assess the brain abnormality and diagnose the condition for treatment. Reducing the formation of the precursor amyloidogenic protein is one of the most successful therapy options for amyloidosis. The slowing of amyloid protein production may result in the removal of amyloid deposits and the restoration of organ dysfunction.

Alzheimer's disease (AD) is known as one of the world's most prominent late life dementia with vast social and economic impacts. It is formed by deposits of protein complexes called Amyloid plaques in the brain. As a specific protein or monomer, $A\beta$ is usually present in the brain. Even then, in Alzheimer's, it accumulates into tangles. Experts claim that these tangles tend to lead to the death of brain cells. This progressive neurodegenerative disorder generally starts slowly. It is necessary to recognize that Alzheimer's disease is the unusual decline of cerebral functions because of the slow progression that characterizes the disease and the delay in early diagnosis, although age is still the major problem.

Image analysis processing is the study of analyzing or addressing medical problems by utilizing various image recognition methods to collect information in an efficacious and productive manner. It has become one of the most important research areas in engineering and medicine. The processes that provide sensory information about the human body are referred to as medical imaging. Medical imaging encompasses several imaging methods like magnetic resonance imaging (MRI), computed tomography (CT), ultrasounds, positron emission tomography (PET), X-ray, and composite methods in the detection and treatment of diseases [5]. These methods are critical for detecting anatomical and functional details about various body organs for diagnostic and testing purposes. Medical imaging has long been used as a biomarker in medical practice.

The area of medical imaging has gained significantly from recent advances in hardware architecture, safe techniques, computing capabilities, and data storage capability. Segmentation, classification, and abnormality recognition in images produced by a variety of clinical imaging methodologies are currently the most common medical image analysis applications. Fig. 9.1 depicts a taxonomy of important medical imaging modalities. The full range of current imaging methods is difficult to depict in a single image, but it does demonstrate the broad utility and volume of clinical imaging data currently produced. Medical image processing aims to help radiologists and clinicians improve the efficiency of the diagnosis and treatment process. Effective medical image processing is used in computer aided detection (CADx) as well as computer aided diagnosis (CAD), making it important in terms of results because it specifically affects clinical diagnosis and care [6]. As a result, in medical image processing, good performance in terms of effectiveness, precision, memory, sensitivity, as well as specificity is critical and attractive.

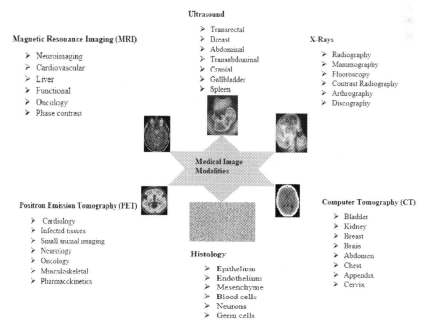

FIGURE 9.1 Typology of medical imaging modalities.

For the purpose of early diagnosis of brain amyloidosis and AD, several diagnostic techniques are improving with time. Certain diagnostic techniques involving medical imaging are detailed below.

- **Structural imaging** is used to determine the shape, location, and volume of brain tissue. Magnetic resonance imaging (MRI) as well as computed tomography (CT) are two structural approaches.

  The brains of the patients with AD shrink dramatically as the disease advances, according to a structural imaging study. Shrinkage in certain brain regions, like the hippocampus, may be an early warning of Alzheimer's, according to a structural imaging study. However, scientists have yet to agree on standardized brain volume measurements that would determine the relevance of a certain degree of shrinkage for any given person at any given time. Tumors, indications of minor or major strokes, injury by severe head trauma, or an accumulation of fluid in the brain can all be discovered using structural imaging.

- **Functional imaging** shows how actively cells in distinct brain areas use sugar or oxygen to reveal well they are performing. Functional imaging approaches comprise functional magnetic resonance imaging (fMRI) as well as positron emission tomography (PET).

  According to a functional imaging study, people with AD have lower brain cell activity in some areas. Studies using fluorodeoxyglucose (FDG)-PET, for example, show that Alzheimer's disease is frequently related to decreased glucose (sugar) utilization in brain regions involved in memory, learning, as well as problem solving. However, as with structural imaging-detected shrinkage, there is currently insufficient data to convert these broad patterns of decreased activity into diagnosis knowledge about specific respective.

- **Molecular imaging** detects cellular or molecular changes connected to specific disorders using highly focused radiotracers. PET, fMRI, as well as single photon emission computed tomography are examples of molecular imaging methods.

  These techniques are among the most active research topics aiming at developing novel ways to diagnose AD in its initial phases. Before AD alters the structure or function of the brain or takes an irreversible toll on memory, thinking, and reasoning, molecular techniques may discover biochemical signs indicating how the illness is progressing. Molecular imaging might potentially provide a novel way to track disease development and improve the effect of next-generation disease modifying therapies. Florbetaben (Neuraceq), Florbetapir (Amyvid), and Flutemetamol (Vizamyl) are three molecular imaging compounds that have been certified for clinical use in the detection of beta-amyloid in the brain.

Imaging plays an important function as a "window on the brain" in the diagnosis and prognosis of brain amyloidosis and AD. These methods have unique benefits and limits, their responsibilities as well as scope are frequently varied and complementary [7].

Its role in diagnosis has evolved to include finding distinctive patterns of structural and functional brain abnormalities in symptomatic patients, hence giving positive evidence for a medical detection of AD. With amyloid imaging, we could see the particular molecular pathology of disorder amyloid aggregation. Imaging has aided in the detection of the frequency of mixed diseases in dementia by finding vascular and non-AD degenerative diseases [7]. Medical imaging is used in certain diagnostic procedures, which are discussed below.

9.2.1 Structural magnetic resonance imaging (sMRI)

MRI photographs provide a significant input in providing valuable knowledge about the brain anatomy as well as defects in the brain soft tissues because of their high resolution and controlled contrast. The abnormality in the brain can be identified through visualization. Protons have the angular momentum that is polarized in the magnetic field, which is used in MRI. This implies that a radiofrequency pulse can alter the total energy of protons, so when the pulse is switched off, the protons will release a radiofrequency signal as they return to their original energy level. "Sequences" may be constructed to be responsive to distinct tissue properties by combining various gradients and pulses.

In general, structural MRI in Alzheimer's disease is used to measure decay and variations in tissue features that generate signal changes on certain patterns, like white matter hyperintensities for T2-weighted MRI due to vascular damage.

Several MR sequences responsive to micro-structural change (e.g., magnetization) have revealed changes in Alzheimer's disease. These patterns are also useful analysis methods, but it is necessary to find a route into the regular clinical treatment of Alzheimer's disease [7]. Developing brain atrophy is a hallmark of neurodegenerative disorder that may be shown in life with MRI (T1-weighted parametric patterns are the best). Dendritic and neuronal losses are regarded to be the main causes of atrophy. Regional MRI volumes (e.g., hippocampus) are strongly associated with neuronal counts at autopsy in studies. The pattern of loss varies between disorders, indicating specific neuronal disease manifestation at the regional level [8,9]. AD is identified by an insidious beginning as well as an unstoppable spread of atrophy, which starts in the middle of the temporal lobe [10]. The entorhinal cortex, followed by the hippocampus, amygdala, and parahippocampus, is the most common location of atrophy [11]. Other limbic lobe components, like the posterior cingulate, are also damaged earlier. The temporal neocortex is then affected, followed by every neocortical brain region in a balanced pattern.

Magnetic resonance imaging has been used as a predictive tool in the diagnosis of AD in several situations, particularly in at-risk individuals with either familial risk of Alzheimer's disease or mild cognitive impairment (MCI). In terms of imaging, investigations have looked at total brain volume, hippocampus, entorhinal cortical volume, as well as variations in these measurements over time as predictors of Alzheimer's disease. Whole-brain volumes are smaller in

Alzheimer's disease than in normal older people on average, and there is strong evidence that the rate of whole-brain shrinkage in AD is twice as fast, 1% per year versus 0.5% per year in normal aging [12].

The use of MRI to assess middle temporal atrophy has been found to have a positive predictive significance for AD. With a sensitivity as well as specificity of 80–85%, visual evaluation can distinguish mild AD from normal aging [13]. It is more difficult to distinguish MCI people who will develop AD soon from those who will not. On MRI, middle temporal atrophy still is a strong indicator of growth, with an accuracy as well as specificity of 50–70% for separating those who would develop AD from those who will not [14]. Structural MRI is not molecularly specific. It cannot identify the histological hallmarks of Alzheimer's disease and so is downstream from the molecular basis. Brain atrophy is a general outcome of neuronal injury, as well as through particular sequences of the loss are associated with different disorders, they are not specific.

9.2.2 Functional magnetic resonance imaging (fMRI)

Functional MRI (fMRI) detects brain functionality by detecting variations in blood flow. This approach is based on the idea that cerebral blood flow and neuronal activity are strongly tied. When a part of the brain is used, blood flow to that part of the brain rises. In aging and early Alzheimer's disease, functional MRI is rapidly being adopted to investigate the functional reliability of brain regions supporting memory and other mental abilities. fMRI is a noninvasive imaging technology that uses fluctuations in the blood oxygen level dependent (BOLD) MR signal to offer an indirect evaluation of neural activity [15]. Increased blood supply to the triggered area is the result of a local increase in metabolism. It is iron in blood hemoglobin that is utilized as a local indication of functional activity, and as an intrinsic T2*-shortage magnet susceptibility-induced intravascular contrast agent. BOLD fMRI is believed to show incorporated neuronal neural transmission via the magnetic resonance signals due to blood flow fluctuations, blood volumes, as well as the blood oxyhemoglobin or deoxyhemoglobin ratios, as shown in Fig. 9.2.

Many of the initial fMRI studies on mild cognitive impairment and Alzheimer's disease were focused on short term memory in the hippocampus and associated structures in the middle temporal lobe. The results were very consistent in patients with clinically confirmed AD and showed a decline in hippocampus activity during new information coding [7]. Due to its better sensitivity to spin-echo sequencing, T2*-weighted patterns are the most utilized in fMRI. An oxidized blood stream includes diamagnetic, oxygenated, and magnetic sensitive hemoglobin. The crucial native T2* in fMRI comparison is regulated by the deoxygenation to oxygenated hemoglobin balance in the blood in a voxel, that depends on local arterial autoregulation. T2* is seen to depend on the deoxygenation of the blood. As deoxyhemoglobin is paramagnetic, the magnetic resonance signal weighted by T2 * is changed [16].

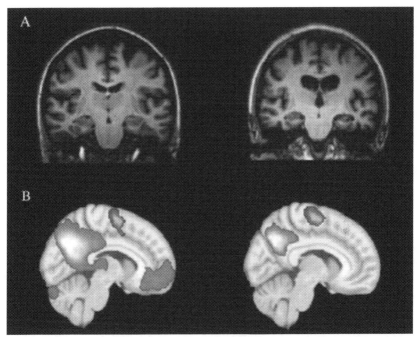

FIGURE 9.2 (A) Structural MRI comparison of a healthy human brain (left) and pathological alterations in Alzheimer's disease (right). (B) Functional MRI of a healthy brain's resting-state network activity (left) against a hypothetical AD brain activation (right). The connection map depicts how brain activity reduces with the disease inside; red/orange (mid gray/light gray in print version) indicates increased linkage, while blue (dark gray in print version) indicates inversely linked activity [17] under CC.

Longitudinal fMRI studies in people with progressive dementia presents many challenges. As these approaches are particularly sensitive to head motion, fMRI is likely to remain highly challenging in studying individuals with more severe cognitive impairment. One of the key benefits of task fMRI activation investigations is wasted if the patients are unable to complete the cognitive task properly.

9.2.3 Computed tomography (CT)

A computed X-ray imaging approach in which a narrow beam of X-rays is targeted at a patient and rapidly rotated around the body, generating signals that are analyzed by the machine's computer to construct cross-sectional images or "slices" of the body is referred to as "computed tomography," or CT. These slices are referred to as tomographic pictures because they carry more information than ordinary X-rays. After the machine's computer collects several successive slices, they may be digitally "stacked" together to create a three-dimensional picture of the patient, allowing for simpler identification and localization of fundamental structures as well as suspected tumors or pathologies. The CT computer utilizes advanced mathematical procedures to create a 2D picture slice of the patient every time the X-ray source completes one full rotation. The thickness of the tissue shown in each imaging slice varies depending on the CT scanner, but it typically ranges between 1 and 10 mm.

A nonenhanced CT of the skull should be enacted on a patient who has a deep neurologic impairment or TIA-like symptoms. CT enables the fast detection of an ICH and the elimination of the key medical differential diagnostic consideration of deep cerebral infarction. The nonenhanced head CT is the recommended imaging modality for the first work-up because it offers important data on the intracranial hemorrhage (ICH) features, such as location, size, shape, as well as extension to the extra-axial regions [18]. A 61-year-old woman reported a one-year history of decreasing mental function. A CT scan of the skull at a local medical facility revealed an enhancing, partially calcified tumor in the left parietal lobe. She had 30 Gy of whole brain radiation after being diagnosed with a brain tumor. Six months later, a CT scan of the head revealed a massive, dense, enhancing left parietal mass with no surrounding swelling [19].

9.2.4 Positron emission tomography (PET)

In contrast to anatomical techniques, PET is a very sensitive non-invasive tool that is well suited for pre-clinical and clinical imaging of cancer biology. Three-dimensional pictures of the concentration and location(s) of the tracer of interest may be rebuilt by a computer utilizing radiolabeled tracers that are delivered in non-pharmacological doses. In the coming decade, PET should play a bigger role in cancer imaging. FDG in the brain PET is used to detect neural behavior. Since the brain's function relies entirely on glucose for energy, the glucose analog FDG is a good indication of cerebral metabolism and may be detected with PET when marked with ^{18}F (half-life 110 min). The majority of the brain's energy budget is devoted to maintaining intrinsic, resting activity, which is mainly maintained in the cortex through glutamaturgic synaptic transmission [20].

Positron emitters require the presence of a nearby nuclear reactor. Commonly utilized positron emitters are ^{11}C, ^{18}F, ^{13}N, as well as ^{15}O. These are radioactive nuclides having short half-lives, see Table 9.1, that can be found in the human body in their non-radioactive form. As a result, such proton emission can be used to label human particles such as glucose (in the form of ^{18}F-labeled 2-deoxy-2-fluoro-D-glucose (FDG)) from Table 9.2. The most commonly used radiolabeled tracer in clinical practice is ^{18}F-FDG, which is readily accessible, has a reasonable half-life, and is reasonably priced. FDG is utilized to determine how much glucose is consumed in different parts of the brain. FDG, like unlabeled glucose, is carried into the cell through a glucose carrier and subsequently phosphorylated into FDG–phosphate. For the sake of research, particularly in simulation experiments. The most often utilized tracers are ^{15}O-labeled tracers, such as, for example, ^{15}O-CO (given through a nasal catheter) is used to evaluate data on the amount of cerebral blood in different parts of the brain. The pharmaceutical sector is increasingly interested in PET's options for responding to queries about environmental medicine methods of action and biodistribution. These can be directly labeled with positron emitters [21].

TABLE 9.1 Positron emitters utilized in PET tracers.

Positron emitters	Half-life	Reference
^{18}F	109.8 min	[21]
^{11}C	20.4 min	
^{15}O	2.05 min	
^{13}N	9.98 min	

TABLE 9.2 Radiopharmaceuticals used in brain PET imaging [21].

Regional cerebral blood flow	$^{15}O\text{-}H_2O$
Regional cerebral blood volume	$^{15}O\text{-}CO$
Energy metabolism	$^{18}F\text{-}FDG$
Protein metabolism	$^{18}F\text{-}DOPA$

PET imaging of β-amyloid (Aβ) in vivo offers a valuable new method for assessing the prognosis, reasons, as well as prospective cure of the disorders in which Aβ contributes. PET visualization of brain Aβ plaquettes along with F-labeled trackers would almost certainly be available in clinical practice to aid AD diagnosis. With the prevalence of Alzheimer's disease because of the population, a long time and the growing focus on initial detection as well as cure, brain amyloid imaging is ready to emerge as a popular procedure. All specialists analyzing PET scans try to understand the dynamic link among amyloid and cognitive impairment, how to better obtain and exhibit photos for amyloid detection, and in what way to identify tracer binding arrangements in Alzheimer's disease and another dementias. PET imaging for amyloidosis is helping to create further reliable treatments providing for improved patient screening for antiamyloid therapy trials and measuring the effects of these therapies on the brain. Both forms of Alzheimer's disease (AD) have mild to regular β-amyloid (Aβ) plaques in the cerebral gray matter, which occur several years before the onset of dementia. As a result, amyloid imaging can be used to confirm or rule out Alzheimer's disease, to differentiate Alzheimer's disease from mild cognitive impairment, and to diagnose Alzheimer's disease sooner [22].

Authors from the University of the Pittsburgh [23,24] modified thioflavin T, a fluorescent dye utilized by pathologists to identify plaquettes in brain tissue samples, to establish the first PET tracker especially for Ab plaquettes. Studies with ^{11}C-PiB, the primary and most commonly investigated PET Aβ ligand, suggest that Aβ imaging can help to diagnose Alzheimer's disease (AD) earlier and differentiate dementia better. In nearly all AD patients, ^{11}C-PiB trials display robust cortical attachment, which correlates well with a reduction in cerebrospinal fluid Aβ42 as well as Alzheimer's disease histopathology [25]. Enhanced ^{11}C-PiB binding has proven to be a strong predictor of development from moderate cognitive dysfunction to Alzheimer's disease [26]. The scientific standards for the diagnosis of possible AD have been updated in light of recent developments in neuroimaging and cerebral spinal fluid examination, allowing for earlier diagnosis and clinical intervention [27]. Although ^{11}C tracers such as ^{11}C-BF2272 and ^{11}C-AZD21383 have been utilized in safety audits, the first published study of an Aβ PET radio compound was in 2002; although this tracer has been reported to recognize both Aβ plaques and tau neurofibrillary tangles (NFT).

As the requirements for diagnosing Alzheimer's disease change, Aβ imaging is expected to play a larger role in care practice, given it is available and inexpensive, and the scans can also be read consistently and reliably outside of academic centers of excellence [22]. Unfortunately, the 20-min radioactive decay half-life of ^{11}C restricts ^{11}C-PiB usage to facilities with an onsite nuclear reactor and ^{11}C nuclear chemistry experience, rendering ^{11}C-PiB PET access limited and price restrictive for regular medical usage [28]. To address such limits, multiple trackers marked with [18]F (half-life, 110 min) were synthesized and tested [29]. This allows for coordinated processing and geographic delivery, which is currently performed globally in the yield of ^{18}F-FDG for clinical usage. In Alzheimer's disease, the ^{18}F tracers often display the dropping of the gray matter as well as white matter limitations, as a result, dropping of the usual white matter arrangement as the primary proof of the cortical amyloid plaque as shown in the images provided in reference [30], but hardly illustrate the strong attachment in the cortical ribbon that is characteristic of a positive ^{11}C-PiB scan. The Food and Drug Administration (FDA) and the European Medicines Agency (EMA) have approved ^{18}F for medical usage.

[11]C-PiB has remained one of the most commonly used tracers for Alzheimer's disease and other dementias, and it has been used as a benchmark for imaging Aβ in vivo [31] ^{11}C-PiB has been shown to bind fibrillar Aβ with high affinity, selectivity, and specificity, while avoiding specific binding to white matter. Fluorine-18-labeled radiotracers have been shown to produce similar results to ^{11}C-PiB [29].

In Aβ imaging, ^{11}C-PiB took the lead. ^{11}C-PiB is a luminescent dye that has proven to have strong bonding and specificity for fibrils Aβ [23,24]. PET experiments using ^{11}C-PiB also illustrated not just a significant variation in ^{11}C-PiB persistence among Alzheimer's disease subjects and age-matched controls [32] but also opposite associations with glucose hypometabolism in certain areas of the brain and reduced cerebrospinal fluid Aβ42 [25]. While Aβ load as measured by ^{11}C-PiB-PET does not correspond with measurement of memory failure in Alzheimer's disease, it does in MCI and healthy older subjects [33].

While the cortical persistence of ^{11}C-PiB appears to be increased in AD, the rate of adhesion is likely variable and has nothing to associate with dementia intensity. The cerebral lobe, cingulate gyrus, prefrontal cortex, striatum, parietal cortex, and lateral temporal cortex have the largest ^{11}C-PiB brain attachment. In most cases, the occipital cortex, sensorimotor cortex, and mesial temporal cortex are unaffected.

The rising rates of positive amyloid scan results in asymptomatic people as they become older has ramifications for the diagnosis precision of these scans for Alzheimer's disease in patients with diagnosed dementia. A positive scan is seen in almost all patients with Alzheimer's disease [22], while a negative scan has been identified in a 91-year-old man with pathological symptoms and cerebral blood flow indicators of Alzheimer's disease but only dissipate plaques at autopsy [34]. However, 12% positive ^{11}C-PiB scans is seen in healthy people at the age of 60 years, 30% of healthy people in the age of 70 years, and 50% of healthy people in the age of 80 years [35]. It is unclear if these numbers would refer to ^{18}F-amyloid ligands as well. The negative prognostic value as well as sensitivity of the ^{11}C-PiB figures is high, but the positive prognostic value and sensitivity decrease with the age. As a consequence, the precision of amyloid imaging for Alzheimer's disease could be over 90% for patients under the age of 70, about 85% for those in their 70s, and 75–80% for those over 80.

The potential to diagnose $A\beta$ accumulation in vivo with PET has advanced the clinical and scientific field of neurodegenerative diseases and enhanced our understanding of the natural history of aging and Alzheimer's disease (AD). ^{18}F tracers, in addition to ^{11}C-PiB, are available, with three of them accepted for clinical use. The uniform uptake value ratio (SUVR), a semiquantitative tool, can be used to determine the burden of $A\beta$ PET. One of the two pathognomonic neuropathologic features of Alzheimer's disease is $A\beta$ plaques. $A\beta$ PET has been integrated into medical trials, whereas it can help with patient selection and measurement of $A\beta$ clearance efficacy. The existence of $A\beta$ can be ruled out using $A\beta$ PET. When $A\beta$ is identified, its prevalence and severity will predict the likelihood of development from normal cognition to moderate cognitive impairment (MCI) and then to Alzheimer's disease dementia [36].

9.3 CNNs in the field of medicine and their basic architecture

CNNs are a type of architecture that uses convolutional filters to perform complex operations efficiently. Image recognition, target identification, facial recognition, fingerprint classification techniques, etc., are the applications of CNN. CNN is an artificial neural network with a unique architecture, as seen in Fig. 9.3. CNN's input data is usually RGB (3 channels) or grayscale images (1 channel). Following the input layer, there are several convolutional or pooling layers (with or without activation functions) [37]. A typical CNN architecture consists of a convolutional layer (feed-forward layer) and a pooling layer, with the network connecting to a fully connected layer after the last pooling layer, there are numerous two-dimensional surfaces with several neurons in each of them. Furthermore, each neuron is thought to function independently [38].

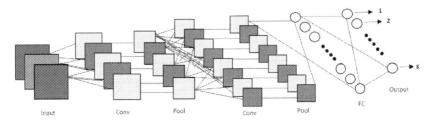

FIGURE 9.3 Basic architecture of CNN.

Every convolution layer produces a feature map of varying size, which is reduced by the pooling layers before being passed to the following layers. The feature extraction module is built into the CNN architecture, and the input data may be compared to a two-dimensional image. CNNs are multi-layer perceptrons that have been biologically influenced. They have a tendency to detect visual features in raw image pixels. Using several layer neurons and common weights in each convolutional layer, such deep neural networks examine just a small portion of the input image, called receptive fields [6]. The complete summary of each portion is given below [39,40].

We need to explain some terminology before we move onto the CNN layers.

- **Kernel:** The kernel is a matrix that slides across the input data, performs a dot product with a sub-region of the input data, and outputs a matrix of dot products, as shown in Fig. 9.4. The kernel in a convolutional neural network is simply a filter that extracts features from pictures. To obtain feature map values, the following formula is often used:

$$O[m, n] = (I, F)[m, n] = \sum_q \sum_r F[q, r]I[m - q, n - r], \qquad (9.1)$$

where I is the input picture, and F denotes our kernel. The output matrix's row and column indexes are denoted by m and n, respectively.

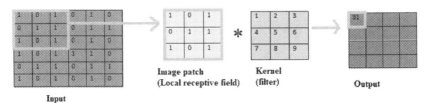

FIGURE 9.4 Application of the filter on the input matrix.

- **Padding:** The size of the picture reduces every time we apply a convolution, and pixels at the image's corners are used just a few times during convolution compared to the center pixels. As a result, we avoid focusing too much on the corners, as this might lead to information loss. To solve these problems, we may add an extra border to the image, as shown in Fig. 9.5. Padded convolution will have the following dimensions:

Input: $m \times m$
Padding: P
Filter size: $k \times k$
Output: $[m + 2P - k + 1] \times [m + 2P - k + 1]$.

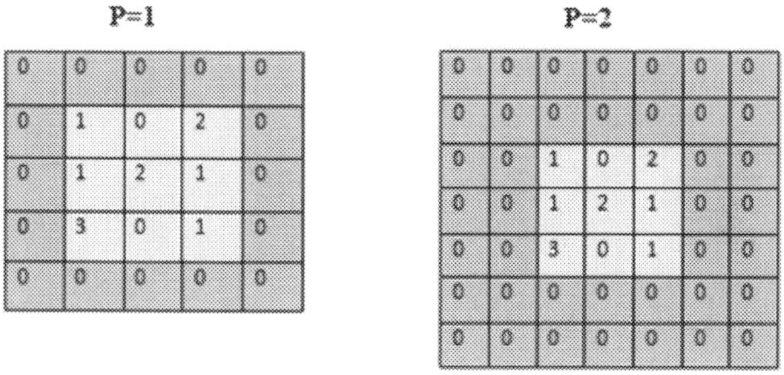

P=1

0	0	0	0	0
0	1	0	2	0
0	1	2	1	0
0	3	0	1	0
0	0	0	0	0

P=2

0	0	0	0	0	0	0
0	0	0	0	0	0	0
0	0	1	0	2	0	0
0	0	1	2	1	0	0
0	0	3	0	1	0	0
0	0	0	0	0	0	0
0	0	0	0	0	0	0

FIGURE 9.5 Padded input matrix.

- **Stride:** The stride is the step taken in the convolution operation. When the size of the stride is increased, the size of the output matrix is decreased. Let us say we choose a stride length of 2. Hence, while convoluting the image, we will take two steps in each the horizontal and vertical direction separately, as shown in Fig. 9.6. Strided convolution will be the following dimension:

Input: $m \times m$
Padding: P
Stride: S
Filter size: $k \times k$
Output: $[(m + 2P - k)/S + 1] \times [(m + 2P - k)/S + 1]$.

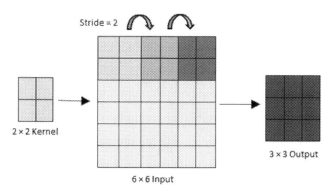

Stride = 2

2 × 2 Kernel

6 × 6 Input

3 × 3 Output

FIGURE 9.6 Strided convolution.

9.3.1 Input layer

The raw data collected for input can be immediately sent into the input layer. There is only one image that is directly inserted into the input layer by its pixel number.

9.3.2 Convolutional layer

Convolution is a mathematical process that takes two inputs (one as the input matrix and the second as convolutional filters).

- Input: $m^{[z-1]}$ with size $(N_h^{[z-1]}, N_w^{[z-1]}, N_c^{[z-1]})$, $m^{[0]}$ being the image in the input;
- Padding: $P^{[z]}$, stride: $S^{[z]}$;
- Number of filters: $N_c^{[z]}$ where every $F^{(n)}$ has the dimension: $(f^{[z]}, f^{[z]}, N_c^{[z-1]})$;
- Bias of the nth convolution: B^z;
- Activation function: ϕ^z;
- Output: $m^{[z]}$ with size $(n_h^{[z]}, N_w^{[z]}, N_c^{[z]})$.

Also, we have:

$$conv(m^z, F^{(n)})_{a,b} = \phi^{[z]}\left(\sum_{p=1}^{N_h^{[z-1]}}\sum_{q=1}^{N_w^{[z-1]}}\sum_{r=1}^{N_c^{[z-1]}} F_{p,q,r}^{(n)} m_{a+p-1,b+q-1,r}^{[z-1]} + B_n^z\right)$$

$$dim(conv(m^{[z-1]}, F^{(n)})) = (N_h^{[z]}, N_w^{[z]}).$$

Thus

$$m^{[z]} = \left[\phi^z(conv(m^{[z-1]}, F^{(1)})), \phi^z(conv(m^{[z-1]}, F^{(2)})), ...,\right.$$
$$\left.\phi^z(conv(m^{[z-1]}, F^{(N_c^{[t]})}))\right]$$

$$dim(m^{[z]}) = (N_h^{[z]}, N_w^{[z]}, N_c^{[z]})$$

$$N_{\frac{h}{w}}^{[z]} = \lfloor\frac{N_{\frac{h}{w}}^{[z-1]} + 2P^{[z]} - f^{[z]}}{S^{[z]}} + 1\rfloor; S > 0$$

$$= N_{\frac{h}{w}}^{[z-1]} + 2P^{[z]} - f^{[z]}; S = 0$$

$$N_c^{[z]} = number\ of\ filters.$$

It is responsible for extracting characteristics from the input picture and is commonly referred to as the up-sampling layer. Every convolutional layer has its convolutional filter, it takes the input data and extracts various characteristics. The number of derived functions increases as the number of convolutional filters in the up-sampling layer increases. The first convolution layer extracts low-level characteristics of the input photos. The higher-level characteristics are provided by the contribution of the second as well as third convolutional layers. For each convolution layer, there are two inputs. The intensity levels of the input data are the first input, and the kernel factors are the second. Weights are another name for the filter coefficients. In the input layer, these are the trainable parameters. Fig. 9.7 depicts the internal design of a convolutional layer.

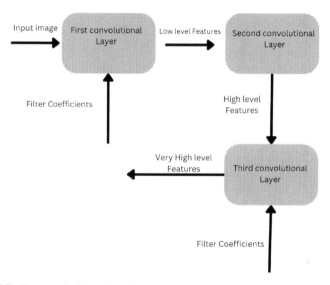

FIGURE 9.7 Framework of convolution layer.

9.3.3 ReLU layer

The term "ReLU layer" refers to the "Rectified Linear Unit" function. After each convolutional layer, this layer uses a simple rectified linear function. This layer's principal goal is to exclude all negative values as well as zeros, as shown in Fig. 9.8. It improves the distinction of the output of every convolution layer, resulting in a finer representation of the characteristics. Here is the formula to calculate it:

$$f(x) = \begin{cases} 0 & \text{if } x < 0 \\ x & \text{if } x \geqslant 0. \end{cases} \tag{9.2}$$

9.3.4 Pooling layer

Each feature map is usually pooled by a pooling layer after passing the input via one or more convolutional layers. The average number is usually calculated using the max-pooling or average pooling process. It is the step of summing up the information and down sampling the image's features, the convolutional

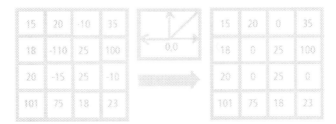

FIGURE 9.8 ReLU operation.

layer comes after that. The more layers of the architecture that are set, the more possible it is that extracting features from input data would allow obvious classification. The max-pooling operation is shown in Fig. 9.9.

$$dim(pooling(image)) =$$
$$\begin{cases} ([\frac{N_h+2P-f}{S}+1], [\frac{N_w+2P-f}{S}+1], N_c) & \text{if} \quad S > 0 \\ (N_h+2P-f, N_w+2P-f, N_c) & \text{if} \quad S = 0. \end{cases} \quad (9.3)$$

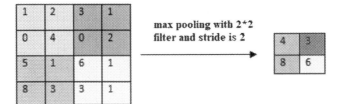

FIGURE 9.9 Max-pooling operation.

9.3.5 Fully connected layers

This is the last and decision making layer for CNN. As input, all of the feature maps are linked together. The nodes or vertices of the neurons in the next layers remain linked to the vertices of the neurons in the preceding layers, however, the vertices of the neurons in every layer remain separated, as shown in Fig. 9.10. This layer combines as well as regularizes the previous convolutions abstracted characteristics to produce a probability for different conditions.

Generally, we have the following equations with regard to the pth node of the qth layer:

$$m_q^{[p]} = \sum_{l=1}^{N_p-1} w_{q,l}^{[p]} c_l^{[p-1]} + B_q^{[p]}$$
$$c_q^{[p]} = \phi^{[p]}(m_q^{[p]}).$$

The input $c^{[p-1]}$ might be the result of a convolution or a pooling layer with the dimensions $(N_h^{[p-1]}, N_w^{[p-1]}, N_c^{[p-1]})$. In order to be able to insert it into the fully connected layer we flatten the tensor to a 1D vector having the dimension $(N_h^{[p-1]} \times N_w^{[p-1]} \times N_c^{[p-1]}, 1)$. Hence,

$$N_p - 1 = N_h^{[p-1]} \times N_w^{[p-1]} \times N_c^{[p-1]}.$$

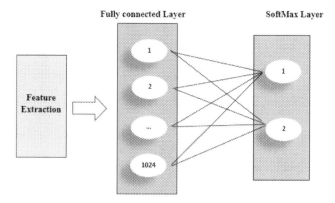

FIGURE 9.10 Fully connected layer.

9.3.6 Output layer

Depending on the conditions, the number of neurons in this layer is determined. The number of the neurons is usually linked to the number of categories to be categorized if classification is needed.

Hence, the whole CNN architecture is given in Fig. 9.11.

9.4 Brain medical imaging and CNN

A Convolutional neural network (CNN) is basically modifications of artificial neural networks (ANNs) that are supported by natural procedures and are designed to recognize various patterns directly from pictures. Sajjad et al. [41] used CNN to tackle the issue of multi-grade brain tumor recognition, inspired by recent successes of CNNs in multiple difficult tasks. Using a fine-tuned CNN platform, Sajjad et al. [41] proposed a novel deep learning system for segmenting and classifying brain tumors into four separate classes. There are three major phases in the proposed system: 1. tumor segmentation, 2. data augmentation, and 3. extraction and classification of deep features. The tumor sectors for both datasets are subdivided in the first phase. Tumor sectors are subdivided using a CNN architecture that has been pre-trained and has layers that specifically support segmentation. The next phase is data augmentation, this involves adjusting

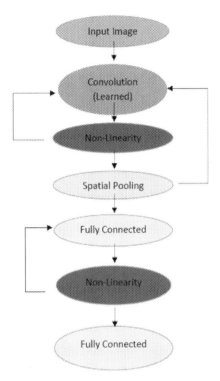

FIGURE 9.11 Layout of CNN.

various parameters to supplement data using various novel and noise invariant strategies.

Medical image processing is the study of analyzing and addressing medical problems by utilizing various image analysis methods to collect information effectively and productively. Machine learning algorithms have been increasingly popular in medical image processing in recent years. Important progress has been made in image recognition, mainly due to the availability of large-scale annotated datasets and the revival of convolutional neural networks (CNNs). As compared to conventional approaches that collect handcrafted features, such machine learning methods are employed to obtain lightweight details for better efficiency of clinical image recognition systems. The importance of human life is demonstrated by advances in medical image processing studies. While the process of collecting and saving digital medical imagery has progressed, analyzing such images has often been difficult and time consuming. As a result, there has always been a pressing need for precise, robust, effective, and user-friendly strategies for early identification and diagnosis of deadly diseases to reduce mortality rates. Brain MRI is a popular diagnostic imaging technique for analyzing and diagnosing a variety of neurological disorders, including brain tumors, cancer, dementia, Alzheimer's disease, epilepsy, and multiple sclerosis.

9.4.1 Background knowledge on machine learning (ML) and deep learning (DL)

In machine learning (ML), a computer is taught how to use its prior experience to solve a problem that has been presented to it. As computer power and memory become more affordable, the idea of applying machine learning to solve problems quicker than humans has gained traction. A vast quantity of data may be processed and analyzed in this way, revealing insights and connections that are not readily apparent to the human eye. To make salient judgments, the machine's intelligent behavior is dependent on several algorithms. On the other hand, Deep Learning (DL) is a sophisticated subfield of machine learning that allows computers to automatically extract, analyze, and interpret meaningful information from raw data by mimicking the way people think and learn. As a collection of neural data-driven approaches, deep learning relies on automated feature engineering procedures. As a result of its automatic learning of characteristics from inputs, it is extremely accurate and performs exceptionally well [42]. An outline of the differences between AI, machine learning, and deep learning (DL) is presented in Fig. 9.12. In machine learning and deep learning, the classifica-

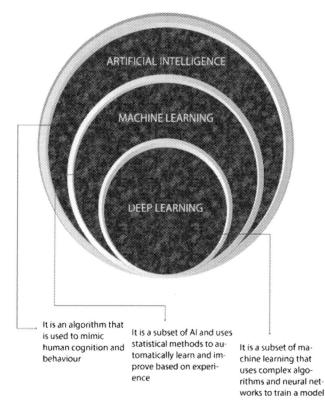

It is an algorithm that is used to mimic human cognition and behaviour

It is a subset of AI and uses statistical methods to automatically learn and improve based on experience

It is a subset of machine learning that uses complex algorithms and neural networks to train a model

FIGURE 9.12 Differences among AI, ML, and DL.

tion algorithm is key to making the appropriate selection. Different classification methods are available in machine learning, and their performance is fairly good. DL is currently replacing ML in most classification applications, despite ML's excellent performance. The principal difference between ML and DL is in how features are extracted for classifiers to work with. Its classification performance is superior to that of ML's classification, which depends on handmade features, because DL features are extracted from many non-linear hidden layers. Consider Fig. 9.13 to better comprehend ML and DL.

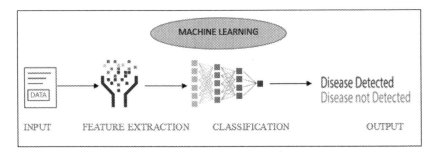

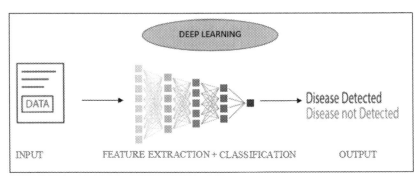

FIGURE 9.13 Differences among ML and DL.

Machine learning techniques are divided into three categories, as shown in Fig. 9.14:

- **Supervised learning:** To learn from a labeled dataset, supervised machine learning algorithms are currently the most widely used approaches for neurodegenerative disease data. A radiologist, for example, is required to label a series of images received from an MRI scan, whereas an expert in neuropathology is required to categorize a set of images collected from postmortem patient samples. Once this 'benchmark' dataset has been labeled, the machine learning algorithm creates a model of the connection between the input features (for example, the size of a brain area on an MRI scan) and the label (for example, a diagnostic category). Based on novel input features, the algorithm may apply this model to new, unlabeled datasets to predict the label for such datasets. For supervised machine learning, gathering huge numbers of correct labels can be a challenge [43].

- **Unsupervised learning:** Algorithms that employ unsupervised machine learning do not require labeled data and are effective for tasks such as grouping data samples into groups and simplifying very complicated datasets. Unsupervised clustering methods, for example, may be used to examine gene expression datasets and discover clusters of patients with common molecular signatures. Latent variable models and other unsupervised clustering techniques can assist in discovering co-expression modules of genes, which are groups of genes likely to be co-regulated or associated with the same biological processes [43].

- **Reinforcement learning:** To obtain the desired outcome in reinforcement learning techniques, a reward or punishment is provided. Examples of algorithmic use include determining a patient's medical history to determine the best drug regimen. Drug–drug interactions that do the algorithm harm would be punished during training, while drugs that improve the disease's course would be rewarded [43–46].

However, supervised and unsupervised learnings are still frequently employed in the field of neurodegenerative disorders.

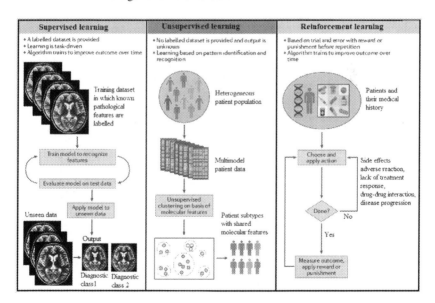

FIGURE 9.14 Categories of machine learning.

Due to the flexibility and power that these approaches provide, they have been widespread and effectively utilized in biology, especially in biomedical research [47], resulting in a broad range of machine learning algorithms. In the United States, more work in medical imaging and "machine learning" is being carried out. It is important to obtain authorization from absolute authorities in order to carry out this task around the globe. An institutional review board must assess the risks and benefits of study participants. Existing data is commonly used, needing previous investigation. In order to share data with their participants, each main investigator in a clinical trial must grant their approval. In the event of a proposed study including the collection of research data, informed consent is required. With the ethical committee's approval, review, and approval, relevant data can be maintained [48]. If the data is open source, then human review is necessary for each image; the quantity of photos and quality may vary depending on the aim and domain. Another stage in making data machine-readable is to structure it. The method of creating medical images for AI development is summarized in Fig. 9.15. These processed image data are used as input data for AI algorithm training.

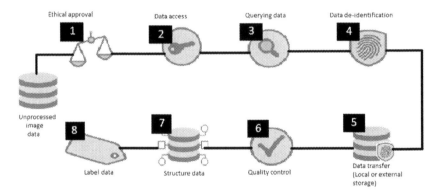

FIGURE 9.15 The procedure for managing medical imaging data.

Language and memory cells in the patient's brain are destroyed in the early stages of Alzheimer's disease, leading to memory loss and reduced capacity to execute daily tasks. Affected individuals lose control over physiological functions as the disorder grows, eventually leading to their death. To diagnose the growth of distinct stages of Alzheimer's disease, radiologists used manual detection techniques in the past. Patient safety might be at risk with these manual methods. Recent techniques based on machine learning and deep learning can detect early stages of Alzheimer's disease automatically [49].

9.4.2 ML-based approaches for AD diagnosis

ML methods for diagnosing patients with different phases of AD are discussed below [49].

Using gene-protein sequences as a potential source of information, Xu et al. [50] studied a computational technique based on SVM-based machine learning. It was concluded that ML-based strategies could be a viable way to predict AD by utilizing gene-coding protein sequence information. In [51], speech processing was used to extract numerous language characteristics for use in a machine learning model. Different feature selection approaches were used to further analyze the extracted linguistic features (semantic, pragmatic, and syntactic) from 242 affected and 242 unaffected AD individuals. The ML classifier is then given the specified characteristics. Using KNN feature selection with an SVM classifier, the proposed ML model obtains the maximum accuracy of 79% in differentiating AD patients from NC patients. Neuropathological abnormalities in patients were used to develop a machine learning prediction model for the early diagnosis of Alzheimer's disease [52]. A neuropathological diagnosis after death was regarded as more definitive than clinical signs. Actually, the proposed model is not clinically applicable, but its accuracy of 77% makes it a step towards precision treatment in AD. Zeng et al. [53] explored a novel switching-delayed PSO-based optimized SVM (SDPSO-SVM) method to distinguish patients with AD from those with NC and MCI. Each of four separate groups, including AD, stable MCI (sMCI), progressive MCI (pMCI), and normal cognitive function (NC), had 92, 82, 95, and 92 participants, respectively. By comparing the suggested technique to many traditional ML algorithms, it was proven that it had high classification accuracy.

For multiway classification of AD and MCI subtypes, the authors of [54] proposed an ML framework with a precise feature selection approach and hierarchical grouping technique. For training and testing, T1 weighted MRI data from four classes namely, AD, cMCI (converted to MCI), MCI (do not convert to MCI), and NC, each with 100 individuals, was collected. The hierarchical grouping procedure transforms a four-way classification into a five-way binary classification. When compared to standard techniques, the suggested feature selection algorithm picks features based on relative significance, resulting in a smaller feature space for each classifier. Further use of the improved classifiers resulted in even higher classification performance. Speech signals generated by the patients themselves were utilized in [55], instead of brain imaging, to distinguish mild AD patients from MCI and NC patients. Combining auditory and linguistic variables can improve the accuracy of classification. Moreover, it is predicted that in the future, speech signal processing will be fully automated, allowing for the automatic identification of AD patients. According to [56], AD, MCI, and NC may be distinguished using 2D textures extracted from T1 weighted MRI scans of 189 AD, 165 MCI-converters, 231 MCI-non-converters, and 227 NC individuals. Using the Rough ROI (RROI) technique, characteristics from the ROIs were extracted, which were subsequently generalized using high-dimensional feature selection techniques. Patients were then classified using machine learning techniques.

A three-layer ANN (input, hidden, and output) was used in [57] to demonstrate its efficacy for the detection of Alzheimer's disease. Of 132 participants, 72 had Alzheimer's disease and 60 did not, according to SPECT brain pictures of cerebral blood flow. Parietal, Ventricular, and Thalamic brain proles were evaluated using 36 numerical values from 12 different regions. Comparing the performance of ANN with that of discriminant analysis (a classic statistical approach), it was observed that ANN was more sensitive and specific in distinguishing AD patients from NC than discriminant analysis. Eke et al. [58] suggested a blood plasma protein-based early detection technique for Alzheimer's disease that is relatively affordable and simpler to obtain. The ADNI database was used to obtain the blood proteome data. The classification was performed using a correlation-based feature subset selection technique. AD was classified using an SVM with a 2-degree polynomial kernel. Rallabandi et al. [59] described a method for predicting AD and MCI early and classifying them from older cognitively normal individuals. From gray matter tissue segmented by anatomical area, the FreeSurfer technique was utilized to generate CT images of various locations. In comparison to other classifiers, non-linear SVM with RBF kernel exhibited superior performance. According to [60], graph theory characteristics from EEG data were utilized to distinguish AD patients from NC patients using an SVM ML method for the first time. However, it is not clear if the graph theory-based model is superior to other forms of EEG analysis or if it can be used to distinguish individuals with AD from those with normal cognition.

Sheng et al. [61] presented a classification technique based on functional features extracted from five main brain areas to correctly identify different stages of AD. In [62], DNA methylation expression proles were obtained from the GEO database, and then integrated genome-wide analysis was done to identify individuals with Alzheimer's disease. SVM, DT, and RF classifiers were used to predict AD. Classifier RF predicts AD more accurately, according to the research. As described in [63], micro-drops of plasma samples were analyzed using a combination of laser-induced breakdown spectroscopy and supervised machine learning (QDA). 31 AD patients and 36 NC patients provided plasma samples for examination. The performance of the suggested system was studied using manually picked characteristics from the different spectra.

9.4.3   DL-based approaches for AD diagnosis

DL methods for diagnosing patients with different phases of AD are discussed below [49]. Islam and Zhang [64] used neuroimaging research with deep CNN to detect distinct phases of Alzheimer's disease, like non-demented, very mild AD, mild AD, and moderate AD, using the axial, coronal, and sagittal planes of an MRI image. Although the accuracy of diagnosing non-demented and very mild stages of dementia was excellent, the accuracy of detecting moderate and mild dementia was poor. Amoroso et al. [65] verified the usage of DNN for diagnosing different phases of AD using MMSE. The International Challenge for Automated Prediction of MCI from MRI Data placed this study third in terms of overall accuracy. DNN's success in this study demonstrates its suitability for future advancements in AD detection systems. Kazemi and Houghten [66] utilized CNN-AlexNet to classify the acquired fMRI data into five categories: NC, significant memory concern, EMCI, LMCI, and AD. AlexNet's identification accuracy was very high because of much preprocessing of the raw data, including the elimination of extraneous tissues, slice timing adjustments, spatial smoothing, high pass filtering, and spatial normalization. In [67], fuses characteristics from 3D patches of MRI and PET images were utilized to study a cascaded deep CNN utilizing the Softmax function to identify AD, MCI, and NC. Not only did the results demonstrate that multimodality is superior to unimodality, but they also showed that deep CNN can identify AD from NC better than an autoencoder. MRI and PET modalities were utilized to distinguish between AD and NC, pMCI and NC, and sMCI and NC in [68]. A unique approach was presented, in which 3D-CNN was utilized to extract the primary features first, and then the FSBi-LSTM was employed to obtain more precise spatial information instead of the generic FC layer. The characteristics were then classified using the SoftMax classifier. To avoid overfitting, the number of filters in the convolution layer was also decreased. A DL method was proposed in [69] that could determine whether a patient had AD, MCI, or neither. The 18F-FDG PET was obtained from the ADNI dataset, with 90% of it being utilized for training and 10% for testing.

A unique strategy to identifying MCI patients who are at a higher risk of progressing MCI to AD was proposed in [70]. MCI to AD conversion and AD vs NC were classified using this proposed approach. It may be used to additional modalities, such as PET, in addition to the imaging modalities mentioned here. Furthermore, the convolutional architecture utilized here makes the system more extensible, since it can be applied to any type of 3D picture collection. Based on dual learning and an Adhoc layer, deep learning is used. As the neural network has fewer parameters, data overfitting is avoided. For testing reasons, they employed a multi-modal feature extractor and 10-fold cross-validation. A deep learning system that uses R-fMRI to identify AD was proposed in [71]. All fMRI and clinical data were used for training and classification. When compared to other current approaches, it has been discovered that the accuracy rose by roughly 25%. The most recent DL object identification methods were used to diagnose AD in [72]. There were three separate approaches utilized, including Faster R-CNN, SSD, and no preprocessing of pictures were required. A simple RNN model has been developed to predict longitudinal AD dementia development using 1677 individuals [73]. It was discovered that the suggested model outperformed the baseline methods in terms of classification. A DL technique utilizing CNN has been investigated, where the OASIS dataset was only utilized for training and the MIRIAD dataset was only used for assessing the model [74]. The findings of this study showed that identifying individuals with MCI was more difficult than identifying patients with AD. The authors of [75] used brain subregions to identify people with Alzheimer's disease. Grey Wolf Optimization was found to be the most promising of the many optimization methods presented here for the correct selection of characteristics. The Siamese CNN model was investigated in [76] for multi-class classification of AD, and it was inspired by Oxford Net. The suggested model's advantage over state-of-the-art models was demonstrated by achieving an exceptional classification accuracy of 99.05%. The authors of [77] used a multi-model DL method with diffusion maps and GM volumes to identify patients with AD and MCI from NC. The authors stated that this was the first study to assess the impact of several scans per person.

Children and adults are at risk of developing a brain tumor, which is among the most deadly and fatal cancers. Earlier detection and brain tumors being classified into various grades are critical for successful treatment. WHO classification codes, brain tumors or tumors in the CNS are further classified into types of malignancy, ranging from class I (benign) to class IV (high malignancy) [78]. Despite many advances in clinical treatment, glioblastomas are still thought to be the most fatal form of tumors with a poor prediction [79]. Histopathology is the main tool for distinguishing grade IV tumors from other grades. The necrosis, microvascular growth, and vascular thrombosis characteristics of class IV tumors can be used to make an initial distinction. These characteristics, on the other hand, are not always obvious and can be difficult to recognize, and pathologists have differing opinions about them [80]. However, there are a few flaws in the histopathological analysis that make glioma diagnosis difficult.

In the area of brain clinical image processing, a variety of work has been undertaken and several workers have contributed to numerous subfields of clinical imaging. The majority of current studies in medical imaging relate to the automated tumor segmentation regions in MRI images. Many scientists have recently proposed various strategies for detecting and segmenting tumor regions in MRI images [81], [82]. Once the tumor has been subdivided in MRI, it must be graded into various levels. Binary classifiers were used in early studies for different groups such as benign and malignant. The authors of [83] suggested a hybrid system that utilizes a genetic algorithm (GA) as well as a support vector machine (SVM) to recognize brain tumors as mild, benign, or malignant. Abdolmaleki et al. [84], used thirteen distinct features to create a three-level back-propagation neural network to differentiate among malignant as well as benign tumors. These characteristics are chosen based on the radiologist's visual perception. When conducting studies on the MRI images of 165 patients, their proposed procedure achieved a precision rate of 91%, as well as 94% for malignant and benign tumors, respectively. Moreover, based on blurred cognitive maps (FCM), Papageorgiou et al. [85] classified brain tumors as higher and lower gliomas. The FCM scoring model was tested on 100 cases and achieved a precision of 90.26% for low grade brain tumors as well as 93.22% for high grade brain tumors.

Scientists also proposed a multi-grading system for brain tumors in addition to binary classifications [41]. For example, Zacharaki et al. [86] described a technique for dividing brain tumors into three types: gliomas, central gliomas, and metastases. The following steps are included in this research: region of interest (ROI) extraction, extraction of feature, and grading. Hsieh et al. [87] suggested a CAD method for measuring the malignancy of gliomas by extracting features from MRI results. Their approach was tested on a dataset of 107 pictures, including 34 images of high grade gliomas and 73 images of lower grade gliomas. Using the global, local, and fused features, the performance accuracy of the CAD device was 76%, 83%, and 88%, respectively. Likewise, Sachdeva et al. [88] proposed a CAD scheme where they derived texture and color features from segmented ROIs and used a genetic algorithm to pick the best features. The precision of GA-ANN as well as GA-SVM was 94.9% and 91.7%, respectively. Several scientists have proposed brain tumor grading approaches based on three different types of brain tumors: meningioma, glioma, as well as pituitary tumor. Cheng et al. [89], for example, suggested a system that combined three related attribute removal methods: the bag of words (BoW) model, the intensity histogram, as well as the co-occurrence grayscale matrix (GLCM). Zia et al. [90] proposed a generic brain tumor classification scheme based on rectangular window picture cropping. Their method used the discrete wavelet transform to remove elements, PCA to reduce dimensionality, and SVM as a classifier.

MR imaging was introduced to easily recognize forms of glioma without the need for brain surgery, which could endanger human lives. CAD or multi-grade brain tumor subdivision schemes [91] were developed to aid radiologists in tumor visualization [92] and tumor type definition. High level characteristics derived from magnetic resonance imaging scans are used to classify various types of classes using deep neural networks, which can aid radiologists to reach early diagnostic judgments and possible care methods. This classification, which has been provided by numerous workers, contains information on the malignancy of tumors at different percentages. Binary sorting, a shortage of datasets for various categories, and poor precision are the main drawbacks of modern brain tumor classification methods. Established procedures classify brain tumors into just two types (benign and malignant), leaving radiologists with a great deal of ambiguity when it comes to further analysis or treatment. The lack of readily accessible evidence is the next big obstacle in brain tumor classification. Furthermore, current methods have not shown satisfactory accuracy to date.

CNN is among the most commonly used deep neural networks in everyday applications. In these networks, the efficiency is usually good, and the feature extraction method is not required manually. However, the high precision comes at a high cost in terms of numerical complexity. The higher number of layers among the input and output layers, as well as the need to change two sets of parameters (one set of filter coefficients as well as another set of weights) in the fully connected network, contribute to CNN's complexity. CNN architecture is similar to LeNet-5 for brain image classification with N classes that accepts a patch made from a 32×32 original 2D clinical picture The network is built up of convolutional, max-pooling, and fully connected layers. Every convolutional layer generates a feature map with varying degrees of detail, which is reduced by the pooling layers before being passed to the following layers. At the output, the fully connected layers produce the desired class prediction, see Fig. 9.16.

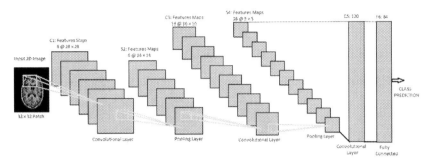

FIGURE 9.16 CNN architecture for class prediction.

The system is influenced by human visuals, which is analogous to a traditional neural network. Inputs are small parts of the image (effective receptive fields). As stated earlier, CNN has several operational layers, including the pooling layer, that is sandwiched among two convolution layers. It aims to monitor overfitting and minimize sample size. Convolution layers gather features, which are then fed to the (FC) layer for classification and other purposes. The fully connected layer follows, with each neuron connected to all neurons in the following layer. Fig. 9.17 depicts the model architecture. The model in this method has five convolutional layers. Filter sizes of 32, 64, and 128, respectively, are used in the first three convolution layers, which are followed by two convolution layers with filter sizes of 64 and 32. Every convolutional layer has a maximum pooling of five layers. The main activation function was originally RELU, but it was subsequently modified to SoftMax to allow binary classification even by the output layer of two nodes. Ultimately, an Adam optimizer and back-propagation were used to train the algorithm [93].

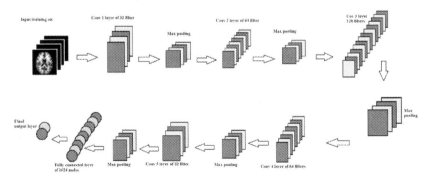

FIGURE 9.17 CNN architecture for brain MRI classification.

References

[1] N.S. Raghavan, L. Dumitrescu, E. Mormino, E.R. Mahoney, A.J. Lee, Y. Gao, M. Bilgel, D. Goldstein, T. Harrison, C.D. Engelman, et al., Association between common variants in rb-

fox1, an rna-binding protein, and brain amyloidosis in early and preclinical Alzheimer disease, JAMA Neurology 77 (10) (2020) 1288–1298.

[2] C. Marino, The role of hsp60 in amyloid beta pathway: Relevance to Alzheimer's disease, 2017.

[3] M. Hoore, S. Khailaie, G. Montaseri, T. Mitra, M. Meyer-Hermann, Mathematical model shows how sleep may affect amyloid-β fibrillization, Biophysical Journal 119 (4) (2020) 862–872.

[4] H. Choi, K.H. Jin, A.D.N. Initiative, et al., Predicting cognitive decline with deep learning of brain metabolism and amyloid imaging, Behavioural Brain Research 344 (2018) 103–109.

[5] M.M. Rahman, B.C. Desai, P. Bhattacharya, Medical image retrieval with probabilistic multiclass support vector machine classifiers and adaptive similarity fusion, Computerized Medical Imaging and Graphics 32 (2) (2008) 95–108.

[6] S.M. Anwar, M. Majid, A. Qayyum, M. Awais, M. Alnowami, M.K. Khan, Medical image analysis using convolutional neural networks: a review, Journal of Medical Systems 42 (11) (2018) 1–13.

[7] K.A. Johnson, N.C. Fox, R.A. Sperling, W.E. Klunk, Brain imaging in Alzheimer disease, Cold Spring Harbor Perspectives in Medicine 2 (4) (2012) a006213.

[8] C. Jack, D. Dickson, J. Parisi, Y. Xu, R. Cha, P. O'brien, S. Edland, G. Smith, B. Boeve, E. Tangalos, et al., Antemortem mri findings correlate with hippocampal neuropathology in typical aging and dementia, Neurology 58 (5) (2002) 750–757.

[9] K. Gosche, J. Mortimer, C. Smith, W. Markesbery, D. Snowdon, Hippocampal volume as an index of Alzheimer neuropathology: findings from the nun study, Neurology 58 (10) (2002) 1476–1482.

[10] R.I. Scahill, J.M. Schott, J.M. Stevens, M.N. Rossor, N.C. Fox, Mapping the evolution of regional atrophy in Alzheimer's disease: unbiased analysis of fluid-registered serial mri, Proceedings of the National Academy of Sciences 99 (7) (2002) 4703–4707.

[11] D. Chan, N.C. Fox, R.I. Scahill, W.R. Crum, J.L. Whitwell, G. Leschziner, A.M. Rossor, J.M. Stevens, L. Cipolotti, M.N. Rossor, Patterns of temporal lobe atrophy in semantic dementia and Alzheimer's disease, Annals of Neurology 49 (4) (2001) 433–442.

[12] A.F. Fotenos, A. Snyder, L. Girton, J. Morris, R. Buckner, Normative estimates of cross-sectional and longitudinal brain volume decline in aging and ad, Neurology 64 (6) (2005) 1032–1039.

[13] E. Burton, R. Barber, E. Mukaetova-Ladinska, J. Robson, R. Perry, E. Jaros, R. Kalaria, J. O'brien, Medial temporal lobe atrophy on mri differentiates Alzheimer's disease from dementia with Lewy bodies and vascular cognitive impairment: a prospective study with pathological verification of diagnosis, Brain 132 (1) (2009) 195–203.

[14] E.S. Korf, L.-O. Wahlund, P.J. Visser, P. Scheltens, Medial temporal lobe atrophy on mri predicts dementia in patients with mild cognitive impairment, Neurology 63 (1) (2004) 94–100.

[15] S. Ogawa, T.-M. Lee, A.S. Nayak, P. Glynn, Oxygenation-sensitive contrast in magnetic resonance image of rodent brain at high magnetic fields, Magnetic Resonance in Medicine 14 (1) (1990) 68–78.

[16] R. Turner, D.L. Bihan, C.T. Moonen, D. Despres, J. Frank, Echo-planar time course mri of cat brain oxygenation changes, Magnetic Resonance in Medicine 22 (1) (1991) 159–166.

[17] M. Pietzuch, A.E. King, D.D. Ward, J.C. Vickers, The influence of genetic factors and cognitive reserve on structural and functional resting-state brain networks in aging and Alzheimer's disease, Frontiers in Aging Neuroscience 11 (2019) 30.

[18] C.P. Chao, A.L. Kotsenas, D.F. Broderick, Cerebral amyloid angiopathy: Ct and mr imaging findings, Radiographics 26 (5) (2006) 1517–1531.

[19] J. Lee, G. Krol, M. Rosenblum, Primary amyloidoma of the brain: Ct and mr presentation, American Journal of Neuroradiology 16 (4) (1995) 712–714.

[20] N.R. Sibson, A. Dhankhar, G. Mason, K. Behar, D. Rothman, R. Shulman, In vivo 13c nmr measurements of cerebral glutamine synthesis as evidence for glutamate–glutamine cycling, Proceedings of the National Academy of Sciences 94 (6) (1997) 2699–2704.

[21] A. Otte, U. Halsband, Brain imaging tools in neurosciences, Journal of Physiology (Paris) 99 (4–6) (2006) 281–292.

[22] C.C. Rowe, K.A. Ellis, M. Rimajova, P. Bourgeat, K.E. Pike, G. Jones, J. Fripp, H. Tochon-Danguy, L. Morandeau, G. O'Keefe, et al., Amyloid imaging results from the Australian imaging, biomarkers and lifestyle (aibl) study of aging, Neurobiology of Aging 31 (8) (2010) 1275–1283.

[23] C.A. Mathis, B.J. Bacskai, S.T. Kajdasz, M.E. McLellan, M.P. Frosch, B.T. Hyman, D.P. Holt, Y. Wang, G.-F. Huang, M.L. Debnath, et al., A lipophilic thioflavin-t derivative for positron emission tomography (pet) imaging of amyloid in brain, Bioorganic & Medicinal Chemistry Letters 12 (3) (2002) 295–298.

[24] W.E. Klunk, Y. Wang, G.-f. Huang, M.L. Debnath, D.P. Holt, L. Shao, R.L. Hamilton, M.D. Ikonomovic, S.T. DeKosky, C.A. Mathis, The binding of 2-(4′-methylaminophenyl) benzothiazole to postmortem brain homogenates is dominated by the amyloid component, Journal of Neuroscience 23 (6) (2003) 2086–2092.

[25] A.M. Fagan, M.A. Mintun, R.H. Mach, S.-Y. Lee, C.S. Dence, A.R. Shah, G.N. LaRossa, M.L. Spinner, W.E. Klunk, C.A. Mathis, et al., Inverse relation between in vivo amyloid imaging load and cerebrospinal fluid aβ42 in humans, Annals of Neurology 59 (3) (2006) 512–519.

[26] A. Nordberg, S.F. Carter, J. Rinne, A. Drzezga, D.J. Brooks, R. Vandenberghe, D. Perani, A. Forsberg, B. Långström, N. Scheinin, et al., A European multicentre pet study of fibrillar amyloid in Alzheimer's disease, European Journal of Nuclear Medicine and Molecular Imaging 40 (1) (2013) 104–114.

[27] G.M. McKhann, D.S. Knopman, H. Chertkow, B.T. Hyman, C.R. Jack Jr, C.H. Kawas, W.E. Klunk, W.J. Koroshetz, J.J. Manly, R. Mayeux, et al., The diagnosis of dementia due to Alzheimer's disease: recommendations from the national institute on aging-Alzheimer's association workgroups on diagnostic guidelines for Alzheimer's disease, Alzheimer's & Dementia 7 (3) (2011) 263–269.

[28] V.L. Villemagne, C.C. Rowe, Amyloid pet ligands for dementia, PET Clinics 5 (1) (2010) 33–53.

[29] R. Vandenberghe, K. Van Laere, A. Ivanoiu, E. Salmon, C. Bastin, E. Triau, S. Hasselbalch, I. Law, A. Andersen, A. Korner, et al., 18f-flutemetamol amyloid imaging in Alzheimer disease and mild cognitive impairment: a phase 2 trial, Annals of Neurology 68 (3) (2010) 319–329.

[30] W. Jagust, Positron emission tomography and magnetic resonance imaging in the diagnosis and prediction of dementia, Alzheimer's & Dementia 2 (1) (2006) 36–42.

[31] K. Shoghi-Jadid, G.W. Small, E.D. Agdeppa, V. Kepe, L.M. Ercoli, P. Siddarth, S. Read, N. Satyamurthy, A. Petric, S.-C. Huang, et al., Localization of neurofibrillary tangles and beta-amyloid plaques in the brains of living patients with Alzheimer disease, The American Journal of Geriatric Psychiatry 10 (1) (2002) 24–35.

[32] C.C. Rowe, S. Ng, U. Ackermann, S.J. Gong, K. Pike, G. Savage, T.F. Cowie, K.L. Dickinson, P. Maruff, D. Darby, et al., Imaging β-amyloid burden in aging and dementia, Neurology 68 (20) (2007) 1718–1725.

[33] K.E. Pike, G. Savage, V.L. Villemagne, S. Ng, S.A. Moss, P. Maruff, C.A. Mathis, W.E. Klunk, C.L. Masters, C.C. Rowe, β-amyloid imaging and memory in non-demented individuals: evidence for preclinical Alzheimer's disease, Brain 130 (11) (2007) 2837–2844.

[34] N.J. Cairns, M.D. Ikonomovic, T. Benzinger, M. Storandt, A.M. Fagan, A.R. Shah, L.T. Reinwald, D. Carter, A. Felton, D.M. Holtzman, et al., Absence of Pittsburgh compound b detection of cerebral amyloid β in a patient with clinical, cognitive, and cerebrospinal fluid markers of Alzheimer disease: a case report, Archives of Neurology 66 (12) (2009) 1557–1562.

[35] M. Mintun, G. Larossa, Y. Sheline, C. Dence, S.Y. Lee, R. Mach, W. Klunk, C. Mathis, S. DeKosky, J. Morris, [11c] pib in a nondemented population: potential antecedent marker of Alzheimer disease, Neurology 67 (3) (2006) 446–452.

[36] N. Krishnadas, V.L. Villemagne, V. Doré, C.C. Rowe, Advances in brain amyloid imaging, in: Seminars in Nuclear Medicine, Elsevier, 2021.

[37] A. Hidaka, T. Kurita, Consecutive dimensionality reduction by canonical correlation analysis for visualization of convolutional neural networks, in: Proceedings of the ISCIE International Symposium on Stochastic Systems Theory and Its Applications, vol. 2017, the ISCIE Symposium on Stochastic Systems Theory and Its Applications, 2017, pp. 160–167.

[38] W.-S. Jeon, S.-Y. Rhee, Fingerprint pattern classification using convolution neural network, International Journal of Fuzzy Logic and Intelligent Systems 17 (3) (2017) 170–176.

[39] H. Gu, Y. Wang, S. Hong, G. Gui, Blind channel identification aided generalized automatic modulation recognition based on deep learning, IEEE Access 7 (2019) 110722–110729.

[40] D.J. Hemanth, J. Anitha, A. Naaji, O. Geman, D.E. Popescu, et al., A modified deep convolutional neural network for abnormal brain image classification, IEEE Access 7 (2018) 4275–4283.

[41] M. Sajjad, S. Khan, K. Muhammad, W. Wu, A. Ullah, S.W. Baik, Multi-grade brain tumor classification using deep cnn with extensive data augmentation, Journal of Computational Science 30 (2019) 174–182.

[42] N.K. Chauhan, K. Singh, A review on conventional machine learning vs deep learning, in: 2018 International Conference on Computing, Power and Communication Technologies (GUCON), IEEE, 2018, pp. 347–352.

[43] M.A. Myszczynska, P.N. Ojamies, A.M. Lacoste, D. Neil, A. Saffari, R. Mead, G.M. Hautbergue, J.D. Holbrook, L. Ferraiuolo, Applications of machine learning to diagnosis and treatment of neurodegenerative diseases, Nature Reviews Neurology 16 (8) (2020) 440–456.

[44] Z. Yu, A. Sohail, T.A. Nofalc, J.M.R. Tavaresd, Explainability of neural network clustering in interpreting the Covid-19 emergency data, Fractals 11 (2021) 1.

[45] Z. Yu, R. Ellahi, A. Nutini, A. Sohail, S.M. Sait, Modeling and simulations of Covid-19 molecular mechanism induced by cytokines storm during Sars-cov2 infection, Journal of Molecular Liquids 327 (2021) 114863.

[46] Z. Yu, R. Arif, M.A. Fahmy, A. Sohail, Self organizing maps for the parametric analysis of Covid-19 seirs delayed model, Chaos, Solitons and Fractals 150 (2021) 111202.

[47] M.W. Libbrecht, W.S. Noble, Machine learning applications in genetics and genomics, Nature Reviews. Genetics 16 (6) (2015) 321–332.

[48] H. Harvey, B. Glocker, A standardised approach for preparing imaging data for machine learning tasks in radiology, in: Artificial Intelligence in Medical Imaging, Springer, 2019, pp. 61–72.

[49] M. Tanveer, B. Richhariya, R. Khan, A. Rashid, P. Khanna, M. Prasad, C. Lin, Machine learning techniques for the diagnosis of Alzheimer's disease: a review, ACM Transactions on Multimedia Computing Communications and Applications 16 (1s) (2020) 1–35.

[50] L. Xu, G. Liang, C. Liao, G.-D. Chen, C.-C. Chang, An efficient classifier for Alzheimer's disease genes identification, Molecules 23 (12) (2018) 3140.

[51] R.B. Ammar, Y.B. Ayed, Speech processing for early Alzheimer disease diagnosis: machine learning based approach, in: 2018 IEEE/ACS 15th International Conference on Computer Systems and Applications (AICCSA), IEEE, 2018, pp. 1–8.

[52] A. Kautzky, R. Seiger, A. Hahn, P. Fischer, W. Krampla, S. Kasper, G.G. Kovacs, R. Lanzenberger, Prediction of autopsy verified neuropathological change of Alzheimer's disease using machine learning and mri, Frontiers in Aging Neuroscience 10 (2018) 406.

[53] N. Zeng, H. Qiu, Z. Wang, W. Liu, H. Zhang, Y. Li, A new switching-delayed-pso-based optimized svm algorithm for diagnosis of Alzheimer's disease, Neurocomputing 320 (2018) 195–202.

[54] D. Yao, V.D. Calhoun, Z. Fu, Y. Du, J. Sui, An ensemble learning system for a 4-way classification of Alzheimer's disease and mild cognitive impairment, Journal of Neuroscience Methods 302 (2018) 75–81.

[55] G. Gosztolya, V. Vincze, L. Tóth, M. Pákáski, J. Kálmán, I. Hoffmann, Identifying mild cognitive impairment and mild Alzheimer's disease based on spontaneous speech using asr and linguistic features, Computer Speech & Language 53 (2019) 181–197.

[56] K. Vaithinathan, L. Parthiban, A.D.N. Initiative, et al., A novel texture extraction technique with t1 weighted mri for the classification of Alzheimer's disease, Journal of Neuroscience Methods 318 (2019) 84–99.

[57] D. Świetlik, J. Białowąs, Application of artificial neural networks to identify Alzheimer's disease using cerebral perfusion spect data, International Journal of Environmental Research and Public Health 16 (7) (2019) 1303.

[58] C.S. Eke, E. Jammeh, X. Li, C. Carroll, S. Pearson, E. Ifeachor, Early detection of Alzheimer's disease with blood plasma proteins using support vector machines, IEEE Journal of Biomedical and Health Informatics 25 (1) (2020) 218–226.

[59] V.S. Rallabandi, K. Tulpule, M. Gattu, A.D.N. Initiative, et al., Automatic classification of cognitively normal, mild cognitive impairment and Alzheimer's disease using structural mri analysis, Informatics in Medicine Unlocked 18 (2020) 100305.

[60] F. Vecchio, F. Miraglia, F. Alù, M. Menna, E. Judica, M. Cotelli, P.M. Rossini, Classification of Alzheimer's disease with respect to physiological aging with innovative eeg biomarkers in a machine learning implementation, Journal of Alzheimer's Disease 75 (4) (2020) 1253–1261.

[61] J. Sheng, M. Shao, Q. Zhang, R. Zhou, L. Wang, Y. Xin, Alzheimer's disease, mild cognitive impairment, and normal aging distinguished by multi-modal parcellation and machine learning, Scientific Reports 10 (1) (2020) 1–10.

[62] J. Ren, B. Zhang, D. Wei, Z. Zhang, Identification of methylated gene biomarkers in patients with Alzheimer's disease based on machine learning, BioMed Research International 2020 (2020).

[63] R. Gaudiuso, E. Ewusi-Annan, W. Xia, N. Melikechi, Diagnosis of Alzheimer's disease using laser-induced breakdown spectroscopy and machine learning, Spectrochimica Acta, Part B: Atomic Spectroscopy 171 (2020) 105931.

[64] J. Islam, Y. Zhang, Early diagnosis of Alzheimer's disease: a neuroimaging study with deep learning architectures, in: Proceedings of the IEEE Conference on Computer Vision and Pattern Recognition Workshops, 2018, pp. 1881–1883.

[65] N. Amoroso, D. Diacono, A. Fanizzi, M. La Rocca, A. Monaco, A. Lombardi, C. Guaragnella, R. Bellotti, S. Tangaro, A.D.N. Initiative, et al., Deep learning reveals Alzheimer's disease onset in mci subjects: results from an international challenge, Journal of Neuroscience Methods 302 (2018) 3–9.

[66] Y. Kazemi, S. Houghten, A deep learning pipeline to classify different stages of Alzheimer's disease from fmri data, in: 2018 IEEE Conference on Computational Intelligence in Bioinformatics and Computational Biology (CIBCB), IEEE, 2018, pp. 1–8.

[67] M. Liu, D. Cheng, K. Wang, Y. Wang, Multi-modality cascaded convolutional neural networks for Alzheimer's disease diagnosis, Neuroinformatics 16 (3) (2018) 295–308.

[68] C. Feng, A. Elazab, P. Yang, T. Wang, F. Zhou, H. Hu, X. Xiao, B. Lei, Deep learning framework for Alzheimer's disease diagnosis via 3d-cnn and fsbi-lstm, IEEE Access 7 (2019) 63605–63618.

[69] Y. Ding, J.H. Sohn, M.G. Kawczynski, H. Trivedi, R. Harnish, N.W. Jenkins, D. Lituiev, T.P. Copeland, M.S. Aboian, C. Mari Aparici, et al., A deep learning model to predict a diagnosis of Alzheimer disease by using 18f-fdg pet of the brain, Radiology 290 (2) (2019) 456–464.

[70] S. Spasov, L. Passamonti, A. Duggento, P. Lio, N. Toschi, A.D.N. Initiative, et al., A parameter-efficient deep learning approach to predict conversion from mild cognitive impairment to Alzheimer's disease, NeuroImage 189 (2019) 276–287.

[71] H. Guo, Y. Zhang, Resting state fmri and improved deep learning algorithm for earlier detection of Alzheimer's disease, IEEE Access 8 (2020) 115383–115392.

[72] J.X. Fong, M.I. Shapiai, Y.Y. Tiew, U. Batool, H. Fauzi, Bypassing mri pre-processing in Alzheimer's disease diagnosis using deep learning detection network, in: 2020 16th IEEE International Colloquium on Signal Processing & Its Applications (CSPA), IEEE, 2020, pp. 219–224.

[73] M. Nguyen, T. He, L. An, D.C. Alexander, J. Feng, B.T. Yeo, A.D.N. Initiative, et al., Predicting Alzheimer's disease progression using deep recurrent neural networks, NeuroImage 222 (2020) 117203.

[74] A. Yiğit, Z. Işik, Applying deep learning models to structural mri for stage prediction of Alzheimer's disease, Turkish Journal of Electrical Engineering & Computer Sciences 28 (1) (2020) 196–210.

[75] D. Chitradevi, S. Prabha, Analysis of brain subregions using optimization techniques and deep learning method in Alzheimer disease, Applied Soft Computing 86 (2020) 105857.

[76] A. Mehmood, M. Maqsood, M. Bashir, Y. Shuyuan, A deep Siamese convolution neural network for multi-class classification of Alzheimer disease, Brain Sciences 10 (2) (2020) 84.

[77] E.N. Marzban, A.M. Eldeib, I.A. Yassine, Y.M. Kadah, A.D.N. Initiative, Alzheimer's disease diagnosis from diffusion tensor images using convolutional neural networks, PLoS ONE 15 (3) (2020) e0230409.

[78] D.N. Louis, A. Perry, G. Reifenberger, A. Von Deimling, D. Figarella-Branger, W.K. Cavenee, H. Ohgaki, O.D. Wiestler, P. Kleihues, D.W. Ellison, The 2016 world health organization classification of tumors of the central nervous system: a summary, Acta Neuropathologica 131 (6) (2016) 803–820.

[79] E.G. Van Meir, C.G. Hadjipanayis, A.D. Norden, H.-K. Shu, P.Y. Wen, J.J. Olson, Exciting new advances in neuro-oncology: the avenue to a cure for malignant glioma, CA: A Cancer Journal for Clinicians 60 (3) (2010) 166–193.

[80] C.L. Nutt, D. Mani, R.A. Betensky, P. Tamayo, J.G. Cairncross, C. Ladd, U. Pohl, C. Hartmann, M.E. McLaughlin, T.T. Batchelor, et al., Gene expression-based classification of malignant gliomas correlates better with survival than histological classification, Cancer Research 63 (7) (2003) 1602–1607.

[81] M. Havaei, A. Davy, D. Warde-Farley, A. Biard, A. Courville, Y. Bengio, C. Pal, P.-M. Jodoin, H. Larochelle, Brain tumor segmentation with deep neural networks, Medical Image Analysis 35 (2017) 18–31.

[82] T. Ateeq, M.N. Majeed, S.M. Anwar, M. Maqsood, Z.-u. Rehman, J.W. Lee, K. Muhammad, S. Wang, S.W. Baik, I. Mehmood, Ensemble-classifiers-assisted detection of cerebral microbleeds in brain mri, Computers & Electrical Engineering 69 (2018) 768–781.

[83] A. Kharrat, K. Gasmi, M.B. Messaoud, N. Benamrane, M. Abid, A hybrid approach for automatic classification of brain mri using genetic algorithm and support vector machine, Leonardo Journal of Sciences 17 (1) (2010) 71–82.

[84] P. Abdolmaleki, F. Mihara, K. Masuda, L.D. Buadu, Neural networks analysis of astrocytic gliomas from mri appearances, Cancer Letters 118 (1) (1997) 69–78.

[85] E. Papageorgiou, P. Spyridonos, D.T. Glotsos, C.D. Stylios, P. Ravazoula, G. Nikiforidis, P.P. Groumpos, Brain tumor characterization using the soft computing technique of fuzzy cognitive maps, Applied Soft Computing 8 (1) (2008) 820–828.

[86] E.I. Zacharaki, S. Wang, S. Chawla, D. Soo Yoo, R. Wolf, E.R. Melhem, C. Davatzikos, Classification of brain tumor type and grade using mri texture and shape in a machine learning scheme, Magnetic Resonance in Medicine 62 (6) (2009) 1609–1618.

[87] K.L.-C. Hsieh, C.-M. Lo, C.-J. Hsiao, Computer-aided grading of gliomas based on local and global mri features, Computer Methods and Programs in Biomedicine 139 (2017) 31–38.

[88] J. Sachdeva, V. Kumar, I. Gupta, N. Khandelwal, C.K. Ahuja, A package-sfercb-"segmentation, feature extraction, reduction and classification analysis by both svm and ann for brain tumors", Applied Soft Computing 47 (2016) 151–167.

[89] J. Cheng, W. Huang, S. Cao, R. Yang, W. Yang, Z. Yun, Z. Wang, Q. Feng, Enhanced performance of brain tumor classification via tumor region augmentation and partition, PLoS ONE 10 (10) (2015) e0140381.

[90] R. Zia, P. Akhtar, A. Aziz, A new rectangular window based image cropping method for generalization of brain neoplasm classification systems, International Journal of Imaging Systems and Technology 28 (3) (2018) 153–162.

[91] G.J. Litjens, J.O. Barentsz, N. Karssemeijer, H.J. Huisman, Clinical evaluation of a computer-aided diagnosis system for determining cancer aggressiveness in prostate mri, European Radiology 25 (11) (2015) 3187–3199.

[92] I. Mehmood, M. Sajjad, K. Muhammad, S.I.A. Shah, A.K. Sangaiah, M. Shoaib, S.W. Baik, An efficient computerized decision support system for the analysis and 3d visualization of brain tumor, Multimedia Tools and Applications 78 (10) (2019) 12723–12748.

[93] N. Chauhan, B.-J. Choi, Performance analysis of classification techniques of human brain mri images, International Journal of Fuzzy Logic and Intelligent Systems 19 (4) (2019) 315–322.

Chapter 10

An intelligent fractal-dimension-based model for brain-tumor MRI analysis

Rakesh Garg, Richa Gupta, and Neha Agarwal

Amity University, Noida, Uttar Pradesh, India

10.1 Introduction

Brain tumor is a kind of disease in which there is an inflammation of cells in the brain that can be cancerous (malignant) or noncancerous (benign). In the United States alone, there are an approximate of 700 000 brief cases of brain tumors reported in 2020, which is further said to be increased to around 850 000 in 2021. Approximately 70% of all the brain tumors are benign, 30% are malignant, 58% of all brain tumors occur in females and approximately 42% of all the brain tumors occur in males. Till now, the imaging techniques of MRIs have been used by doctors to confirm the severity of this disease, so using this technology we were able to build a model by utilizing different CNN architectures that would help in the diagnosis of the disease that a person is suffering from as true or false. The computer-aided diagnosis systems have shown the potential for improving diagnostic accuracy. Throughout the study, different models are analyzed over the different domains. The MRI images are an essential mode of confirmation of the presence of a brain tumor. A fractal analysis of MRI of brain tumors is a new dimension to study the texture present in the image. A fractal can be defined as a mathematical set that shows a self-similar pattern, independent of the scale of the display. The self-similarly in the pattern is analogous to the zooming/magnifying function, used in computer graphics to see details of the image. Many people perceive fractals as beautiful images with an infinite pattern. However, the underlying concept behind fractals is more than just an image. The growing attraction of fractals has encouraged scientists and researchers around the world to explore the different applications of fractal geometry in the real world. Earth science, image processing, biology, mathematical modeling, physiology, and medicine are among the few examples of the fields where fractals have successfully been used for the advancement of society [1,2]. Evolution of fractals and computers over time has a significant role in this. Increasing computational capability with time allows deeper investigations into the vast number

Intelligent Fractal-Based Image Analysis. https://doi.org/10.1016/B978-0-44-318468-0.00018-0

Copyright © 2024 Elsevier Inc. All rights are reserved, including those for text and data mining, AI training, and similar technologies.

of applications of fractal geometry. Texture analysis of the image is a promising application of fractal geometry.

The texture of an image can be determined and compared by fractal dimension [3]. The fractal analysis is used for texture segmentation for clinical diagnostic purposes. Fractal-geometry-based texture segmentation was initially proposed by Keller et al. [4]. Later, many other algorithms were proposed based on fractal geometry [5,6]. The medical field is not the only area where fractal-based texture segmentation is used. The food industry, agriculture, and environmental science are other examples that have benefited from this study [7–11]. Images containing large constant-intensity regions and fewer irregularities are difficult to analyze. Such kinds of images are better studied using multifractals. Various studies have been conducted for feature extraction based on multifractals [12,13].

Fractal features are widely used to characterize 1D or 2D signals in medicine. Fractal dimension is a fractal signature used to study different patterns present in medical imagery [14,15]. Mammography, bone imaging, and brain imaging are explored and these benefit from using fractal geometry. Fractal dimension is used commonly for analyzing in the field of brain imaging. A study of brain-imaging-based fractal features is mainly conducted by magnetic resonance imaging (MRI). An early study carried by Cook et al. [16] found that the fractal dimension is not discriminative on healthy brain and brain with irregularities. This work was done using the box-counting method to estimate fractal dimension. A binarization process of an image in box-counting methods causes loss of valuable information. Later, fractal dimension was proposed based on morphological operators [17,18]. Iftekharuddin et al. [19] proposed a fractal-analysis-based tumor-detection algorithm. In this chapter, fractal dimension and segmentation are used to analyze MRI scans of brain tumors. Here, a CNN-Sequential Model is trained based on the dataset of 2000 images. The performance of existing models is not satisfactory due to restrictions in the dataset, so it is the need of the hour to use a larger dataset. In this chapter, different models such as InceptionV3, VGG-16, RESNET-50, Xception VGG-19, and Efficient NetB0 are built on the improved dataset.

The chapter is organized as follows: Sections 10.2 and 10.3 give an overview of fractals and tumor detection from MRI images, respectively. Section 10.4 discloses the proposed algorithm. Simulation results and discussion are presented in Section 10.5. The paper is concluded in Section 10.6.

10.2 Background

Conventional geometry describes the dimension of the geometrical object in integer space only. For example, the point has zero dimension and lines are one-dimensional and so on. In contrast to conventional geometrical objects, for natural objects it is not possible to give dimensions in integer space. A fractal curve will have a dimension between one and two. This is another exciting feature of fractals in that they have noninteger dimension. Mandelbrot defined fractal sets formally as a set having Hausdorff dimension greater than the topological dimension. Fractal geometry is used in various fields of industries involving application images. Application in image processing often required calculating the fractal dimension associated with it.

10.2.1 Fractal dimension

Fractal dimension is a means of comparison between two fractals. Formally, FD can be defined as follows. Assume a complete metric space is denoted as (X, d) and A is the nonempty compact subset of X. For $\varepsilon > 0$ let function $B(x, \varepsilon)$ represent the closed ball of radius ε and center at point x. Let $N(A, \varepsilon)$ be the function denoting the minimum number of closed balls required to cover set A, if there exists D called the fractal dimension of set A as shown in Eq. (10.1):

$$D = \lim_{\varepsilon \to 0} \left\{ \frac{Ln(N(A, \varepsilon))}{Ln(1/\varepsilon)} \right\}. \tag{10.1}$$

There are many methods to estimate fractal dimension (FD) that can broadly classify into three categories: the box-counting method, fractal Brownian motion, and the area method.

The box-counting method [20] is the most frequently used method for evaluating FD. The fractal dimension is defined according to Eq. (10.2) by covering the signal using a box of length r:

$$FD = -\lim \frac{\log(N(r))}{\log(r)}, \tag{10.2}$$

where N is the number of boxes required to cover a binary signal and limit parameter r tends to zero.

The Box-Counting Theorem [21]

Assume set $A \in h(R^m)$, where Euclidean space is used. Cover space R^m by consecutive square boxes. Every box is of the same side length of $(1/2^n)$. Let function $N_n(A)$ denote the required number of square boxes that intersect the attractor. The fractal dimension of A is denoted by D if Eq. (10.3) is satisfied:

$$D = \lim_{n \to \infty} \frac{\ln(N_n(A))}{\ln(2^n)}. \tag{10.3}$$

Although it is the most famous method for estimation of FD, it suffers from box-size sensitivity [22]. Also, the signal must be in binary form before box covering. Chaudhuri and Sarkar [23] proposed an alternative to a box-counting method to overcome some of the limitations of BC. Box-size sensitivity is handled by using a different size of the box and the number of the box is calculated as the difference between the maximum and minimum gray level in any box. Another class of fractal dimension is based on the fact that most of the physical models are fractal Brownian functions [24]. These functions are statistically affine, and their fractal dimension is not altered by linear transformation and scaling. There are two algorithms named Variogram [25] and power spectrum, which estimate the FD of two-dimensional images of the fractal Brownian function. Finally, a class of methods to compute FD is the area-measurement method. This method uses various structuring elements like a triangle, erosion of different size, and the estimated area of the intensity surface of the signal for any scale. There are three algorithms commonly used to estimate FD under the area-measurement category: the Isarithm method, the triangular prism method, and the blanket method estimate FD based on area measurement.

Fractal dimension is an integral part of the fractal model. The randomness associated with natural images makes them eligible for fractal analysis. Fractal analysis is explored in depth to observe the change in the fractal dimension in correspondence to the change in the texture of the image. The following section presents the study of MRI images using FD and segmentation to extract ROI.

10.2.2 Tumor detection and MRI images

The human body consists of various types of cells. Sometimes, cells lose the capacity to control their development. Cells replicate repeatedly in an inappropriate manner. The additional cell's structure like this is a mass of tissue called a tumor. Metastatic brain tumors start as a disease somewhere else in the physique and move to the cerebrum. There are more than 120 separate sorts of cerebrum tumors. The reason for brain tumors is obscure.

MRI is an efficient method that provides complete information about the targeted brain tumor. MRI of the brain facilitates adequate diagnosis, treatment, and monitoring of the disease. Presently, medical-field specialists appraise the size and position of the tumor by inspecting the MRI images. It is an extremely time-consuming process and error prone.

Basics of tumor detection from MRI images

Tumor identification in the current situation is made physically by radiologists without the utilization of a wholly computerized or semirobotized framework. This might be ascribed to the absence of such advanced frameworks that can bring about a fast and precise method.

In a mechanized framework, the picture is sustained as the data to the framework. Such a framework might be of incredible utilization for a therapeutic examination of a brain tumor. Henceforth, there is a substantial need to create such new frameworks and enhance the existing ones. The following algorithm is an attempt to analyze the texture based on fractal features to extract a region of interest.

Segmentation-based fractal-texture analysis (SFTA) extraction algorithm

SFTA [26] is composed of two major processes, named decomposition and extraction. The algorithm partitions a grayscale image into a set of binary images. It uses two threshold binary decomposition algorithms. Following decomposition in the second process, the feature vector is formed using the image's fractal dimension, mean value, etc.

Decomposition algorithm

1. Input grayscale image $f(x, y)$ with grayscale varying from 1 to L.
2. Compute the set of threshold T using a multilevel Otsu algorithm as follows. For m threshold $\{n_1, n_2 \ldots n_{m-1}\}$ divide the image into m classes. The optimal thresholds are chosen by minimizing interclass variance according to Eqs. (10.4) and (10.5):

$$\{n_1, n_2, ..., n_{m-1}\} = Arg \max\{\sigma_b^2(n_1, n_2, ..., n_{m-1})\} \tag{10.4}$$

$$\sigma_b^2 = \sum_{k=1}^{m} w_k (\mu_k - \mu_t)^2 \tag{10.5}$$

$$\text{where } w_k = \sum p_i, \ \mu_k = \sum \frac{i p_i}{w_k}.$$

Parameter w_k is the zeroth-order cumulative moment of the kth class C_k and μ_t is the mean of the whole image.
3. Repeat step two for each image.
4. Select a pair of the threshold from set T and apply decomposition using Eq. (10.6):

$$f_b(x, y) = \begin{cases} 1 & \text{if } t_u < f(x, y) < t_p \\ 0 & otherwise \end{cases}. \tag{10.6}$$

Feature-vector extraction

5. Compute region boundary of binary image f_b using Eq. (10.7):

$$\Delta(x, y) = \begin{cases} 1 & \exists(x', y') \in N_8 \wedge f_b\left(x', y'\right) = 0 \wedge f_b(x, y) = 1 \\ 0 & otherwise \end{cases}, \tag{10.7}$$

where N_8 is 8 connected neighbors, is the border image.
6. Compute the fractal dimension of each image using the box-counting method.
7. Compute the mean gray level and pixel count of each binary image.
8. Form a feature vector as {FD, mean, pixel count}.

The dimension of the feature vector is equal to the number of binary images multiplied by three. The algorithm is successfully used to study the anomalies present in human lungs. In the following section, a fractal dimension and segmentation-based detection algorithm is proposed.

10.3 Dataset

The dataset that was developed for the discovery of Brain Tumor was a 2D form of the MRI scans. It consisted of nearly 3300 images of MRI of the Brain that were split into different categories, i.e., Glioma Tumor, Meningioma Tumor, Pituitary Tumor and No tumor at all. The dataset was already split into double sets, i.e., Training Set and Testing Set. As shown in Table 10.1, the training set had 2070 images, whereas the testing set had around 400 images.

TABLE 10.1 Brain-tumor dataset.

Dataset - 4	Images	Training Set	Testing Set	Source 5 [27]
Glioma Tumor	926	826	100	926
Meningioma Tumor	937	822	115	937
Pituitary Tumor	901	827	74	901
No Tumor	500	395	105	500
Total	3264	2870	394	3264

The dataset had around 926 images belonging to the Glioma Tumor, out of which 826 images were a part of the training set and a 100 images were in the testing set. The dataset had around 937 images belonging to the Meningioma Tumor, out of which 822 images were a part of the training set and 115 images were in the testing set. The dataset had around 901 images belonging to the Pituitary Tumor out of which 827 images were a part of the training set and 74 images were in the testing set. The dataset had around 500 images belonging to the No Tumor out of which 395 images were a part of the training set and a 105 images were in the testing set.

10.4 Proposed algorithm

The model for the recognition of the Brain Tumor was based on the dataset comprising of 2D pictures of the MRI scans. The models trained throughout the study were CNN (using sequential model), and various Transfer Learning models comprising of RESNET-50, INCEPTION V3, VGG19, VGG16, XCEPTION, EfficientNetB0, RESNET101V2, DeenseNet-201, and Inception-ResnetV2. The models were trained for 30 epochs each with early stopping being configured to each model.

During the process of training for Transfer Learning Models, the input shape of the pictures was taken as 224 × 224 pixels and each matrix had 3 dimensions associated with it. The top layer of every model was not included, and the layers were restricted from being trained to reduce the parameters being trained onto, thus helping to reduce the training time of the model. Then, a flattened layer was added with a dropout layer of 0.5 and then a dense layer to finally end the classification task. The loss function entertained was binary crossentropy and the optimizer function was Adam. Moreover, a decrease in the learning rate of the layers was utilized with a factor of 0.3 and a level of patience of 2 epochs that signifies that the learning rate pf they would tend to model would decrease by a factor of 0.3 after it encounters two epochs with the same validation accuracy and when the accuracy increases, the model's weights are saved in a separate HDF5 file. Furthermore, early stopping was utilized monitoring the validation accuracy through 5 epochs that signifies that in the situation of the model not being able to give better scores (or a flat line) in the validation accuracy, there is a very low chance that it will yield better results in the coming epochs. Hereby, to diminish the consumption of the resources and to save time, the model training stops after 5 consecutive epochs.

Step 1. Preprocessing of image

The preprocessing of the image was carried out to enhance the input image and make it better for further processing by removing noise, smoothing the edges, or sharpening them as seemed fit. Some filtering and contrasting steps were performed. Noise removal: Apply Gaussian and Butterworth lowpass filters to remove the noise in the images. However, the application of filters blurs the edges in the image. Additionally, the median filter was utilized to evaluate the noise. Contrast enhancement: Histogram equalization and contrast stretching were applied to the test image to upgrade the grayscale of the image.

Step 2. Image segmentation

After the preprocessing, segmentation of the image was performed. Segmentation iteratively divides an image into its constituent regions. This was carried out until the region of interest (ROI) was isolated. Thresholding is utilized to extract an object from its background by assigning an intensity value that is known as the threshold value for every pixel such that each pixel is either classified as an object point or a background point. Here, Otsu's thresholding method is used as a thresholding process.

Step 3. Otsu's thresholding method

Otsu's thresholding method chooses the threshold to minimize the intraclass variance between the thresholded black and white pixels. In Otsu's method, we try to minimize the intraclass variance that is defined as the sum of variance of two classes using Eqs. (10.8) and (10.9):

$$\sigma^2 = w_1\sigma_1^2 + w_2\sigma_2^2, \tag{10.8}$$

where w_x is the class weight given by:

$$w_x = \sum_{a}^{b} p(i), \tag{10.9}$$

where parameter p, is the class occurrence probability. It is calculated as the division of the total number of pixels in the image by the number of pixels in the class. Otsu's technique coverts grayscale images to binary images by setting a threshold to divide the pixels into two classes. Usually, this method is used to divide a histogram into two classes that reduces the intraclass variance of the data contained within the class.

Step 4. K-mean clustering

K-mean clustering is the most popular method used for image segmentation and classification. In clustering, method data is divided into some groups. The K-mean clustering algorithm works in two parts. Initially, k centroids are calculated, and then a grouping of the data points is done using the minimum Euclidean distance. K-mean clustering is an iterative process, which means centroids are calculated again, and regrouping is performed for newly formed centroids. The following algorithm is to make clusters of the images of size $m \times n$.

Step 1. Assume k clusters and their respective centroids.

Step 2. Estimate distance between the centroid and each pixel of an image. Eq. (10.10) is estimated between the two data points using the Euclidean distance metric:

$$dis = \| f(x, y) - c_k \| . \qquad (10.10)$$

Step 3. Reform the clusters for each data point based on the minimum-distance value to the centroid achieved from the above equation.

Step 4. Find the change in location of centroid using Eq. (10.11):

$$c_k = \frac{1}{k} \sum_{y \in c_k} \sum_{x \in c_k} f(x, y). \qquad (10.11)$$

Step 5. Repeat from step 2 until the threshold value is within limits.

Step 6. Form image from the clusters obtained.

The K-mean clustering algorithm is straightforward and easy to implement as an unsupervised learning algorithm. It suffers from variation in the result based on the initial choice of the centroid. Also, computational complexity increases as the number of data points, or the number of clusters formed, increases. Fuzzy classification is applied to aggregate the components into a fuzzy set (elements that have a level of enrolment) whose participation capacity was characterized by the truth value of the fuzzy propositional function. Binary images differentiate the foreground (brain) from the background.

Step 5. Feature vector

A fractal feature is estimated on the binary image obtained after segmentation and classification. Fractal output and the segmented image are used to detect the result.

The experiment results are discussed in the next section.

10.5 Results and experiment

The experiment is conducted using a digital image as a two-dimensional function or array $f(x, y)$. Different measures like mean, standard deviation, entropy, and smoothness are used to analyze the image for the various problem domains. The MRI scan of healthy brain and an embedded tumor can be downloaded from http://www.med.harvard.edu/aanlib. In Table 10.2 fractal dimension, box size, and the number of boxes are used to characterize and analyze the image. The size of the box is varied from 1×1 to 128×128. The number of boxes required to cover is represented using n. A comparison of a fractal image called an Apollonian gasket is conducted with an MRI scan of the brain. A box-counting method is used for the comparison of two images. The aim of comparison is to show the difference between the grayscale image (MRI) and the fractal image with respect to the fractal dimension of each. A box-counting method is used to estimate the FD of both the images.

TABLE 10.2 Comparison of fractal dimension vs. box size and some boxes.

Fractal image	FD	1.6727	1.5219	1.5607	1.7118	1.8087	1.8817	1.7711	1.8314	1.5025	1.58 2
	r	1	2	4	8	16	32	64	128	256	512 1024
	n	44 000	27 466	11 786	4265	1386	421	121	37	12	4 1
MRI image	FD	1.8523	1.8644	1.8231	1.6415	1.0588	1.5849	1	1		
	r	1	2	4	8	16	32	64	128	256	
	n	3629	1005	276	78	25	12	4	2	1	

Fig. 10.1 is a comparative analysis of the fractal image and MRI scan of the brain by fractal dimension and box size. According to the plot (b) for a large box size, a lower number of boxes is required to cover the fractal curve. Plots (c) and (d) are a comparison between the box and fractal dimension for fractal curve and MRI scan. It is noted that the method does not perform well for the images with a higher dimension.

An alternative to the box-counting method is a differential box-counting method. The objective is to address the issues presented by the BC method. Fig. 10.2 compares the fractal dimension estimation using the box-counting method and differential box-counting method for an MRI scan. The image is sliced using two threshold decomposition algorithms, as discussed in Section 10.3.

The outcome of the differential box-counting methods is better for varying threshold values. Based on the results of Fig. 10.3 the following experiment is conducted.

The result of the proposed algorithm's step 3 is showcased in Table 10.3. The comparative analysis of threshold set T using a multilevel Otsu algorithm for MRI image is summarized in Table 10.3. The threshold is varied from 2 to 8 for the MRI image to determine the optimum value for the classification. The algorithm aims to find an optimum threshold that minimizes interclass variance. The range of threshold returns varied from 0.0039 to 0.5117.

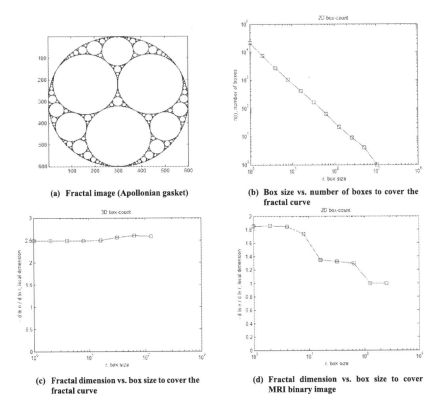

(a) Fractal image (Apollonian gasket)

(b) Box size vs. number of boxes to cover the fractal curve

(c) Fractal dimension vs. box size to cover the fractal curve

(d) Fractal dimension vs. box size to cover MRI binary image

FIGURE 10.1 Comparative analysis of fractal image and MRI scan of a brain.

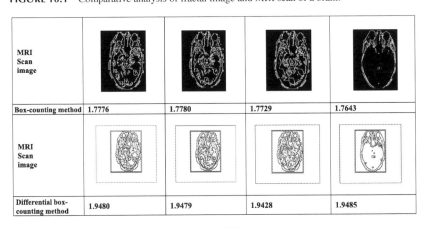

MRI Scan image				
Box-counting method	1.7776	1.7780	1.7729	1.7643
MRI Scan image				
Differential box-counting method	1.9480	1.9479	1.9428	1.9485

FIGURE 10.2 Estimation of fractal dimension of MRI scans.

The number of thresholds and computation cost presents a tradeoff. The experiment is conducted using a set of six threshold values to binaries the MRI image. Fig. 10.4 is a visual display of binary decomposition using a multilevel

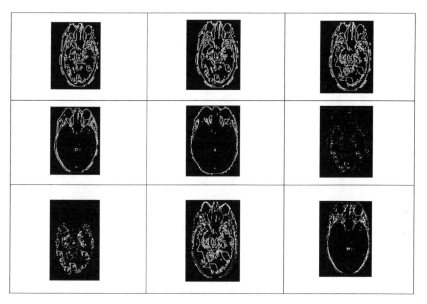

FIGURE 10.3 Border images of MRI scan of brain using threshold binary decomposition.

TABLE 10.3 Comparative analysis of threshold set T.

Number of thresholds	2	4	6	8
Set of the threshold for MRI Scan of Brain	0.0039, 0.5039	0 0.0039 0.5039 0.5078	0 0.0039 0.0078 0.5039 0 0.5078	0 0 0.0039 0.0078 0.5039 0 0.5078 0.5117

Otsu algorithm for several thresholds, 6. The next row shows the result of the fractally transformed image using the box-counting algorithm.

A segmentation of binary images is conducted using the SFTA algorithm. The segmentation using SFTA returns a feature vector of dimension one by six multiplied by the number of the threshold. For example, for different thresholds equal to 6 feature vectors is of dimension 1×36. The following metric is a feature vector extracted using the SFTA algorithm. The vector is formed using a fractal dimension, mean value and pixel count of each binary image. Fig. 10.4 shows the feature vector.

Feature-vector extraction is used further to extract regions of interest. It is worth noting that the fractal dimension along with the SFTA algorithm gives better accuracy in extracting ROI. An example of extraction is shown in Fig. 10.5.

$$\begin{bmatrix} NaN & NaN & 0 & 0.0015 & 0.1256 & 4.652 & 0.0015 & 0.1291 & 4.7250 \\ 0.0015 & 0.1323 & 4.6710 & 0.0014 & 0.1827 & 2.3710 & 0.0013 & 0.2128 & 1.7470 \\ 0.0011 & 0.2550 & 0.4890 & 0.0011 & 0.2550 & 0.4890 & 0.0015 & 0.0750 & 3.7730 \\ 0.0011 & 0.0804 & 0.5440 & 0.0013 & 0.1028 & 1.4500 & 0.0015 & 0.1149 & 3.9080 \\ 0.0013 & 0.1655 & 1.8290 & 0.0014 & 0.2124 & 2.1520 & NaN & NaN & 0 \end{bmatrix}$$

FIGURE 10.4 SFTA-based feature vector.

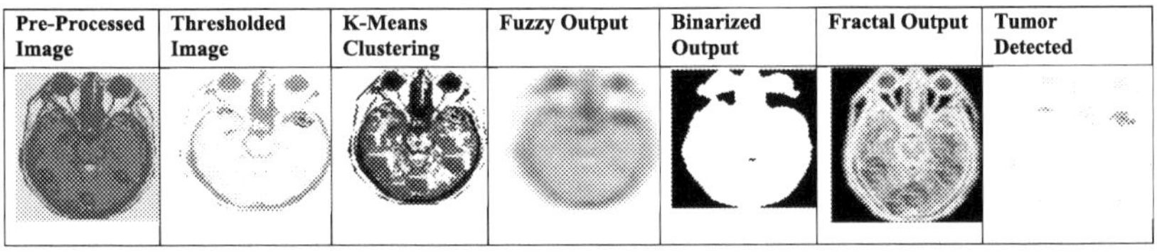

Pre-Processed Image	Thresholded Image	K-Means Clustering	Fuzzy Output	Binarized Output	Fractal Output	Tumor Detected

FIGURE 10.5 A visual display of the result of tumor detection using an algorithm proposed in Section 10.4.

The study of the brain-tumor detection model is developed using Python with a Flask framework. The image-classification models developed using TensorFlow and different algorithms and their corresponding outcomes are presented in Table 10.4. Also, the following Libraries are used to conduct the experiment Numpy (v-1.18.4), Pandas (v-1.0.3), Flask (v-1.1.2), Sci-Kit Learn (v-0.23.0), Tensorflow(v-2.2), and Keras. The analysis used the TensorFlow library as the platform on which several models were trained. The selection of the library is due to its simplicity and the amount of flexibility it offers. Also, the study utilized the "Keras" library in TensorFlow for its Deep-Learning Models. In the first section of study, we built the CNN-Sequential Model regarding the Dataset2, where the training and testing precision of this model were 60 and 80%, respectively. This model did not give the best result, so we had to find and work on the bigger dataset. Dataset2 was built on several Convolutional

TABLE 10.4 Cumulative results for all the models that have been trained for brain tumors.

	Cumulative Results for Brain Tumor						
Model / Metrics	Data Source	Training Accuracy	Training Loss	Testing Accuracy	Testing Loss	Epochs	Training Time (Seconds)
CNN - Sequential	Data 2	0.61	0.85	0.81	0.52	22.00	42.83
InceptionV3	Data 1	0.97	0.07	0.93	0.26	12.00	174.29
VGG-16	Data 1	0.94	0.15	0.91	0.25	15.00	277.39
RESNET-50	Data 1	0.96	0.07	0.93	0.24	11.00	178.30
Xception	Data 1	0.96	0.08	0.92	0.31	8.00	280.35
VGG-19	Data 1	0.89	0.28	0.88	0.30	15.00	633.33
EfficientNetB0	Data 1	0.96	0.10	0.91	0.23	15.00	291.93

Neural-Network models such as InceptionV3, VGG-16, RESNET-50, Xception, VGG-19, and Efficient NetB0.

Table 10.4 exhibits the results of different models. It is observed from the test result that the Inception V3 and Xception model have nearly same training accuracy of 96%, training loss of 6%, and testing accuracy of 92% with the testing loss of 26%.

Data 1 4800 Images of all kinds of tumors (2D)
Data 2 2000 MRI Scans of Brain Tumors

Figs. 10.6 and 10.7 are comparative analysis of accuracy and loss function. It is observed from Figs. 10.6 and 10.7 that the second-best performance was obtained using the model of RESNET-50. The training accuracy of 96%, training loss of 7%, and testing accuracy of 93% are noted using RESNET-50. Fig. 10.7 show the highest loss function is associated with CNN-sequential model.

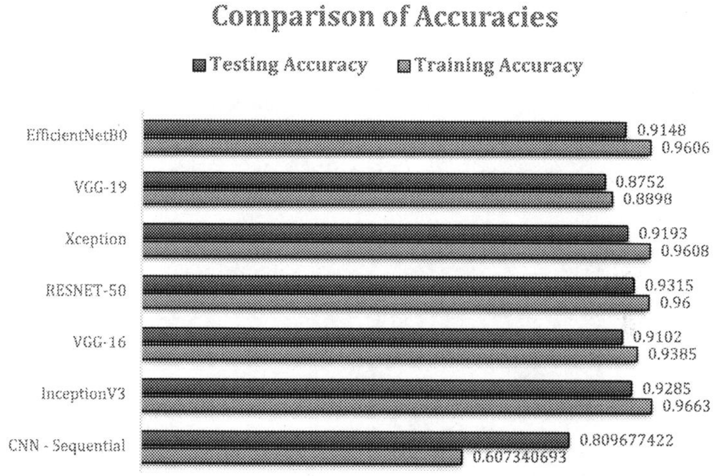

Comparison of Accuracies

■ Testing Accuracy   ■ Training Accuracy

Model	Testing Accuracy	Training Accuracy
EfficientNetB0	0.9148	0.9606
VGG-19	0.8752	0.8898
Xception	0.9193	0.9608
RESNET-50	0.9315	0.96
VGG-16	0.9102	0.9385
InceptionV3	0.9285	0.9663
CNN - Sequential	0.809677422	0.607340693

FIGURE 10.6 Comparison of accuracies of models trained for brain tumors.

Figs. 10.8, 10.9, and 10.10 are image-classification model screenshots depicting the process's intermediate steps.

Fig. 10.9 shows the uploading of MRI scan and corresponding probability of the presence of a pituitary brain tumor in the image.

Fig. 10.10 shows the final result of MRI scan along with the probability of the presence of a tumor.

Out of these, the third best performing model was the Xception model with training accuracy of 96%, training loss of 8%, and testing accuracy of 91%.

10.6 Conclusions

In this chapter, the fractal analysis of the MRI image was carried out. It was observed that segmentation algorithms and classification methods combined with

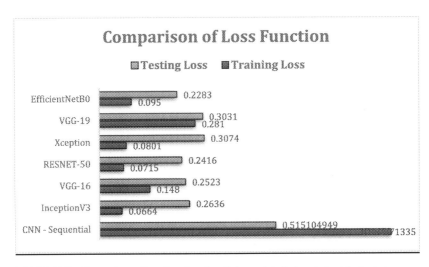

FIGURE 10.7 Comparison of loss functions for all the models that are trained for brain tumors.

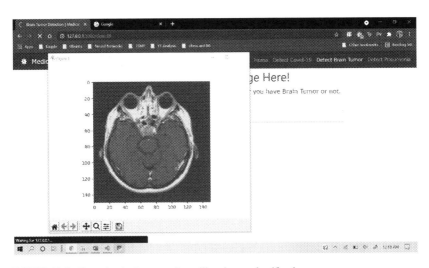

FIGURE 10.8 Detecting brain tumor that utilizes image classification.

morphological operators extract the ROI. Although the threshold approach produced the results faster it was computationally expensive. Some operators like open, spur, dilate, and close proved helpful in extracting the brain tumor from the MRI brain images. It is observed that color images should be converted to the grayscale image before segmentation and feature extraction. The fractal dimension using BC and the differential box-counting method were used to observe the structure of the MRI scan. However, there are other methods available to estimate the fractal dimension of natural images with a higher dimension. This

FIGURE 10.9 Upload of MRI Scan on the application.

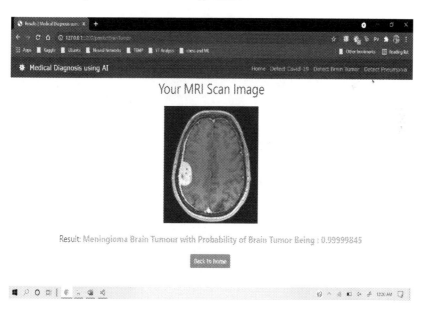

FIGURE 10.10 Result of the MRI Scan.

increases the scope to explore the various methods of fractal-dimension methods and corresponding application fields. Usually, the doctors spend around 14–15 min to analyze the CT scan and with the help of machine learning we can save this time and can predict whether a patient is infected or not in a few seconds.

The study for Brain Tumor was conducted by making use of the MRI scan dataset publicly available on the internet. First, the CNN-Sequential model built on the Dataset2. The training and testing accuracy of the model were reported 60 and 80%, respectively. As the existing model did not give a satisfactory result this led to work on the bigger dataset. Dataset2 was built on several convolutional neural-network models such as InceptionV3, VGG-16, RESNET-50, Xception VGG-19, and Efficient NetB0. The results of different models out of which Inception V3 and Xception model were very close, with Inception giving a training accuracy of 96%, training loss of 6%, and testing accuracy of 92% with the testing loss of 26%. The second-best performing model was the RESNET-50 with a training accuracy of 96%, training loss of 7%, and testing accuracy of 93%. Out of these, the third best performing model was the Xception model with a training accuracy of 96%, training loss of 8%, and testing accuracy of 91%.

References

[1] J. Qin, W. Pan, X. Xiang, Y. Tan, G. Hou, A biological image classification method based on improved CNN, Ecol. Inform. 58 (Jul. 2020) 101093, https://doi.org/10.1016/J.ECOINF.2020. 101093.

[2] J. Wang, H. Zhang, P. Han, C. Liu, Y. Xu, Pixel re-representations for better classification of images, Pattern Recognit. Lett. (2020), https://doi.org/10.1016/j.patrec.2020.04.027.

[3] P. Shanmugavadivu, V. Sivakumar, Fractal dimension based texture analysis of digital images, Proc. Eng. 38 (2012) 2981–2986, https://doi.org/10.1016/j.proeng.2012.06.348.

[4] J.M. Keller, S. Chen, R.M. Crownover, Texture description and segmentation through fractal geometry, Comput. Vis. Graph. Image Process. 45 (2) (1989) 150–166, https://doi.org/10. 1016/0734-189X(89)90130-8.

[5] W.Y. Hsu, C.C. Lin, M.S. Ju, Y.N. Sun, Wavelet-based fractal features with active segment selection: application to single-trial EEG data, J. Neurosci. Methods 163 (1) (2007) 145–160, https://doi.org/10.1016/j.jneumeth.2007.02.004.

[6] A. Kikuchi, N. Unno, T. Horikoshi, T. Shimizu, S. Kozuma, Y. Taketani, Changes in fractal features of fetal heart rate during pregnancy, Early Hum. Dev. 81 (8) (2005) 655–661, https:// doi.org/10.1016/j.earlhumdev.2005.01.009.

[7] G. Taraschi, J.B. Florindo, Computing fractal descriptors of texture images using sliding boxes: an application to the identification of Brazilian plant species, Phys. A, Stat. Mech. Appl. 545 (2020), https://doi.org/10.1016/j.physa.2019.123651.

[8] U. Gonzales-Barron, F. Butler, Fractal texture analysis of bread crumb digital images, Eur. Food Res. Technol. 226 (4) (2008) 721–729, https://doi.org/10.1007/s00217-007-0582-3.

[9] M. Navid, S.S.F. Hamidpour, F. Khajeh-Khalili, M. Alidoosti, A novel method to infrared thermal images vessel extraction based on fractal dimension, Infrared Phys. Technol. 107 (2020), https://doi.org/10.1016/j.infrared.2020.103297.

[10] N.A. Valous, F. Mendoza, D.W. Sun, P. Allen, Texture appearance characterization of presliced pork ham images using fractal metrics: Fourier analysis dimension and lacunarity, Food Res. Int. 42 (3) (2009) 353–362, https://doi.org/10.1016/j.foodres.2008.12.012.

[11] R. Lopes, P. Dubois, I. Bhouri, M.H. Bedoui, S. Maouche, N. Betrouni, Local fractal and multifractal features for volumic texture characterization, Pattern Recognit. 44 (8) (2011) 1690–1697, https://doi.org/10.1016/j.patcog.2011.02.017.

[12] M.A. Mohammed, B. Al-Khateeb, A.N. Rashid, D.A. Ibrahim, M.K. Abd Ghani, S.A. Mostafa, Neural network and multi-fractal dimension features for breast cancer classification from ultrasound images, Comput. Electr. Eng. 70 (2018) 871–882, https://doi.org/10.1016/j. compeleceng.2018.01.033.

[13] Y. Karaca, M. Moonis, D. Baleanu, Fractal and multifractional-based predictive optimization model for stroke subtypes' classification, Chaos Solitons Fractals 136 (2020), https://doi.org/10.1016/j.chaos.2020.109820.

[14] D. Oprić, et al., Fractal analysis tools for early assessment of liver inflammation induced by chronic consumption of linseed, palm and sunflower oils, Biomed. Signal Process. Control 61 (2020), https://doi.org/10.1016/j.bspc.2020.101959.

[15] K.K. Al-Nassrawy, D. Al-Shammary, A.K. Idrees, High performance fractal compression for EEG health network traffic, Proc. Comput. Sci. 167 (2020) 1240–1249, https://doi.org/10.1016/j.procs.2020.03.439.

[16] M.J. Cook, S.L. Free, M.R.A. Manford, D.R. Fish, S.D. Shorvon, J.M. Stevens, Fractal description of cerebral cortical patterns in frontal lobe epilepsy, Eur. Neurol. 35 (6) (1995) 327–335, https://doi.org/10.1159/000117155.

[17] E. Kalmanti, T.G. Maris, Fractal dimension as an index of brain cortical changes throughout life, In Vivo (Brooklyn) 21 (4) (2007) 641–646.

[18] T.G. Smith, K. Brauer, A. Reichenbach, Quantitative phylogenetic constancy of cerebellar Purkinje cell morphological complexity, J. Comp. Neurol. 331 (3) (1993) 402–406, https://doi.org/10.1002/cne.903310309.

[19] K.M. Iftekharuddin, W. Jia, R. Marsh, Fractal analysis of tumor in brain MR images, Mach. Vis. Appl. 13 (5–6) (2003) 352–362, https://doi.org/10.1007/s00138-002-0087-9.

[20] D.A. Russell, J.D. Hanson, E. Ott, Dimension of strange attractors, Phys. Rev. Lett. 45 (14) (1980) 1175–1178, https://doi.org/10.1103/PhysRevLett.45.1175.

[21] M.F. Barnsley, Fractals everywhere, 2000.

[22] F. Normant, C. Tricot, Method for evaluating the fractal dimension of curves using convex hulls, Phys. Rev. A 43 (12) (1991) 6518–6525, https://doi.org/10.1103/PhysRevA.43.6518.

[23] B.B.B. Chaudhuri, N. Sarkar, Texture segmentation using fractal dimension, IEEE Trans. Pattern Anal. Mach. Intell. 17 (1) (1995) 72–77, https://doi.org/10.1109/34.368149.

[24] A.P. Pentland, Fractal-based description of natural scenes, IEEE Trans. Pattern Anal. Mach. Intell. PAMI-6 (6) (Nov. 1984) 661–674, https://doi.org/10.1109/TPAMI.1984.4767591.

[25] M.F. Goodchild, Fractals and the accuracy of geographical measures, J. Int. Assoc. Math. Geol. 12 (2) (1980) 85–98, https://doi.org/10.1007/BF01035241.

[26] A.F. Costa, G. Humpire-Mamani, A.J.M. Traina, An efficient algorithm for fractal analysis of textures, in: 2012 25th SIBGRAPI Conference on Graphics, Patterns and Images, IEEE, 2012, August, pp. 39–46.

[27] R. Hashemzehi, S.J.S. Mahdavi, M. Kheirabadi, S.R. Kamel, Detection of brain tumors from MRI images base on deep learning using hybrid model CNN and NADE, Biocybern. Biomed. Eng. (Jun. 2020), https://doi.org/10.1016/J.BBE.2020.06.001.

Chapter 11

Fractal dimension analysis using hybrid RDBC and IDBC for gray scale images

Surbhi Vijh[a], Sumit Kumar[b], and Mukesh Saraswat[c]

[a]School of Engineering and Technology, Sharda University, Greater Noida, India, [b]ASET, Amity University, Noida, Uttar Pradesh, India, [c]Jaypee Institute of Information Technology, Noida, Uttar Pradesh, India

11.1   Introduction

Nature often displays patterns called "fractals," which tend to repeat indefinitely over adequate time. Fractals can be seen everywhere in nature, ranging from mountains to clouds, rivers and their tributaries to snowflakes, and the human nervous system to strawberries [1]. Fractals are used in a variety of domains, including geology, computational geometry, computer vision, computer animation, and fractal antennas. Sierpinski Triangle, Cantor Set, Mandelbrot Set, and other synthetic fractals are some examples.

Fractal images are self-similar and independent of scale; therefore, when an image is zoomed, the structure of the magnified area stays the same and holds the comparable property of the original image. This is because it is invariant under bi-Lipschitz transformation, which signifies that the fractal dimension is immune to geometrical deformations. Mandelbrot proposed fractal geometry, characterizing the form and look of an image to scale size [2]. A collection of dots or pixels in the proper order with regular or uneven pixel intensities could be used to depict an image. Peitgen et al. [3] show that an image may be divided into several boxes, and as and when all the boxes are put in the right order, the ideal image results. According to Harrington et al. [4], when a picture is divided or zoomed in, the scaling behavior varies exponentially as a function of the dimension. The quantity of partitions for a 1D line is inversely related to the power of 1. Similarly, if a 2D image or plane is divided into different sections, the number of divisions is equal to the square of the length, while for a 3D object, the number of divisions is equal to the cube of the length. Furthermore, Marusina et al. [5] demonstrated that there is a sufficient level of complexity in the distribution of small objects on medical images that are randomly arranged or have fuzzy contours. These things could include, for example, the development of

Copyright © 2024 Elsevier Inc. All rights are reserved, including those for text and data mining, AI training, and similar technologies.

tiny foci. Mohammedzadeh et al. [6], proposed that using fractal analysis to detect the existence of tiny focal lesions in medical imaging can be crucial.

Based on the box-counting technique, a study conducted by Sarkar et al. [7] compared four different box-counting approaches proposed by Pentland et al. [8], Peleg et al. [9], Gagnepain et al. [10], and Keller et al. [11]. Further, another approach is shown referred to as, the differential box-counting approach, addressing the shortcomings of the above algorithms. While Sarkar et al. [7] concluded that a satisfactory result was obtained from the approach suggested by Keller et al. [11], the investigations presented the scale of the superiority of the differential box-counting approach. Nayak et al. [12] demonstrated that triangle box partition is used to partition a square grid into asymmetric triangular boxes in order to improve the efficiency of the approach by reducing fit error and concurrently supplying a more precise box count. By using a practical concept concerning the upper and lower bounds for improved FD evaluation, Jin et al. [13] established the Relative DBC technique. Another revised approach of the DBC was given analogously by Biswas et al. [14] by utilizing a parallel algorithm technique for an effective estimation. Moreover, changing the DBC by shifting the box in the spatial plane was another method that Chen et al. [15] reported as being similar to RDBC. Li et al. [16] introduced Improved Box Counting (IBC), which differs from traditional approaches in three key areas: a suitable choice of box height, an accurate estimation of the box number, and the intensity of surface partition. Then, Clarke et al. [17] suggested a similar approach to determine the fractal dimensions of an image, called the Triangular prism surface area method. The Variation technique was introduced by Dubuc et al. [18] using the idea of an e-neighborhood. To estimate FD, Voss et al. [19] introduced probability into the fractal theory and developed a method that Keller et al. [11] later refined. Fuzzy logic was employed by Castillo et al. [20] to identify FD. Furthermore, Graph edges were employed by Jalan et al. [21] to estimate a network FD.

Fractal geometry has emerged as the most well liked and practical method for analyzing the roughness of digital images. The image intensity surface can be thought of as a fractal object, and the mathematical characteristics of these objects are assessed by taking into consideration their fractal dimension. The real world may have the most complicated and disturbed items that are impossible to quantify using Euclidean geometry. The idea of fractal dimension was created to assess the roughness of complicated objects, and it can be used in a variety of image analysis applications, including the measurement of shape and classification, segmentation and texture analysis, and other fields of graphics and image analysis. Self-resemblance is the primary characteristic of fractal dimensions, and it is followed by several other theories that apply to a large class of fractals in terms of both grayscale and color images.

This chapter discusses the fractal dimensional analysis of grayscale images and has been divided into sections as described below. Section 11.1 is referred to as the introduction while Section 11.2 discusses the elaborate literature review.

Section 11.3 shows the proposed approach for fractal dimension, Section 11.4 presents the experimental result analysis and finally, Section 11.5 elaborates on the conclusions.

11.2 Literature review

11.2.1 Preliminaries

11.2.1.1 Fractal dimensions (FD)

Elements that occur in nature are typically exceedingly complex and irregular, making it challenging for Euclidean geometry for an accurate explanation. Fractals are mathematical structures that can be used to represent complex and inconsistent objects. Furthermore, the fractal property is seen as a trustworthy indicator because it is an innate characteristic of such items [22]. The fractal-based analysis depends on the estimation of fractal dimensions. Fractal dimension, or FD, is a crucial metric for fractals that describes their irregularity and complexity. Additionally, there is a significant association between the FD of an image surface and how rough it appears to a human [23].

11.2.1.2 Estimation of fractal dimensions

Most fractal-based approaches directly or indirectly depict texture using fractal dimensions. The measurement of fractal dimension has been proposed using several different approaches. For instance, the covering-blanket strategy applies Mandelbrot's concept of the δ-blanket technique to surface area computation, while the box-counting approach forecasts FD over the total number of boxes encompassing an image for every box dimension, and the fractal Brownian motion approach predicts FD by using the Fourier power spectrum in the image intensity surface [2,10,8].

11.2.1.3 Differential box counting (DBC)

If the limit exists, the following generic description of fractal dimension for a bounded set A is provided as shown in Eq. (11.1):

$$D = \lim_{r \to 0} \frac{\log(N_r)}{\log(1/r)}, \tag{11.1}$$

where, N_r is the absolute minimum number of unique copies of A in scale r required to cover A [18]. An image of size $M \times M$ can be thought of as a surface $Z = f(x, y)$ in three-dimensional (3D) Euclidean space, where (x, y) are the coordinates that denote the position of a pixel in the picture plane, and Z denotes the corresponding gray level. The xy-plane is divided into blocks of non-overlapping size $s \times s$, in which s is an integer with $1 \leq s \leq M/2$. Each block has a column of boxes with sizes $s \times s \times s'$, where s' is the height of the box. If there are G total gray levels, then $s' = \lfloor s \times G/M \rfloor$ [7]. Boxes present in each block are numbered $1, 2, 3 \ldots$. Let the (i, j)th block's pixels with the lowest and highest gray levels fall inside the boxes marked with the numbers k and l, respectively. The number of boxes required to cover the (i, j)th block is therefore determined using Eq. (11.2):

$$N_r (i, j) = l - k + 1. \tag{11.2}$$

This equation supplies mathematical logic for a single block, by defining the ratio of the sides s over the dimension of the image M has been represented by r, a scalar quantity. Therefore to cover the entirety of the image A, the number of boxes needed can be represented by Eq. (11.3):

$$Y_r = \sum_{i,j} y_r (i, j). \tag{11.3}$$

The fractal dimension of the image can be calculated from the regression Eq. (11.4):

$$\log(Y_r) = C + D \log(1/r), \tag{11.4}$$

where D represents the box-counting fractal dimension, and C is an arbitrary constant for every image.

11.2.1.4 Relative differential box counting (RDBC)

By adopting the same minimum and maximum intensity points on the matrix and considering the scale limit such as the lower and upper limits of scale ranges for precise FD estimation of texture images, Jin et al. [13] introduced an enhanced version of DBC termed Relative DBC (RDBC). This method uses the metric property formula devised by Pentland et al. [8] shown in Eq. (11.5):

$$P = N_r \times r^D. \tag{11.5}$$

Here, P denotes the bounded set that is to be measured. For a certain value of r, the grids are partitioned into the dimensions $r \times r$. Post this assumption, the value of Y_r is calculated differently from in Eq. (11.3). Jin et al. propose the following preliminary for a single grid $g_r (i, j)$ for the calculation of Y_r:

$$y_r (i, j) = u_r (i, j) - b_r (i, j), \tag{11.6}$$

where $b_r (i, j)$ and $u_r (i, j)$ are the minimum and maximum gray levels. Using Eq. (11.6), the modified formula for the calculation of Y_r is:

$$N_r = \sum_{i,j} \left\lceil \frac{k \times y_r(i, j)}{r} \right\rceil. \tag{11.7}$$

Here, $\lfloor x \rfloor$ denotes the ceiling of x, which is the smallest integer value $\geq x$. k is the z-coordinate modification coefficient defined as $k = M/G$. The FD value is 2 when $Y_r = 0$.

11.2.1.5 Improved differential box counting (IDBC)

For deciding the FD of a gray scale image, Lai et al. [24] proposed improving the DBC approach. As a result, the difference between the boxes where the highest and lowest intensity value occur falls [16]. Therefore their proposed modified version of Eq. (11.8):

$$y_r = \begin{cases} \left\lfloor \frac{I_{max}-I_{min}+1}{s'} \right\rfloor, \; when \; I_{max} \neq I_{min} \\ \overline{\qquad y_r = 1, \; when \; I_{max} = I_{min} \qquad} \end{cases}. \qquad (11.8)$$

Here, I_{min} and I_{max} are the (i, j)th block's minimum and maximum gray levels, respectively. The formula for calculating Y_r remains the same as in Eq. (11.3).

11.2.2 Issues in the various box-counting methods

Numerous modified ways are proposed to address the difficulties with the differential box-counting approach, but the updated approaches lacked good generality. When the observed image's total gray scales are less than 256, the standard differential box-counting procedures may result in errors when computing the dimensions of the image [25]. Due to the same degree of roughness, when a new image is created from an existing one by adding or subtracting a fixed value, their fractal dimensions tend to match. When smooth pictures are used to estimate FD using DBC, RDBC, and IDBC, which are created by adopting the smallest difference between the minimum and maximum intensity points, the results are erroneous. To put it more precisely, these approaches cannot produce reliable results when the highest and minimum intensity points fall inside a grid of boxes in the z-direction. Also, another significant flaw in both DBC and IDBC is that they produce significantly more computing error, which was calculated by deducting the fractal dimension of linearly modified images from the fractal dimension of the original images from the Brodatz database [27]. Furthermore, High-dimensional space makes the situation worse [17,26]. Moreover, the box scales used to measure fractals provide another problem that has a significant impact on the precision of FD estimation [28,29]. The demonstration states that various box scale selections frequently result in varying values for the computed FDs.

11.3 Proposed methodology

The examination of RDBC and IDBC methods shows some certain limitations based on drawbacks or discussed issues. Therefore the adjustment coefficient is proposed in the method (RIDBC) referred to as a hybrid of RDBC and IDBC. As the average brightness magnitude on a grid determines all the intensity points along with the lowest and highest points within the grid, the box's respective average value will be high or nearly towards the upper bound. Thus to determine more precise results and achieve the accuracy level of the first box, the adjustment coefficient is subtracted to compute the roughness on the surface appropriately. This enables us to solve the mentioned challenges and problems. The proposed RIDBC is depicted in Fig. 11.1.

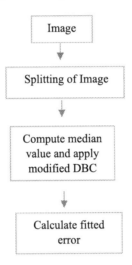

FIGURE 11.1 Flow process.

In this proposed method (RIDBC), an image of size $T * T$ is considered, which is further split into size $Y * Y$, where Y depicts the size of the box of integer format lying between the range of 2 to $T/2$. With (x, y) showing 2D spatial space and the third coordinate Z depicting gray level G, the image can be visualized in 3D temporal space. To compute the proposed approach, we determine the Median value and the lowest gray level for each subsequent box in order to assess the minimum possible count of boxes in Z coordinates utilizing DBC, RDBC, and IDBC. For more effective results, the subtraction of the median and minimum intensity value of each box is computed so that the adjustment coefficient can be placed in a minimum box for precise result analysis. However, if $Y' = Y * f/h$ is less than 1 then y_r should be larger than 1. Thus y_r should be referred to as a box whose median is equivalent to minimum intensity values. Then, $y_r(i, j)$ can be estimated as follows using Eqs. (11.9) and (11.10), respectively.

$$Q = \text{ceil}(\text{Median}(i, j) - \text{Min}(i, j)/Y') \tag{11.9}$$

$$y_r(i, j) = \begin{cases} Q(i, j) & if \ I_{median} - I_{min} \\ 1 & otherwise \end{cases} . \tag{11.10}$$

11.3.1 Algorithm steps

a. Splitting an image into the size of a box as $Y * Y$.

b. Consider the column of boxes with a scale of $Y * Y * Y'$ beginning the data point with the block's smallest amount of gray level. Evaluate $y_r(i, j)$ for each block by applying Eq. (11.10).

c. The estimated number of bins y_r encompassing the entire image surface is computed for various Y scales using Eq. (11.2).

d. Map the least square method for fitting the $\log(y_r)$ versus $\log(r)$.

11.4 Experimental results

This section discusses the exploratory findings and evaluates the effectiveness of the suggested approach. In this experimental investigation, three pre-specified selected methods (DBC, RDBC, and IDBC) are implemented and compared with the proposed method. The RIDBC method is experimented and validated on four different images, i.e., Brodatz images, smooth texture using a linear transformation, smooth synthetic textured images, and rough synthetic textured images.

(a) Texture-based real images

The experimental analysis is performed on a set of 10 real images collected from the Brodatz repository for analysis and visualization purposes, as depicted in Fig. 11.2. The corresponding fractal dimension and error fit for different selected methods (DBC, RDBC, and IDBC) and the proposed approach is shown in Table 11.1. Furthermore, the linear transformation is applied for smoothening the images using Eq. (11.11). The resultant transformed matrix evaluated from the original image $Q(i, j)$ is referred to as QTA and is computed as follows:

$$QTA(i, j) = \text{round}(\frac{Q(i, j) - \min(i, j)}{\max(i, j) - \min(i, j)} * W) + (G/2 - 2). \quad (11.11)$$

Here, minimum and maximum frequency points of the original image are depicted as $\min(i, j)$ and $\max(i, j)$, G shows the gray level, and W depicts the weighted coefficient varying from the range from 1 to 50.

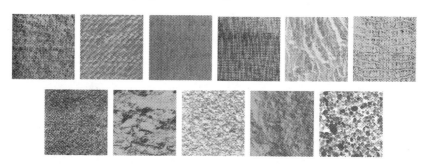

FIGURE 11.2 Images collected from Brodatz repository.

TABLE 11.1 Estimated FDs and fitted error.

Estimated FDs and fitted errors of the Brodatz texture images

Image	Fractal dimension				Error Fit			
	DBC	RDBC	IDBC	RIDBC	DBC	RDBC	IDBC	RIDBC
1	2.436221	2.487428	2.487428	2.487428	0.066992	0.066992	0.057993	0.0579928
2	2.431272	2.4139	2.547826	2.547826	0.066992	0.060993	0.061993	0.0599925
3	2.618122	2.612668	2.647715	2.581358	0.065992	0.064992	0.057994	0.0629922
4	2.605396	2.60479	2.591256	2.607921	0.055993	0.055994	0.040995	0.0549932
5	2.538332	2.568733	2.576308	2.605598	0.066992	0.062993	0.048995	0.0599925
6	2.310476	2.331484	2.335726	2.418142	0.066925	0.066925	0.057935	0.0579348
7	2.29775	2.33815	2.362188	2.296235	0.066925	0.060932	0.061931	0.0599326
8	2.451876	2.439756	2.450664	2.54419	0.065926	0.064928	0.057936	0.0629293
9	2.434302	2.433494	2.445614	2.529141	0.055938	0.055938	0.040955	0.0549384
10	2.421273	2.471874	2.582873	2.511062	0.066925	0.06293	0.048946	0.0599323

Furthermore, the error fit can be evaluated using Eq. (11.12). The computational errors are calculated by deducting the spatial pattern of the existing Brodatz repository visuals from the fractal dimension of the linear processed image:

$$EQ = \frac{1}{t}\sqrt{\sum_{i=1}^{t}\frac{rx_i + v - u_i}{1 + r^2}}. \quad (11.12)$$

The estimation of FD is performed on the basis of W, the observation shows that W is directly proportional to FD. As the value of W decreases, the intensity variation that occurs corresponding to FD also decreases. As previously stated, as the value of W increases, the acquired FD also increases, whereas methods such as DBC and RDBC fail to approximate accurate FD. In comparison to other methods, the proposed approach fractal dimension result analysis and error fit in Table 11.1 shows that it provides more consistent and precise outcomes with a varying range of W. The bar graph visualization of FD and error fit for Brodatz images are shown in Fig. 11.3.

(b) Transformed synthetic textured images

The exploration and investigation are performed on transformed images accumulated from the original image. The transformation is achieved by applying the linear function through Eq. (11.12), along with the weighted coefficient having $W = 10$. The transformed images are shown in Fig. 11.4. The visualization shows that most of the images contain roughness, they are not entirely smooth. It is stated that fractal dimension is required to be 2 in the case of smoother images and nearly 3 for maximum rougher images. The experimental result shows that both DBC and RDBC provide precise estimates due to an under-counting issue, however, IDBC generates FD as 2 for the images presented in Table 11.2, showing that it is not able to identify the pointed variation between the median

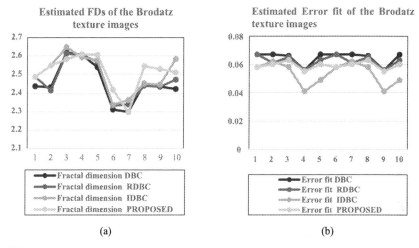

(a)   (b)

FIGURE 11.3   (a) Visualization of evaluated FDs and (b) error fitted.

FIGURE 11.4   Transformed synthetic textured images.

TABLE 11.2 Estimated FDs and fitted error.

Estimated FDs and fitted errors of the transformed images							
Image	Fractal dimension				Error fit		
	DBC	RDBC	IDBC	Proposed	DBC	RDBC	IDBC	Proposed
1	2.121168	2.094672	2	2.107218	0.065644	0.093812	0	0.03201
2	1.987776	1.948608	2	2.04561	0.065644	0.085828	0	0.01067
3	2.065296	2.11293	2	2.161686	0.064665	0.06986	0	0.03977
4	2.09202	2.194836	2	2.192898	0.054867	0.066866	0	0.04268
5	2.04663	2.109972	2	2.16852	0.065644	0.061602	0	0.04365
6	1.96605	1.982574	2	2.11089	0.065579	0.065448	0	0.0097
7	1.939836	1.92423	2	2.069682	0.065579	0.059587	0	0.0291
8	1.948506	1.935756	2	2.075598	0.0646	0.140718	0	0.02134
9	1.929534	1.955544	2	2.062338	0.054813	0.054704	0	0.02425
10	1.831308	1.898424	2	2.043672	0.065579	0.08982	0	0.0194

and minimum intensity of fall inside the grid of boxes in the z-direction in relation to shifting operation in the spatial dimension. Its error fit was recorded as zero for all cases and it cannot be taken into account due to the identical FD for all image data.

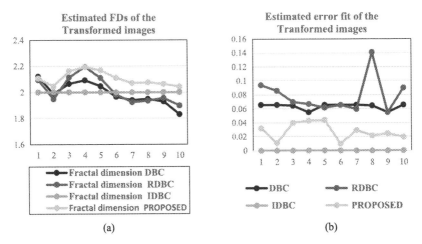

(a) (b)

FIGURE 11.5 (a) Visualization of evaluated FDs and (b) error fitted.

Thus it can be concluded that different methods such as DBC, RDBC, and IDBC do not provide appropriate results as compared to the proposed approach for fractal dimensional analysis, as shown in Fig. 11.5. The outcomes show that the proposed method obtains more accurate estimation and less error fit along with minimum computational error, as shown in Table 11.3. The computational variation is listed in the bar graph and is shown in Fig. 11.6.

TABLE 11.3 Obtained computational error.

Estimated computational error acquired from Brodatz images

Images	Computational error			
	DBC	RDBC	IDBC	RIDBC
1	0.36927	0.04851	0.44451	0.38412
2	0.45342	0.47421	0.51678	0.40194
3	0.56133	0.50985	0.61479	0.43164
4	0.52272	0.42273	0.55935	0.42768
5	0.50094	0.46926	0.5445	0.44847
6	0.48708	0.47223	0.30888	0.3861
7	0.34353	0.36729	0.33462	0.20097
8	0.51975	0.52371	0.42174	0.48411
9	0.49401	0.50589	0.41679	0.46431
10	0.50094	0.52272	0.55143	0.45936

Estimation of computational error

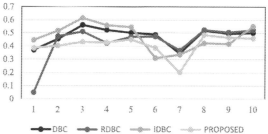

FIGURE 11.6   Visualization of computational error.

(c) Smooth texture image

In this experimental work, the 3 smooth texture-based images are considered, as depicted in Fig. 11.7. As stated, FD is 3 for smoother images and close to 3 for the most severely rough images. Therefore the result shows that the proposed method provides an FD value of 2 for a smoother image depicting the promising outcome in comparison to existing methods, as shown in Table 11.4.

FIGURE 11.7   Smooth texture images.

TABLE 11.4 Estimated FDs of smooth images.			
Estimated FDs of three Smooth images			
Image	Fractal dimension		
	DBC	RDBC	IDBC	RIDBC
1	2	2	2	2
2	2	2	2	2
3	2	2	2	2

(d) Synthetic textured images

In this setup, 10 synthetic images with a maximum gray scale of 255 are experimented and displayed in Fig. 11.8. Each image is formulated from real Brodatz images by increasing each intensity points in a manner such that at the lth, the highest gray level should not be greater than 255. Thus it can be concluded that intensity surfaces are only moved in the z direction up a few gray values from their original position. Therefore either there is an increment or decrement by a constant value, then FD remains the same as both have an equal level of roughness, as shown in Table 11.5.

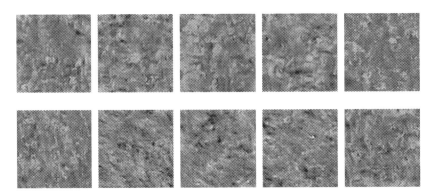

FIGURE 11.8 Synthetic textured images determined with different gray level shift.

TABLE 11.5 Estimated FDs and fitted error.

	Generated synthetic natural textured images			
Image	Fractal dimension			
	DBC	RDBC	IDBC	RIDBC
1	2.30236692	2.388925851	2.380925851	2.388925851
2	2.334993629	2.31830956	2.420904	2.44693209
3	2.514444369	2.509206347	2.542865486	2.597913622
4	2.502222318	2.501640316	2.488642262	2.546473284
5	2.437814053	2.467011173	2.474286203	2.502416319
6	2.21898115	2.239157234	2.24323125	2.322383577
7	2.2067591	2.24555926	2.268645355	2.25304094
8	2.35478171	2.343141662	2.353617706	2.443440076
9	2.337903641	2.337127638	2.348767686	2.428987016
10	2.325390589	2.37398779	2.480591229	2.495162394

(e) Test on synthetic rougher images

The analysis is performed on 5 synthetic rough images. In reference to the minimum and maximum intensity values of the image. In this for all the considered images, the minimum value varies from 5 to 25 and the maximum intensity value is referred to as 256, respectively. From the experimental analysis in Table 11.6, it can be seen that the proposed method can determine the greater roughness degree that tends to decrease with minimum intensity value as compared to other existing methods, as depicted in Fig. 11.9.

TABLE 11.6 Estimated FDs of synthetic Rougher images.

Computed FD of synthetic images presented

Image	Fractal dimension			
	DBC	RDBC	IDBC	RIDBC
1	2.76284	3	2.857111	2.866711
2	2.801992	3	2.905085	2.936319
3	2.733324	3	2.981439	2.997496
4	2.666782	3	2.986371	3.055768
5	2.925377	3	2.969143	3.0029
6	2.662777	3	2.691878	2.78686
7	2.648111	3	2.722374	2.703649
8	2.825738	3	2.824341	2.932128
9	2.805484	3	2.818521	2.914784
10	2.790469	3	2.976709	2.994195

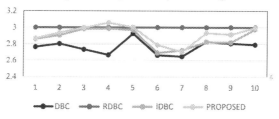

FIGURE 11.9 Visualization of FDs for synthetic Rougher images.

11.5 Conclusion

Fractal images are the magnitude and identical images referred to as fractal geometrical properties in which the structure of the magnified area stays the same and holds the comparable property of the original image for pattern recognition. In this chapter, an effective and enhanced differential box-counting technique is presented to achieve a precise and accurate evaluation of the fractal dimension of images. The proposed approach provides promising results in comparison to existing methods such as DBC, RDBC, and IDBC, respectively. The method is applied to different categories of images. The appropriate result outcomes are acquired by adding or removing constant values from each intensity value. Moreover, the method tends to be robust and efficiently obtains the fractal dimension with less computational error.

References

[1] A. Husain, J. Reddy, D. Bisht, M. Sajid, Fractal dimension of coastline of Australia, Scientific Reports 11 (1) (2021) 1–10.

[2] B.B. Mandelbrot, B.B. Mandelbrot, The Fractal Geometry of Nature, vol. 1, WH Freeman, New York, 1982.

[3] H.O. Peitgen, H. Jürgens, D. Saupe, M.J. Feigenbaum, Chaos and Fractals: New Frontiers of Science, vol. 7, Springer, New York, 1992.

[4] S. Harrington, Computer Graphics: A Programming Approach, 2nd edition, McGraw-Hill, New York, 1987, pp. 109–112.

[5] M.Y. Marusina, A.P. Volgareva, V.S. Sizikov, Noise suppression in the task of distinguishing the contours and segmentation of tomographic images, Journal of Optical Technology 82 (10) (2015) 673–677.

[6] Z. Mohammadzadeh, R. Safdari, M. Ghazisaeidi, S. Davoodi, Z. Azadmanjir, Advances in optimal detection of cancer by image processing; experience with lung and breast cancers, Asian Pacific Journal of Cancer Prevention 16 (14) (2015) 5613–5618.

[7] N. Sarkar, B.B. Chaudhuri, An efficient differential box-counting approach to compute fractal dimension of image, IEEE Transactions on Systems, Man and Cybernetics 24 (1) (1994) 115–120.

[8] A.P. Pentland, Fractal-based description of natural scenes, IEEE Transactions on Pattern Analysis and Machine Intelligence 6 (1984) 661–674.

[9] S. Peleg, J. Naor, R. Hartley, D. Avnir, Multiple resolution texture analysis and classification, IEEE Transactions on Pattern Analysis and Machine Intelligence 4 (1984) 518–523.

[10] J.J. Gagnepain, C. Roques-Carmes, Fractal approach to two-dimensional and three-dimensional surface roughness, Wear 109 (1–4) (1986) 119–126.

[11] J.M. Keller, S. Chen, R.M. Crownover, Texture description and segmentation through fractal geometry, Computer Vision, Graphics, and Image Processing 45 (2) (1989) 150–166.

[12] S.R. Nayak, J. Mishra, A modified triangle box-counting with precision in error fit, Journal of Information & Optimization Sciences 39 (1) (2018) 113–128.

[13] X.C. Jin, S.H. Ong, A practical method for estimating fractal dimension, Pattern Recognition Letters 16 (5) (1995) 457–464.

[14] M.K. Biswas, T. Ghose, S. Guha, P.K. Biswas, Fractal dimension estimation for texture images: a parallel approach, Pattern Recognition Letters 19 (3–4) (1998) 309–313.

[15] W.S. Chen, S.Y. Yuan, C.M. Hsieh, Two algorithms to estimate fractal dimension of gray-level images, Optical Engineering 42 (8) (2003) 2452–2464.

[16] J. Li, Q. Du, C. Sun, An improved box-counting method for image fractal dimension estimation, Pattern Recognition 42 (11) (2009) 2460–2469.

[17] K.C. Clarke, Computation of the fractal dimension of topographic surfaces using the triangular prism surface area method, Computers & Geosciences 12 (5) (1986) 713–722.

[18] B. Dubuc, C. Roques-Carmes, C. Tricot, S.W. Zucker, The variation method: a technique to estimate the fractal dimension of surfaces, in: Visual Communications and Image Processing II, vol. 845, SPIE, October 1987, pp. 241–248.

[19] R.F. Voss, Random fractals: characterization and measurement, in: Scaling Phenomena in Disordered Systems, Springer, Boston, MA, 1991, pp. 1–11.

[20] O. Castillo, P. Melin, A new method for fuzzy estimation of the fractal dimension and its applications to time series analysis and pattern recognition, in: PeachFuzz 2000. 19th International Conference of the North American Fuzzy Information Processing Society-NAFIPS (Cat. No. 00TH8500), IEEE, 2000, July, pp. 451–455.

[21] S. Jalan, A. Yadav, C. Sarkar, S. Boccaletti, Unveiling the multi-fractal structure of complex networks, Chaos, Solitons and Fractals 97 (2017) 11–14.

[22] H. Potlapalli, R.C. Luo, Fractal-based classification of natural textures, IEEE Transactions on Industrial Electronics 45 (1) (1998) 142–150.

[23] Y. Li, Fractal dimension estimation for color texture images, Journal of Mathematical Imaging and Vision 62 (1) (2020) 37–53.

[24] K. Lai, C. Li, T. He, L. Chen, K. Yu, W. Zhou, Study on an improved differential box-counting approach for gray-level variation of images, in: 2016 10th International Conference on Sensing Technology (ICST), IEEE, 2016, November, pp. 451–455.

[25] T. Ai, R. Zhang, H.W. Zhou, J.L. Pei, Box-counting methods to directly estimate the fractal dimension of a rock surface, Applied Surface Science 314 (2014) 610–621.

[26] Y.U. Liu, L. Chen, H. Wang, L. Jiang, Y. Zhang, J. Zhao, et al., An improved differential box-counting method to estimate fractal dimensions of gray-level images, Journal of Visual Communication and Image Representation 25 (5) (2014) 1102–1111.

[27] S.R. Nayak, J. Mishra, G. Palai, A modified approach to estimate fractal dimension of gray scale images, Optik 161 (2018) 136–145.

[28] S. Buczkowski, S. Kyriacos, F. Nekka, L. Cartilier, The modified box-counting method: analysis of some characteristic parameters, Pattern Recognition 31 (4) (1998) 411–418.

[29] K. Foroutan-pour, P. Dutilleul, D.L. Smith, Advances in the implementation of the box-counting method of fractal dimension estimation, Applied Mathematics and Computation 105 (2–3) (1999) 195–210.

Chapter 12

Preliminary analysis and survey of retinal disease diagnosis through identification and segmentation of bright and dark lesions

Jaskirat Kaur[a], Deepti Mittal[b], Ramanpreet Kaur[c], and Gagandeep[d]

[a]*Department of Electronics and Communication Engineering, Punjab Engineering College (Deemed to be University), Chandigarh, India,* [b]*Department of Electrical and Instrumentation Engineering, Thapar Institute of Engineering and Technology, Patiala, Punjab, India,* [c]*Department of Electronics and Communication Engineering, Chandigarh Engineering College, Mohali, Punjab, India,* [d]*Department of Computer Science Engineering, Chandigarh Engineering College, Mohali, Punjab, India*

12.1 Introduction

Retinal diseases pose significant health related issues having major incidence in developing countries. Retinal diseases mainly comprise of systemic and non-systemic diseases. Systemic retinal diseases are those that spread in the whole body and affect more than one organ at a time, whereas non-systemic retinal diseases are those that reside only in the retina. The present work is centered on the detailed study of various retinal lesions related to systemic retinal abnormalities. Systemic retinal abnormalities have adverse effect on various parts of body such as eyes, liver, kidneys, heart, the neural system, but the first to be affected is the retina, and hence the vision [1]. Additionally, prolonged cases of retinal diseases are the significant cause of severe vision loss or total blindness [2]. Vision loss due to systemic retinal diseases accounts for almost 15% of blindness cases worldwide [1]. According to the WHO, globally at least 2.2 billion have vision impairment, of whom at least 1 billion have a vision impairment that could have been prevented [2]. A survey conducted by the International Diabetes Federation, found that the increase in prevalence of retinal diseases affects half a billion only due to diabetes and is projected to affect a billion people by 2045 [4]. The rapid increase in the rate of obesity, inactive lifestyle, lack of physical movement are the main aspects of increased prevalence of systemic retinal diseases. Systemic retinal diseases are the seventh leading cause of death worldwide and

Copyright © 2024 Elsevier Inc. All rights reserved, including those for text and data mining, AI training, and similar technologies.

India tops the list by contributing 49% (72 million) of worlds retinal diseases cases and this figure is expected to almost double by 2025 [5].

Retinal diseases are chronic in nature and are mainly depicted by the incidence of one or more retinal lesions like dark and bright retinal lesions. Bright lesions are mainly categorized into exudates and cotton wool spots and dark lesions into micro-aneurysms and hemorrhages. Systemic retinal diseases are broadly categorized into non-proliferative retinal diseases and proliferative retinal diseases. Non-proliferative retinal diseases advance from a mild stage to moderate and then severe stage [6]. Mild non-proliferative retinal diseases are depicted by the presence of one retinal lesion, whereas moderate stage is depicted by the incidence of two to four retinal lesions present in not more than two quadrants of retinal fundus image. However, the severe stage of retinal disease is identified by the presence of multiple types of retinal lesions in more than two quadrants of a retinal fundus image. The proliferative stage of retinal diseases is an advanced stage depicted by the growth of new retinal blood vasculature in different regions of a retinal fundus image, and it may lead to total blindness. Also, no treatment is effective at this stage of retinal disease. Patients depicting non-proliferative stage of retinal diseases generally exhibit no characteristic symptoms and, in most cases, sight is unaffected until a proliferative stage of retinal disease is attained. In brief, vision loss due to systemic retinal diseases is stage dependent and that further is depicted by the presence of retinal lesions in a retinal fundus image. It is therefore important to detect the stage of retinal diseases. One of the key challenges in the process of diagnosing retinal diseases is the accurate detection and segmentation of retinal lesions. Thus the current study is focused on the preliminary study of the retinal lesions' classification on retinal fundus images for the diagnosis of retinal diseases.

The rest of the chapter is organized as follows. Sections 12.2 and 12.3 describe the retinal imaging modalities used for the identification and diagnosis of retinal diseases. Section 12.4 details the eye anatomy and the type of retinal diseases that exist. Section 12.5 describes the needs and challenges in the various steps of the retinal disease detection and diagnosis, followed by state-of-the art in Section 12.6 and conclusions in Section 12.7, respectively.

12.2 Retinal imaging modalities

Retinal imaging modalities such as retinal fundus imaging (FI), fundus fluorescein angiography (FFA), and optical coherence tomography (OCT) are the commonly used diagnostic techniques by expert ophthalmologists to detect non-proliferative retinal diseases and progressive retinal disease cases. There is no perfect retinal imaging modality for all ophthalmological applications and needs. In addition, each imaging modality is limited by its mechanism and type of image being produced. Medical imaging modalities that do not enter the skin are termed as non-invasive. The non-invasive medical imaging modalities like retinal fundus imaging and OCT play a significant role in the diagnosis and treatment of retinal abnormalities by identifying the type of retinal lesion, its size, shape, and number. The challenges with imaging modalities in extraction of clinically important information related to healthy structures and retinal lesions have received enormous attention. Fundus fluorescein angiography has been the gold standard to understand, confirm diagnosis, and treat retinal abnormalities [7]. However, being an invasive procedure, it is quite difficult to be accepted by the patients because of the associated side effects such as vomiting and the risk of developing different allergies [8]. Due to this non-invasive retinal imaging methods such as FI and OCT are widely used for retinal disease diagnosis during mass screening and to gather information about the particular abnormality.

The diagnostic clarification related to retinal abnormalities is a frequent problem in the clinical routine. For imaging the retina, most hospitals use fundus imaging, FFA, or OCT. Presently, there is no consensus concerning the optimal strategy for imaging the retina. All modalities have their advantages and limitations. The choice of which modality to use is identified based on the expert ophthalmologist, the availability of the equipment, the condition of the patient, and the experience of the ophthalmologist. To ensure the best treatment of the patient, preferably all lesions related to particular abnormality should be detected. Therefore choosing the modality with the most benefits and fewest limitations is essential. With the brief introduction of FFA, OCT, and fundus imaging modalities in the following paragraphs, some comparative statements are made to obtain a better understanding of the preferences among imaging modalities in order to detect non-proliferative and progressive diabetic retinopathy.

Fundus fluorescein angiography (FFA) is a procedure to image the retina using a fundus camera by injecting a fluorescent dye. Sodium fluorescein dye is injected into the systemic circulation following which the retina is lit with blue light at a wavelength of 490 nm. The final FFA image is obtained by snapping the fluorescent green light that is emitted by the dye. FFA produces a high contrast image of the retina and is mainly used to analyze pathologies related to choroid and blood circulation that reside in the retina. However, FFA, being an invasive technique, has a wide range of complications. The most common reactions are transient nausea, vomiting, pain, and redness in the eyes after the procedure.

Optical coherence tomography (OCT) is a procedure that uses coherent light to capture 2D and 3D retinal images from optical scattering media. It is based on the concept of interferometry to generate a cross-sectional view of the retina. Optical coherence tomography is useful in diagnosing retinal conditions related to macula. Lesions present in macula are easier to capture using OCT than in the other parts of retina. It is primarily helpful in identifying diabetic macular oedema, macular hole, age-related macular degeneration. OCT is non-invasive and easy to interpret, but OCT does not identify the changes in blood vessels.

Therefore this imaging modality is unable to document a disease with bleeding in the retina such as diabetic retinopathy.

Fundus imaging is a very popular and primarily used imaging modality. Almost all the ophthalmologists in the world have fundus cameras that provide high resolution color fundus photographs. Fundus imaging is done by using a digital fundus camera to capture the image of the true retinal fundus. Fundus cameras are non-invasive, easy to use, widely available, cost effective, and good at documenting the front view of the retina where all the retinal lesions and anatomical structures are clearly visible. The images produced by fundus photography are the 2D representation of 3D semi-transparent retinal tissues. They capture 30° to 50° views of the retina.

In summary, FFA, OCT, and fundus photography imaging modalities of retinal imaging have their own advantages and disadvantages. Fundus photography is an attractive medical imaging technique in general because of its unique set of virtues including (i) low cost, making it feasible for a patient from a low socio-economic status, (ii) versatility and widespread availability, making its clinical relevance high worldwide, (iii) initial imaging modality due to its relative simplicity in comparison to other imaging modalities, (iv) risk-free system, making it the first preference among other imaging modalities for screening of retinal abnormalities, and (v) fast and portable, making it convenient to use. In continuation, the fundus images are easy to study once the observer is trained. As a result, fundus imaging is an extremely useful means of diagnostic imaging and the most widely employed imaging method for screening purposes and detection of retinal lesions related to systemic retinal diseases. Therefore retinal fundus images are utilized in the present work to design computer-aided methods to diagnose non-proliferative retinal diseases and proliferative retinal diseases.

12.3 Fundus imaging

Fundus imaging is a procedure in which a 2D representation of the 3D retinal semi-transparent tissues of the back portion of the eye, i.e., the fundus, is projected onto the imaging plane. This projection is attained on the screen of a fundus machine by the image intensities that depict the expanse of light reflected from retina. Fundus photographs thus obtained record the details of the retina, that converts the optical visuals/images in front of our eyes into electrical signals for the brain to interpret. Fundus photography, a highly useful means of diagnostic imaging, has a wide array of clinical applications. The basis of its operation, as shown in Fig. 12.1, is the projection of light into the eye through the cornea followed by the reflection, processing, and display of image intensities reflected from retinal structures within the eye [3]. Ophthalmologists, with the aid of these retinal fundus images, take the follow-up, diagnose, and treat various retinal diseases such as diabetic retinopathy, age-related macular degeneration, hypertensive retinopathy, etc. In order to interpret a fundus image of the retina, it is essential to understand how a fundus imaging system works and how reflected light is converted into an image.

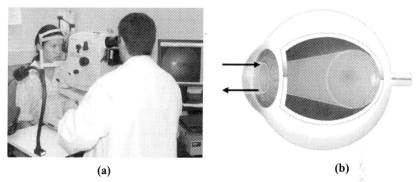

(a)   (b)

FIGURE 12.1   (a) Retinal imaging with fundus camera, (b) path of transmitted and reflected light from fundus camera [3].

12.3.1   Fundus image formation

The fundus cameras comprises a complex microscope connected to a flash enabled camera. The design of the fundus camera is based on the working of an indirect ophthalmoscope. The observation illumination produced from either the observing lamp or the automated flash is projected on a circular mirror using a set of filters. This set of mirrors replicates the illumination into a sequence of lenses that focus the illumination. A cover on the topmost lens forms the doughnut shaped illumination. The doughnut shaped illumination is replicated onto a circular mirror with a central aperture to form a ring shape before exiting the camera through the objective lens, and proceeds into the eye through the cornea. The resulting retinal image formed from the reflected light exits the cornea through the central, un-illuminated hole of the doughnut formed by the illumination system. As the paths of the projection of light into the retina and reception of light from the retina are not dependent, there are negligible reflections of the illuminating source that are captured in the image. The resulting image formed is directed towards the telescopic eye piece attached to the fundus machine. Afterwards, when a switch is pushed to capture an image, the mirror intersects the path of the lighting system and light from the bulb in allowed to enter the eye. Concurrently, a mirror is directed opposite the observing telescope, which then sends the light onto the acquisition medium, i.e., film or a digital CD. The produced image on the film is a straight, magnified interpretation of the fundus described by an angle of view termed as field of view (FOV). Generally, an angle of 30° to 50° of the retinal area (FOV) is used to create the retinal fundus image 2.5 times larger than the original retinal fundus image. Wide angle fundus cameras capture images between 45° and 145° that offer comparatively low magnification of retinal area and minimizes the image to half of the original size, whereas narrow angle fundus cameras capture images at FOV of 20° or less, providing five times magnification.

Fundus photography has three modes of examination:

(i) Color fundus photography: In this mode, the retina is illumined by white light and observed in full color. Generally, this mode is the primary mode to capture retinal fundus images during the first round of diagnosis.

(ii) Red-free fundus photography: The illuminating light is filtered to improve the contrast of retinal lesion structures with respect to the background. The filtering process removes red colors from the colored retinal fundus images.

(iii) Stereo fundus photography: Retinal fundus image intensities in this mode depict the amount of mirrored light from more than two angles of view to create a 3D picture. The 3D image generated by this process is used to obtain better surface characteristics than the above-mentioned photography modes.

12.4 Eye anatomy and retinal diseases

The anatomical view of a human eye is demonstrated in Fig. 12.2 [3]. The retina is a multilayered sensory tissue that lines the back of the eye, senses light and produces electrical impulses. Such electrical impulses transmit to the brain from the optic nerve that are subsequently converted into images through photoreceptors. Photoreceptors are composed of two types: rods and cones. Rods can sense variations in contrast even at low illumination but cannot sense any variation in color, whereas, cones are very sensitive to changes in color. Rods and cones reside mainly in the macula, the region that is accountable for day vision. The central region of macula is termed as fovea, where the human eye is able to differentiate visual details with high precision. All these photoreceptors are attached to the brain via a bundle of optic nerves. The circular region where the optic nerve exits the pupil is called the blind spot due to the absence of photoreceptors, also termed the optic disk. Retinal blood vasculature having a tree like structure supplies nutrients to the retina. Blood vessels have less reflectance as compared to other retinal structures; thus, they appear darker with respect to the background of fundus images. The retinal measures approximately 72% of a sphere about 22 mm in diameter, 0.56 mm thick near the optic disk, and thinnest at the center of fovea in adult humans.

The retinal image serves as a window to assess many systemic abnormalities and brain-related abnormalities. The structures that can be visualized on a retinal fundus image are optic disk, peripheral retina, blood vessels, and retinal lesions if present. Visually extractable features on retinal fundus images to detect diabetic retinopathy are based on the presence of type, size, shape, and number of lesions. Thus retinal fundus images are used to investigate and monitor the progression of abnormalities that affect the eye. Experienced ophthalmologists

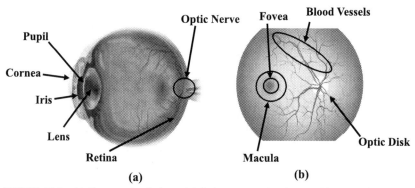

FIGURE 12.2 (a) Cross-sectional view of right human eye, (b) photographic view of the retina [3].

visualize these features in retinal fundus images to characterize various stages of retinal diseases and normal retinas.

12.4.1 Normal retina

On retinal fundus images, the appearance of a normal retina is homogeneous and clear providing a clean view of the fundus, as shown in Fig. 12.3 of the right eye. This image has no signs of any abnormality or pathology. There are no lesions, scars, or pigmentary changes in the retinal fundus. The normal retina has a uniform appearance along with the presence of main distinguishing structures such as blood vessels, optic disk, macula, and fovea. These main distinguishing structures are termed landmarks or anatomical structures. The network of blood vessels is the collection of dark red, finely elongated structures. Blood vessels merge into the highest light circular region known as the optic disk. The blood vessels in this fundus image are normal in course and caliber. Approximately two and half disk diameters to the right of the optic disk, a dark circular region, is the fovea, which is at the center of the area known as macula. These retinal structures like the optic disk and blood vessels, can provide valuable information about the health of the retina. Such retinal components, such as the optic disk and blood vessels, may provide useful details on retina health. This is why it is important to analyze and detect changes in their morphology. However, the prime indicator of the occurrence of a particular retinal abnormality or diseases is the appearance of lesions.

12.4.2 Retinal lesions associated with various retinal diseases

Retinal lesions are recognized visually as alterations of the normal appearance of the retinal fundus image. Lesions appear with various attributes, such as intensities, shape, color, size, etc., in retinal fundus images. They can be divided

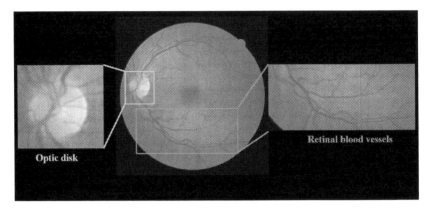

FIGURE 12.3 Appearance of a normal retina on a fundus image.

mainly into bright and dark (red) retinal lesions. Brief descriptions about these retinal lesions are given below:

Dark lesions

Dark lesions are mainly categorized into micro-aneurysms and haemorrhages.

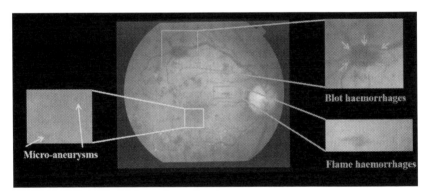

FIGURE 12.4 A fundus image having typical appearance of dark lesions.

Micro-aneurysms

Micro-aneurysms are the first sign of systemic retinal disease. They are dilated, aneurismal retinal vessels that appear as small red (mid gray in print version) dots detached from blood vessels in colored retinal fundus images, as shown in Fig. 12.4. These lesions may pump fluid and blood through the eye, resulting in vision-threatening lesions such as exudates and haemorrhages. These lesions are the smallest detectable retinal lesions and the local expansions of the weakened retinal capillaries. Micro-aneurysms are of variable size ranging from 10 to 100 μm. Some type of blood vessel-related abnormality or high blood pressure can lead to a retinal micro-aneurysm, but the most common cause is diabetes mellitus. Micro-aneurysms alone are unlikely to cause external symptoms and need no treatment. Most micro-aneurysms are reversible if preventive measures such as healthy lifestyle, management of diabetes, etc., are taken timely.

Haemorrhages

These are distinguished by a dark red color and are of many forms such as: "dot", "blot", and "flame" haemorrhages. Dot haemorrhages are tiny red objects and are usually referred to as to micro-aneurysms. Hemorrhages of blots and fires lead to blood loss in deeper layers of the retina. These manifest as large and dark retinal lesions with abnormal contours on the fundus images [9]. Fig. 12.4 shows an example image highlighting different types of haemorrhages inside the marked bound regions of rectangular shape. A retinal haemorrhage can be caused by diabetes, hypertension or blockages in the retinal vein. They are a more serious retinal disease because they are associated with the onset of chronic diabetic retinopathy. Some of the common symptoms of haemorrhages are (i) visual disturbances producing unclear vision, (ii) presence of blind spots in vision, (ii) pain in the eye.

The detection and differentiation of micro-aneurysms and haemorrhages can be complex for the expert ophthalmologist as they are characterized by their weak contrast with the background. Also, these lesions have similar attributes in terms of color and texture to anatomical structures like blood vessels.

Bright lesions

Bright lesions are mainly categorized as exudates and cotton wool spots.

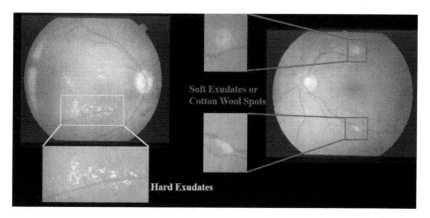

FIGURE 12.5 A fundus image having typical appearance of bright lesions.

Exudates

Retinal exudates are masses of fatty matter molded in the retina from lipids and protein leakage from micro-aneurysms [10,11]. The leakage of lipids and protein is due to increased pressure in the walls of micro-aneurysms that are thinner than normal capillaries. In retinal fundus images, these lesions appear as bright yellow (light gray in print version) crystalline granules having sharper definition, as shown in Fig. 12.5. Exudates can have varying shapes, sizes, and locations and are the hallmark of the progressive diabetic retinopathy. Their scale may range from a few pixels to as wide as an input image, appearing as a circle. Raising the exudate scale and positions means increasing the frequency degree of the abnormality. The presence of hard exudates results in various degrees of reduced vision, depending on the location of the exudates. Exudates may cause rapid vision impairment if not treated in time. In order to prevent the further development of exudates, retinal areas comprising exudates are exposed to a laser for photocoagulation by physicians.

Cotton wool spots

Cotton wool spots (or soft exudates) are small, swollen micro-infarcts that appear because of obstructed blood vessels resulting in the impaired blood supply to that area. Furthermore, the decreased blood flow injures the nerve fibers in that location resulting in blood circulation in local capillaries. They appear on fundus images as yellowish or white fluffy patches with blurred edges, as depicted in Fig. 12.5. Usually, their local spread is less than 1/3 the optic disk area in diameter and are generally found throughout the retina. Cotton wool spots alone do not induce vision problems but are strongly associated with disorders influencing the development off the eye, as in the case of systemic retinal diseases. They also typically appear with other retinal abnormalities that cause significant symptoms and have long-term implications. Cotton wool spots sometimes vanish on their own, although any scattered lack of vision can be irreversible.

The design of computer-aided retinal disease diagnostic methods is done by extracting diagnostic information related to systemic retinal diseases from retinal fundus images by collectively identifying the presence of dark lesions, such as micro-aneurysms and haemorrhages, bright lesions, such as exudates and cotton wool spots, and healthy structures, such as blood vessels and the optic disk. The computer-aided method is generally developed by using the following steps: (i) Retinal image enhancement, (ii) characterization of healthy structures and retinal lesions, and (iii) computer-aided classification and grading method. The following sections highlight the need of these steps in detail.

12.5 Current challenges and needs in computer aided retinal disease analysis

12.5.1 Need and challenges in computer aided retinal disease detection method

Retinal fundus imaging is one of the valuable and non-invasive tools for imaging the retina. However, due to its qualitative, subjective, and experience-based nature, fundus images can be affected by imaging conditions such as illumination and machine settings. The visualization criteria for distinguishing various retinal lesions are slightly confusing and extremely based upon the ophthalmologist's experience. This frequently results in ambiguity in the diagnostic procedure and reduces its objectivity and reproducibility. The visualization of retinal lesions is a tedious task with retinal fundus images because of their varying sizes, shapes, contrast with respect to the background, etc. Moreover, overlapping characteristics of anatomical structures and retinal lesions confuses the ophthalmologists when trying to differentiate them. In practice, an ophthalmologist analyzing a huge number of retinal fundus images can have visual tiredness, resulting in the likelihood of spurious responses. It is therefore important to develop a computer-aided system for the identification of retinal diseases that ophthalmologists may use as a non-invasive diagnostic tool to support their findings on the basis of the characteristic visual appearance of retinal lesions on fundus images. Retinal disease diagnosis with the assistance of an efficient and reliable computer-aided detection method may reduce the frequency of further examinations.

12.5.2 Need and challenges in retinal image enhancement

It is extremely necessary to specifically distinguish all the typical signs of an abnormality found in retinal fundus images to achieve good diagnostic precision. Although artifacts and blur arise during retinal fundus image capture, the interpretation and analysis of retinal fundus images is severely hampered. Artifacts and blur are the prime factors that degrade the contrast resolution and obstruct the meaningful information present in retinal fundus images. Such objects are added due to lighting variations that are often responsible for differences in the non-uniform strength. Additionally, the blur; camera-dependent variables reduce the contrast resolution more and give the picture a noticeable fuzziness. An experienced ophthalmologist's detection of retinal abnormality is dependent on the conception of different structures in a retinal fundus image. Subsequently, the blurred contours of different retinal structures are extremely unwanted when accurately detecting lesions. Therefore there is a need to develop a methodology to enhance the visual quality of retinal fundus images in order to enhance the representation of the concept of retinal structures, the identification of lesions present in the retinal fundus picture must be effective. Therefore enhancement methods are primarily designed with the intention to improve visualization of the details available in retinal fundus images in order to provide accurate detection of anatomical structures and retinal lesions.

12.5.3 Need and challenges in characterization of anatomical structures and lesions

Fundus images can be influenced by qualitative and subjective variations that may result in a vague investigation and detection of different retinal lesions.

Furthermore, the visual criteria for differentiating retinal lesions and anatomical structures are rather subjective and highly dependent upon the ophthalmologist's experience. This often affects the diagnostic procedure by limiting its objectivity. Additionally, the visualization of retinal lesions and anatomical structures with retinal fundus images is not an easy task because of their varying sizes and limited contrast with respect to the background. Therefore there is a need to characterize anatomical structures and retinal lesions quantitatively in order to develop a computer-aided diagnostic method to support ophthalmologists in their decision making.

12.5.3.1 Segmentation of retinal blood vasculature

Precise segmentation of blood vasculature, is the first step in extracting diagnostic information for initial identification of diabetic retinopathy. However, challenges are faced during blood vessel segmentation from various types of pathological retinal fundus images: (i) the occurrence of artifacts due to poor background lighting during image acquisition; (ii) fuzzy appearance due to inappropriate focus; (iii) large variety of thicknesses and tortuosity; (iv) overlying of blood vasculature and lesions due to same intensities; and (iv) possibility of lost evidence due to tiredness and due to the large amounts of data being analyzed. Therefore there is a need to develop an efficient blood vasculature segmentation method to precisely segment and then remove blood vessels from retinal fundus images. Therefore blood vasculature segmentation methods for the detection of retinal diseases are designed to eliminate efficiently the blood vasculature from retinal fundus images.

12.5.3.2 Detection of the optic disk

The accurate detection of the optic disk center and approximate removal of the optic disk region can help in reducing the spurious responses during diabetic retinopathy diagnosis and screening. Challenges faced in the automated detection of optic disk are: (i) the same features of the optic disk and bright lesions in a retinal fundus image in terms of color, strength, and contrast, (ii) partial acquisition of the optic disk, (iii) hidden by major blood vessels crossings over the optic disk boundary, and (iv) ambiguous contour of the optic disk. Therefore there is a need to precisely locate and segment the optic disk to avoid spurious responses during diagnosis of non-proliferative and progressive retinal diseases.

12.5.3.3 Segmentation of retinal lesions

Retinal lesions are primary visual indicators of various stages and types of retinal diseases. Therefore retinal lesion identification and segmentation is an important step in diagnosing various retinal diseases. However, the obstacles encountered in the identification and precise segmentation of retinal lesions are: (i) the existence of artifacts due to inappropriate lighting; (ii) blurry presence of lesions due to inappropriate concentration; (iii) ambiguity in definition of the class, extent, magnitude and precise contour of the lesions due to related geometry and diversity in structures; (iv) resemblance in the characteristics of healthy parts of retina comprising of the optic disk and blood vasculature to those of bright and dark lesions, respectively; and (v) likelihood of missing evidence due to human tiredness, while analyzing a huge number of retinal images. Therefore there is a need to reduce the significances of these restrictions in the precise computer-aided segmentation of retinal lesions. Thus for the detection of retinal diseases, a generalized lesion segmentation method is designed by introducing a novel adaptive image quantization and dynamic decision thresholding-based method. Additionally, for the detection of non-proliferative retinal diseases several dark and bright lesions are segmented regardless of related heterogeneity, bright and faint edges in different retinal fundus images.

12.5.4   Need and challenges in computer aided classification and grading method

Experienced ophthalmologists with retinal photographs envision numerous shape, size, and texture-based features to distinguish retinal lesions and then determine the frequency degree of a retinal disease. The grading of the level of non-proliferative retinal disease needs the comprehensive and quantitative authentication of all types of retinal lesions existing in the fundus image. Nevertheless, the skill and expertise of the ophthalmologist greatly depends on the manual labeling and segmentation, accuracy of valuation of lesions and relevant parameters. Therefore the confusion resides in: (i) reading correct contours owing to their varying forms and strengths leading to contradictions; and (ii) missing the probability of retinal lesions with a limited number of pixels. In fact, each patient's evaluation is wearisome manually, and is a repetitive operation. In fact, the high risk of tests and the lack of specialists discourage many people receiving proper and timely treatment. It is also important to develop a computer-aided tool for the diagnosis of severity rates to support ophthalmologists, which will reduce the costs related with specialist graders and remove the uncertainty related to manual evaluation. Furthermore, the detection and classification of various retinal lesions are useful not only for the detection for a particular disease but also for preparation for treatment. The specialists assess the exact location of laser-exposed lesions for diagnosis. Thus an efficient retinal lesion classification and non-proliferative retinal disease grading method is required for fast and accurate detection of non-proliferative retinal diseases.

12.6   State-of-the-art for retinal disease diagnosis

The studies reported so far, in the computer-aided detection and grading of systemic retinal diseases, focus mainly on either segmentation of anatomical structures or the segmentation of particular retinal lesions related to systemic retinal disease such as diabetic retinopathy [12–16]. Most of the methods use a particular type of bright or dark lesions or combinations of them to diagnose retinal disease [17–19]. In practice, expert ophthalmologists consider the information related to various combinations of lesions such as their shape, size, and location to draw conclusions. Hence, there is a need to develop a more practical, accurate, and precise computer-aided diagnostic method by blending information related to retinal lesions and anatomical structures present in retinal fundus images. Thus a solution is needed in terms of a computer aided diagnostic method by introducing objectivity in segmentation of retinal lesions and anatomical structures regardless of associated heterogeneity, bright and faint edges in different retinal fundus images. The computer-aided non-proliferative retinal disease detection method is desirable with an extensive set of geometrical-, color-, and texture-based features for mathematically interpreting the visualization details of ophthalmologists. Further, the classification methods designed to date reveal that there have been a limited number of trials to design a computer-aided method that can provide classification among five retinal lesion classes, i.e., No DR, EXU, CWS, MA, and HEM, followed by grading of a particular retinal disease. Therefore a significant computer-aided detection method is required to classify these retinal lesion classes and determine the severity level of non-proliferative retinal disease effectively. Retinal lesions with proper and clear boundaries are easily identified by the ophthalmologists. However, the overlapping retinal lesions with anatomical structures or among themselves may result in ambiguous diagnosis and confliction even for the expert ophthalmologists. Thus the techniques designed must include retinal images of varying attributes to provide a generalized computer-aided detection method for retinal disease diagnosis.

Further, the literature review also revealed that the neural network (NN) classifier had been used in various state-of-the-art methods [20–22] for the classification of retinal lesions, whereas the multi-step classifiers have never been used to develop computer aided detection methods for retinal lesions despite the fact that multi-step classifiers are favored in many other medical applications such as liver disease detection [23], brain tumor detection [24], breast cancer detection [25], retinal image database [26], etc. Also, in the past few years a number of deep learning strategies have been proposed for the detection and diagnosis of various retinal diseases using either fundus imaging or optical coherence tomography [27–29]. Therefore the computer aided detection and grading of retinal diseases methods are being designed using multistep classifiers.

The literature review also reveals that most of the computer-aided diagnostic methods have utilized open-source benchmark databases and the limitation in using a particular open-source benchmark database is that it remains specific to only one type of retinal lesion [30–35]. Therefore few researchers have used various benchmark databases available publicly to provide an efficient computer-aided diagnostic solution. Furthermore, images in open-source benchmark databases are of superior quality to the images encountered by ophthalmologists in their clinical practice. Therefore there is also a need to provide a practical solution that should work well with clinical retinal images. Thus it is a prerequisite to develop a composite database by combining publicly available retinal images datasets and clinically acquired images to provide a generalized computer-aided method for the diagnosis of retinal abnormality.

12.7 Conclusions

The present study was conducted to provide a general background to design a computer-aided retinal disease diagnostic method. Detailed understanding of various retinal lesions related to systemic retinal diseases and their characteristic visual appearance is also provided in this chapter. This preliminary analysis and survey is designed to understand the mathematical interpretation of various retinal anatomical structures (landmarks) and lesions to detect various retinal diseases and grade severity level of non-proliferative and proliferative retinal diseases from retinal fundus images. Furthermore, the need of: (i) computer-aided retinal disease detection; (ii) retinal image enhancement; (iii) characterization of landmark (healthy) structures; (iv) characterization of retinal lesions; and (v) computer-aided retinal disease classification and grading method have also been discussed. Lastly, the present status of the retinal disease diagnosis method using various retinal imaging modalities has also been briefly described.

References

[1] W. Sarah, R. Gojka, G. Anders, S. Richard, K. Hilary, Global prevalence of diabetes: estimates for the year 2000 and projection for 2030, Diabetes Care 27 (5) (2004) 1047–1053.

[2] Available: https://www.who.int/news-room/fact-sheets/detail/blindness-and-visual-impairment. (Accessed 11 June 2020).

[3] Available: https://ophthalmology.med.ubc.ca/. (Accessed 21 June 2020).

[4] International Diabetes Federation, Diabetes complications, [Online]. Available: https://www.idf.org/aboutdiabetes/what-is-diabetes/complications.html. (Accessed 11 June 2020).

[5] Available: https://www.firstpost.com/india/diabetes-is-indias-fastest-growing-disease-72-million-cases-recorded-in-2017-figure-expected-to-nearly-double-by-2025-4435203.html. (Accessed 3 December 2017).

[6] J. Chu, Y. Ali, Diabetic retinopathy: a review, Drug Development Research 2 (4) (1989) 226–237.

[7] D.A. Salz, A.J. Witkin, Identifying and monitoring diabetic retinopathy, [Online]. Available: http://retinatoday.com/2016/03/identifying-and-monitoring-diabetic-retinopathy/. (Accessed 5 December 2017).

[8] RxList-Fluorescite Side Effects Center, [Online]. Available: https://www.rxlist.com/fluorescite-side-effects-drug-center.htm. (Accessed 3 January 2015).

[9] A. Sjølie, J. Stephenson, S. Aldington, E. Kohner, H. Janka, L. Stevens, J. Fuller, Retinopathy and vision loss in insulin-dependent diabetes in Europe. The EURODIAB IDDM complications study, Ophthalmology 2 (1997) 252–260.

[10] P. Frith, R. Gray, S. MacLennan, P. Ambler, The Eye in Clinical Practice, 2nd ed., Blackwell Science Ltd., London, 2001.

[11] J.G. Hollyfield, R.E. Anderson, M.M. LaVail (Eds.), Retinal Degenerative Diseases, Laser Photocoagulation: Ocular Research and Therapy in Diabetic Retinopathy, vol. 572, Springer, 2006, pp. 195–200.

[12] P.M. Rokade, R.R. Manza, Automatic detection of hard exudates in retinal images using Haar wavelet transform, International Journal of Application or Innovation Engineering and Management 4 (5) (2015) 402–410.

[13] B. Zhang, L. Zhang, L. Zhang, F. Karray, Retinal vessel extraction by matched filter with first-order derivative of Gaussian, Computers in Biology and Medicine 40 (4) (2010) 438–445.

[14] D. Welfer, J. Scharcanski, D.R. Marinho, A coarse-to-fine strategy for automatically detecting exudates in color eye fundus images, Computerized Medical Imaging and Graphics 34 (3) (2010) 228–235.

[15] X. Zhang, et al., Exudate detection in color retinal images for mass screening of diabetic retinopathy, Medical Image Analysis 18 (7) (2014) 1026–1043.

[16] Q. Liu, et al., A location-to-segmentation strategy for automatic exudate segmentation in colour retinal fundus images, Computerized Medical Imaging and Graphics 55 (2017) 78–86.

[17] Habib, et al., Incorporating spatial information for microaneurysm detection in retinal images, Advances in Science, Technology and Engineering Systems Journal 2 (3) (2017) 642–649.

[18] M. Usman Akram, S. Khalid, A. Tariq, S.A. Khan, F. Azam, Detection and classification of retinal lesions for grading of diabetic retinopathy, Computers in Biology and Medicine 45 (1) (2014) 161–171.

[19] I.N. Figueiredo, S. Kumar, C.M. Oliveira, J.D. Ramos, B. Engquist, Automated lesion detectors in retinal fundus images, Computers in Biology and Medicine 66 (2015) 47–65.

[20] D. Santhi, D. Manimegalai, S. Parvathi, S. Karkuzhali, Segmentation and classification of bright lesions to diagnose diabetic retinopathy in retinal images, Biomedical Engineering 61 (4) (2016) 443–453.

[21] N.M. Salem, A.K. Nandi, Segmentation of retinal blood vessels using scale-space features and K-nearest neighbour classifier, in: International Conference on Acoustics, Speech, Signal Processing, 2006, pp. 1001–1004.

[22] M. Garcia, C. Valverde, M.I. Lopez, J. Poza, R. Hornero, Comparison of logistic regression and neural network classifiers in the detection of hard exudates in retinal images, in: 35th Annual International Conference of the IEEE Engineering in Medicine and Biology Society (EMBC), 2013, pp. 5891–5894.

[23] D. Mittal, V. Kumar, S.C. Saxena, N. Khandelwal, N. Kalra, Neural network based focal liver lesion diagnosis using ultrasound images, Computerized Medical Imaging and Graphics 35 (4) (2011) 315–323.

[24] K. Kharat, P. Kulkarni, M. Nagori, Brain tumor classification using neural network based methods, International Journal of Computer Science and Informatics 1 (4) (2012) 85–90.

[25] I.I. Esener, S. Ergin, T. Yuksel, A new feature ensemble with a multistage classification scheme for breast cancer diagnosis, Journal of Healthcare Engineering 2017 (2017).

[26] J. Kaur, D. Mittal, Construction of benchmark retinal image database for diabetic retinopathy analysis, Part H, Journal of Engineering in Medicine 234 (9) (2020).

[27] Jo-Hsuan Wu, Tin Yan Alvin Liu, Application of deep learning to retinal-image-based oculomics for evaluation of systemic health: a review, Journal of Clinical Medicine 12 (1) (2023) 152, https://doi.org/10.3390/jcm12010152.

[28] B. Goutam, M.F. Hashmi, Z.W. Geem, N.D. Bokde, A comprehensive review of deep learning strategies in retinal disease diagnosis using fundus images, IEEE Access 10 (2022) 57796–57823, https://doi.org/10.1109/ACCESS.2022.3178372.

[29] R.T. Yanagihara, C.S. Lee, D.S.W. Ting, A.Y. Lee, Methodological challenges of deep learning in optical coherence tomography for retinal diseases: a review, Translational Vision Science & Technology 9 (2) (2020) 17–19.

[30] B. van Staal, J.J. Abramoff, M.D. Niemeijer, M. Viergever, M.A. Ginneken, DRIVE: digital retinal images for vessel extraction, IEEE Transactions on Medical Imaging (2004) 501–509.

[31] T. Kauppi, et al., DIARETDB1: standard diabetic retinopathy database, http://www.it.lut.fi/project/imageret/diaretdb1/. (Accessed 10 January 2022), 2007.

[32] Messidor, Methods to evaluate segmentation and indexing techniques in the field of retinal ophthalmology, Adcis, 2004.

[33] Available: http://www.adcis.net/en/Download-Third-Party/Messidor.html. (Accessed 13 January 2022).

[34] ADCIS, e-optha: a color fundus image database, ADCIS, http://www.adcis.net/en/Download-Third-Party/E-Ophtha.html. (Accessed 8 January 2022).

[35] M. Goldbaum, STructured Analysis of the Retina (STARE), http://www.ces.clemson.edu/~ahoover/stare/. (Accessed 10 December 2021), 2000.

Index

G

H

Printed in the United States
by Baker & Taylor Publisher Services